Subsurface Ventilation *and* Environmental Engineering

Subsurface Ventilation

and

Environmental Engineering

Malcolm J. McPherson

The Massey Professor of Mining Engineering,
Virginia Polytechnic Institute and State University
and
President,
Mine Ventilation Services,
Incorporated
USA

SPRINGER-SCIENCE+BUSINESS MEDIA, B.V.

First edition 1993

Originally published by Chapman & Hall in 1993
Softcover reprint of the hardcover 1st edition 1993

Typeset in 10/12 pt Bembo by Thomson Press (India) Limited, New Delhi, India

ISBN 978-94-010-4677-0 ISBN 978-94-011-1550-6 (eBook)
DOI 10.1007/978-94-011-1550-6

A catalogue record for this book is available from the British Library

Library of Congress Cataloging-in-Publication Data

McPherson, Malcolm J.
Subsurface ventilation and environmental engineering / Malcolm J. McPherson.—1st ed.
p. cm.
Includes bibliographical references and index.
1. Mine ventilation. 2. Air—Pollution. 3. Mine fires. 4. Mine explosions. 5. Underground areas—heating and ventilation. 6. Underground areas—Fires and fire prevention. 7. Explosions.
I. Title.
TN301.M37 1992
622'.42—dc20 92–6185
CIP

This work has been undertaken in fulfilment of a long-standing promise to my former teacher, mentor and dear friend

Professor Frederick Baden Hinsley

The book is dedicated to his memory

Contents

PART FOUR

Heat and Humidity

14

Psychrometry: the study of moisture in air

14.1 INTRODUCTION

Around the surface of the earth, air that is not affected by any local source of pollution has a composition that is remarkably constant. Air analyses are usually carried out on samples from which all traces of water vapour have been removed. The moisture-free composition of air is given on both a volume and mass basis in Table 14.1. Argon forms the largest fraction, by far, of the monatomic gases. The value of the molecular weight given is, in fact, that for argon.

However, there is another gas present in the free atmosphere, water vapour, that is rather different to the others in that its concentration varies widely from place to place and with time. This is because the pressures and temperatures that exist within the blanket of air that shrouds our globe also encompass the ranges over which water may exist in the gaseous, liquid or solid forms—hence the appearance of clouds, rain, snow and ice. Evaporation of water, mainly from the oceans, coupled with wind action, produces and transports water vapour through the atmosphere. Increases in pressure or, more effectively, decreases in temperature may result in condensation of the water vapour to form clouds which, in turn, can produce droplets large enough to be precipitated as rain, snow or hail.

On other planets with very different atmospheres and gravitational fields, similar phase changes occur in other gases. Because of its variable concentration within the earth's atmosphere, airborne water has become the subject of a special study, **psychrometry**.

Changes of phase are particularly important within the confines of closed environments, including subsurface ventilation systems. Ice to liquid and ice to vapour phase changes occur in mines located in cold climates and, particularly, if situated in permafrost. However, the vast majority of humidity variations that occur in underground airflows are caused by the evaporation of liquid water or the condensation of water vapour. This chapter concentrates on the phase changes between liquid water and water vapour. Virtually all mines produce water from the strata and/or dust

Table 14.1 Composition of dry air

Gas	*Volume (%)*	*Mass (%)*	*Molecular weight*
Nitrogen	78.03	75.46	28.015
Oxygen	20.99	23.19	32.000
Carbon dioxide	0.03	0.05	44.003
Hydrogen	0.01	0.0007	2.016
Monatomic gases	0.94	1.30	39.943
	100.00	100.00	
Equivalent molecular weight of dry air			28.966

suppression techniques. Even with the hygroscopic minerals of evaporite mines, the water vapour content in return airways is normally higher than that in the intakes.

Subsurface environmental engineers have a particular interest in psychrometry for two reasons. First, if we are to comprehend fully the thermodynamic processes that occur in ventilation circuits then variations in humidity must be taken into account. For example, strata heat may be emitted into a wet airway without there being a corresponding increase in air temperature. This could occur if all the added heat were utilized in exciting some of the water molecules until their kinetic energy exceeded the attractive forces of other molecules in the liquid water. They would then escape through the liquid–air surface and become airborne as a gas. The process of evaporation increases the energy content of the air–vapour mixture. This may be termed a **latent** (or hidden) rise in the heat content of the air as there is no commensurate increase in temperature and, hence, no indication on an ordinary thermometer.

Alternatively, if there were no liquid water present, then the strata heat would be directed immediately to the airstream, causing a temperature rise of the air that would be sensed by a thermometer. This is an increase in the **sensible** heat of the air.

These examples illustrate that if we are to predict quantitatively the climatic effects of strata heat, water inflows, machines or air coolers, then we need to have methods of analysis that take humidity into account.

The second reason for the study of psychrometry is the effect of heat and humidity on the human body. This is examined in detail in Chapter 17. However, for the time being, we will concentrate on developing means of quantifying the psychrometric relationships that enable predictions to be made of temperature and other climatic variables in the environment.

14.2 BASIC RELATIONSHIPS

Most of the psychrometric equations that are used in practice are based on the premise that air is a mixture of perfect gases and that the air itself behaves as a perfect gas.

Within the ranges of temperatures and pressures that are reasonable for human tolerance, this assumption gives rise to acceptable accuracy. The majority of this chapter assumes perfect gas laws. However, in certain areas of some underground facilities (including possible future mining scenarios), the atmosphere will require control, but not necessarily within physiologically acceptable ranges. For this reason, more accurate relationships are included that take some account of deviations from the perfect gas laws.

14.2.1 Basis of measurement

A question that should be settled before embarking on a quest for psychrometric relationships is how best to express the quantity of water vapour contained within a given airstream. As we saw in Chapter 3, it is preferable to conduct our analyses on a mass (kg) basis rather than volume (m^3), as variations in pressure and temperature cause the volume of the air to change as it progresses through a ventilation system. That choice relied on the assumption that the mass flow of air remained constant along a single airway. Now we are faced with a different situation. The addition of water vapour to an airstream through evaporative processes, or its removal by condensation, result in the mass flow of the air–vapour mixture no longer being constant.

Within the mixture, molecules of water vapour coexist with, and occupy the same volume as the nitrogen, oxygen and other gases that constitute the air. We can assume, however, that in the absence of chemical reactions or the addition of other gases, it is only the concentration of water vapour that varies, as a result of evaporation and condensation. The mass flow of the rest of the air remains constant.

It is a convenient, although somewhat artificial, device to consider the air to be divided into a fixed mass of 'dry air' and an associated but variable mass of water vapour. For most purposes, we can then refer to the moisture content in terms of grams or kilograms of water vapour per kilogram of 'dry air'. Occasionally, we may use the alternative measure of grams or kilograms of vapour per kilogram of the real mixture of air and vapour. Throughout this chapter we shall use the term 'air' to mean the actual mixture of air and water vapour, and 'dry air' for that fraction which does not include the water vapour. Hence, a moisture content of 0.02 kg/(kg dry air) means that in each 1.02 kg of air, 0.02 kg are water vapour and 1 kg is 'dry air'.

14.2.2 Moisture content (specific humidity) of air

In order to quantify the mass of water vapour associated with each kilogram of 'dry air', let us conduct an imaginary experiment.

Suppose we have a closed vessel of volume V m^3 containing 1 kg of perfectly dry air at a pressure P_a Pa and temperature T K. If we inject X kg of water vapour at the same temperature, the pressure within the vessel will increase to

$$P = P_a + e \quad \text{Pa}$$

where e is the partial pressure exerted by the water vapour (Dalton's law of partial pressures for perfect gases). Both the air and the water vapour occupy the same volume, V, and are at the same temperature, T.

The problem is to determine the mass of water vapour, X, if we can measure nothing more than the initial and final pressures, P_a and P.

From the general gas law, we have the following

1. For the X kg of water vapour,

$$eV = XR_vT \quad \text{J} \tag{14.1}$$

   where R_v = gas constant for water vapour (461.50 J/(kg K)), and
2. for the original 1 kg of dry air,

$$P_aV = 1 \times R_aT \quad \text{J} \tag{14.2}$$

   where R_a = gas constant for dry air (287.04 J/(kg K)).

   Dividing equation (14.1) by (14.2) gives

$$\frac{e}{P_a} = \frac{R_v}{R_a}X$$

However, the absolute (barometric) pressure in the vessel is

$$P = P_a + e$$

Hence,

$$X = \frac{R_a}{R_v}\frac{e}{P - e} \tag{14.3}$$

Inserting the values of the gas constants gives

$$X = 0.622\frac{e}{P - e} \quad \frac{\text{kg}}{\text{kg dry air}} \tag{14.4}$$

This gives the moisture content of the vessel in kg per kg of dry air, provided that the partial pressure of the water vapour, e, can be evaluated. This is, of course, simply the difference betwen the initial and final absolute pressures. However, e can be determined independently, as we shall see a little later in the chapter.

14.2.3 Saturation vapour pressure

Returning to our experiment, suppose we continue to inject water vapour. The partial pressure of water vapour (and, hence, the absolute pressure) will continue to rise—but only to a certain limiting value. The partial pressure of the dry air fraction, P_a, will remain constant. If we insist on forcing yet more vapour into the system while keeping the temperature constant, then the excess will condense and collect as liquid water on the sides and bottom of the vessel. (It is, in fact, possible to achieve a condition of supersaturation in the laboratory but that will not occur in natural atmospheres and is not considered here.)

When the system refuses to accept any more water vapour then we say, rather loosely, that the air has become saturated. In fact, it is not the air but the space that has become saturated. If we were to repeat the experiment starting with the vessel evacuated and containing no air then exactly the same amount of vapour could be injected before condensation commenced, provided that the temperature remained the same.

The pressure, e_s, exerted by the water vapour at saturation conditions depends only on temperature and not on the presence of any other gases. The relationship between saturation vapour pressure and temperature for water has been determined not only experimentally but also through thermodynamic reasoning by a number of authorities. The tables of Goff and Gratch produced in 1945 are still regarded as a standard. The simplest analytical equation is known as the Clausius–Clapeyron equation:

$$\frac{1}{e_s}\frac{de_s}{dT} = \frac{L}{R_v T^2} \quad \mathrm{K}^{-1} \tag{14.5}$$

where the latent heat of evaporation, L(J/kg), is the heat required to evaporate 1 kg of water.

The derivation of the Clausius–Clapeyron equation is given in the appendix at the end of the chapter. This equation is based on the perfect gas laws and must, therefore, be regarded as an approximation. Furthermore, the latent heat of evaporation, L, is not constant. The higher the initial temperature of the liquid water then the less will be the additional heat required to evaporate it. The relationship between latent heat of evaporation and temperature is near linear. The equation

$$L = (2502.5 - 2.386t)1000 \quad \mathrm{J/kg} \tag{14.6}$$

(where temperature, t, is in degrees Centigrade) is accurate to within 0.02% over the range 0 to 60 °C.

The equation may also be written as

$$L = (3154.2 - 2.386\,T)1000 \quad \mathrm{J/kg} \tag{14.7}$$

where the temperature T is in kelvins.

The Clausius–Clapeyron equation can be integrated over any given interval:

$$\int_1^2 \frac{de_s}{e_s} = \frac{1}{R_v}\int_1^2 \frac{a - bT}{T^2}\,dT$$

where $a = 3\,154\,200$ J/kg and $b = 2386$ J/(kg K) from equation (14.7). Integrating gives

$$\ln(e_{s2}/e_{s1}) = \frac{1}{R_v}\left[a\left(\frac{1}{T_1} - \frac{1}{T_2}\right) + b\ln(T_1/T_2)\right]$$

giving

$$e_{s2} = e_{s1} \exp\left\{\frac{1}{R_v}\left[a\frac{T_2 - T_1}{T_1 T_2} + b\ln(T_1/T_2)\right]\right\} \quad (14.8)$$

If the saturation vapour pressure, e_{s1}, at any given temperature, T_1, is known, then equation (14.8) allows the saturation vapour pressure, e_{s2}, at any other temperature, T_2, to be calculated. At 100 °C, the saturation vapour pressure is one standard atmosphere (101.324 kPa) by definition (section 2.2.3). Hence, this may be used as a starting point for the integration.

There are two problems with equation (14.8). First, it is cumbersome for rapid calculation and, secondly, the assumption of perfect gas behaviour in its derivation introduces some uncertainty. The latter is not serious. Integrating down from a vapour pressure of 19.925 kPa at 60 °C using equation (14.8) gives a maximum error of only 0.52% over the range of 0 to 60 °C. Nevertheless, we can do much better and simplify the format of the relationship at the same time.

Let us take temperature T_1 to be the freezing point of water, 273.15 K or 0 °C. Then the temperature T_2 may be expressed in degrees Centigrade as

$$t = T_2 - 273.15 \quad ^\circ\text{C}$$

or

$$T_2 = t + 273.15 \quad \text{K}$$

Furthermore, the logarithmic term in equation (14.8) can be rewritten as

$$\ln\left(\frac{T_1}{T_2}\right) = \ln\left(1 + \frac{T_1 - T_2}{T_2}\right) = \ln\left(1 - \frac{t}{T_2}\right)$$

Expanding from the logarithmic series for the condition where t is much less than T_2 gives the term as approximately $-t/T_2$.

Substituting into equation (14.8), simplifying and gathering constants together gives the form

$$e_{s2} = A\exp\left(\frac{Bt}{C + t}\right) \quad \text{Pa} \quad (14.9)$$

where A, B and C are constants. Curve fitting from the standard tables of Goff and Gratch over the range 0 to 60 °C gives the widely used equation

$$e_{s2} = 610.6\exp\left(\frac{17.27\,t}{237.3 + t}\right) \quad \text{Pa} \quad (14.10)$$

This is accurate to within 0.06% over the given range.

For the more usual mining range of 10 to 40 °C the curve-fitting procedure gives

$$e_{s2} = 610.162\exp\left(\frac{17.291\,t}{237.481 + t}\right) \quad \text{Pa} \quad (14.11)$$

having an excellent accuracy of within 0.01% for that range.

Example Find the saturation vapour pressure at a temperature of 30 °C.

Solution

1. Inserting $t = 30$ °C into equation (14.10) gives $e_s = 4241.7$ Pa or 4.2417 kPa. This is in error by 0.026% when compared with the value of 4.2428 kPa given by the Goff and Gratch tables.
2. Using equation (14.11) gives $e_s = 4.2431$ kPa, having an error of only 0.007%

14.2.4 Gas constant and specific heat (thermal capacity) of unsaturated air

From the gas laws for 1 kg of dry air and X kg of associated water vapour, we have

$$P_a V = 1 R_a T$$

and

$$eV = X R_v T$$

Adding these two equations gives

$$(P_a + e)V = (R_a + X R_v)T$$

or

$$PV = (R_a + X R_v)T \quad \text{J/kg} \tag{14.12}$$

Alternatively, we can treat the $1 + X$ kg of air–vapour mixture as a perfect gas having an equivalent gas constant of R_m. Then

$$PV = (1 + X) R_m T \quad \text{J/kg} \tag{14.13}$$

Equating (14.12) and (14.13) gives

$$R_m = \frac{R_a + X R_v}{1 + X} \quad \text{J/(kg K)} \tag{14.14}$$

Hence, we have shown that the gas constant for the moist air is given simply by adding the gas constants for dry air, R_a, and water vapour, R_v, in proportion to the relative masses of the two components.

The equivalent gas constant for moist air can also be expressed in terms of pressures by substituting for X from equation (14.4), leading to

$$R_m = 287.04 \frac{P}{P - 0.378e} \quad \text{J/(kg K)} \tag{14.15}$$

Similarly, the equivalent specific heat (thermal capacity) of moist unsaturated air can be found by adding the specific heats of the two components in proportion to their masses. The specific heat at constant pressure becomes

$$C_{pm} = \frac{C_{pa} + X C_{pv}}{1 + X} \quad \text{J/(kg K)} \tag{14.16}$$

where C_{pa} = specific heat of dry air at constant pressure (1005 J/(kg K)) and C_{pv} = specific heat at constant pressure for water vapour (1884 J/(kg K)). Also, the specific heat of moist air at constant volume is given as

$$C_{vm} = \frac{C_{va} + XC_{vv}}{1 + X} \quad \text{J/(kg K)} \tag{14.17}$$

where C_{va} = specific heat of dry air at constant volume (718 J/(kg K)) and C_{vv} = specific heat of water vapour at constant volume (1422 J/(kg K)).

14.2.5 Specific volume and density of unsaturated air

The actual specific volume of moist unsaturated air, V_m, can be calculated from the general gas law for 1 kg of the air–vapour mixture:

$$V_m = R_m \frac{T}{P}$$

Substituting for R_m from equations (14.14) and (14.15) gives

$$V_m = \frac{R_a + XR_v}{1 + X} \frac{T}{P} \quad \text{m}^3\text{/(kg of moist air)} \tag{14.18}$$

and

$$V_m = 287.04 \frac{T}{P - 0.378e} \quad \text{m}^3\text{/(kg of moist air)} \tag{14.19}$$

In these equations, T is in kelvins and P in pascals.

As we saw in section 14.2.1, it is more convenient, for most purposes, to conduct our analyses on the basis of 1 kg of 'dry air' rather than a kilogram of the true mixture. The **apparent** specific volume, based on 1 kg of dry air, is simply

$$V_m(\text{apparent}) = 287.04 \frac{T}{P - e} \quad \text{m}^3\text{/(kg dry air)} \tag{14.20}$$

The actual density of the moist air, ρ_m, is the reciprocal of the actual specific volume:

$$\rho_m = \frac{1 + X}{R_a + XR_v} \frac{P}{T} \quad \text{kg moist air/m}^3 \tag{14.21}$$

or

$$\rho_m = \frac{P - 0.378e}{287.04T} \quad \text{kg moist air/m}^3 \tag{14.22}$$

A useful approximate formula to calculate the effect of moisture on air density can be derived from equation (14.21) giving

$$\rho_m = \frac{P}{R_a T}(1 - 0.608X) \quad \text{kg moist air/m}^3 \tag{14.23}$$

This equation makes clear that the density of air decreases as its moisture content rises.

The **apparent** density, based on 1 kg of dry air, is the reciprocal of equation (14.20):

$$\rho_m(\text{apparent}) = \frac{P-e}{287.04T} \quad \text{kg dry air/m}^3 \qquad (14.24)$$

14.2.6 Relative humidity and percentage humidity

Relative humidity is widely used by heating and ventilating engineers as an important factor governing comfort of personnel in surface buildings. In open surface areas or in subsurface facilities, its use as a physiological parameter is very limited and can be misleading, owing to the wider range of air temperatures that may be encountered (see, also Chapter 17). On the other hand, the concept of relative humidity, rh, is a convenient way of expressing the degree of saturation of a space. It is defined as

$$\text{rh} = \frac{e}{e_{sd}} \times 100\% \qquad (14.25)$$

where e_{sd} = saturation vapour pressure at the air (dry bulb) temperature.

This definition indicates that relative humidity is the ratio of the prevailing vapour pressure to that which would exist if the space were saturated at the same temperature. A similar concept is that of **percentage humidity**, ph, defined as

$$\text{ph} = \frac{X}{X_{sd}} \times 100\% \qquad (14.26)$$

where X_{sd} is the moisture content (kg/(kg dry air)) that would exist if the space were saturated at the same dry bulb temperature.

Percentage humidity is approximately equal to relative humidity over the normal atmospheric range. Substituting from equation (14.4) gives

$$\text{ph} = \frac{X}{X_{sd}} = \frac{e}{P-e}\,\frac{P-e_{sd}}{e_{sd}} \times 100$$

$$= \text{rh} \times \frac{P-e_{sd}}{P-e}$$

As P is much larger than e or e_{sd} over the normal atmospheric range, the numerator and denominator in the final expression are near equal.

Example An airstream of temperature 20° and barometric pressure 100 kPa is found to have an actual vapour pressure of 1.5 kPa. Determine

1. the moisture content of the air, X,
2. the gas constant, R_m,
3. the specific heat, C_{pm},
4. the actual and apparent specific volumes,

5. the actual and apparent densities,
6. the relative humidity, rh, and
7. the percentage humidity

Solution

1. *Moisture content, X.*

$$X = 0.622 \frac{1.5}{100 - 1.5}$$
$$= 0.009\,472\ \text{kg vapour/(kg dry air) or } 9.472\ \text{g/(kg dry air)}$$

2. *Gas constant.* From equation (14.15)

$$R_m = \frac{287.04 \times 100}{100 - (0.378 \times 1.5)}$$
$$= 288.68\ \text{J/(kg K)}$$

3. *Specific heat, C_{pm}.* From equation (14.16)

$$C_{pm} = \frac{1005 + 0.009\,472 \times 1884}{1 + 0.009\,472}$$
$$= 1013\ \text{J/(kg K)}$$

4. *Specific volumes.* From equation (14.19), the actual specific volume is

$$V_m = \frac{287.04(273.15 + 20)}{1000(100 - 0.378 \times 1.5)}$$
$$= 0.8463\ \ \text{m}^3\text{/(kg moist air)}$$

(the 1000 in the denominator is necessary to convert the pressures from kPa to Pa). The apparent specific volume from equation (14.20) is

$$V_m(\text{apparent}) = \frac{287.04(273.15 + 20)}{1000(100 - 1.5)}$$
$$= 0.8543\ \text{m}^3\text{/(kg dry air)}$$

5. *Air densities.* Actual density is given by equation (14.22) or the reciprocal of actual specific volume:

$$\rho_m = \frac{1}{V_m} = \frac{1}{0.8463}$$
$$= 1.1816\ \text{kg moist air/m}^3$$

Similarly, the apparent density is given as

$$\rho_m(\text{apparent}) = \frac{1}{V_m(\text{apparent})}$$

$$= \frac{1}{0.8543} = 1.1705\,\text{kg dry air/m}^3$$

6. *Relative humidity, rh.* From equation (14.10), the saturation vapour pressure at a temperature of 20 °C is

$$e_{sd} = 610.6 \exp\left(\frac{17{\cdot}27 \times 20}{237.3 + 20}\right)$$

$$= 2337.5\,\text{Pa} \quad \text{or} \quad 2.3375\,\text{kPa}$$

Then, from equation (14.25)

$$rh = \frac{e}{e_{sd}} \times 100 = \frac{1.5}{2.3375} \times 100$$

$$= 64.17\%$$

7. *Percentage humidity, ph.* If the space were saturated at 20 °C, then the saturation vapour pressure would be 2.3375 kPa. Equation (14.4) gives the corresponding saturation moisture content to be

$$X_{sd} = 0.622 \frac{2.3375}{100 - 2.3375}$$

$$= 0.014\,887\,\text{kg/(kg dry air)}$$

and equation (14.26) gives

$$ph = \frac{X}{X_{sd}} \times 100$$

$$= \frac{0.009\,472}{0.014\,887} \times 100 = 63.62\%$$

14.3 THE MEASUREMENT OF WATER VAPOUR IN AIR

There are a number of methods of measuring the humidity of air. Numerous instruments are available commercially. These may be divided into five types.

14.3.1 Chemical methods

This involves passing a metered volume of air through a hygroscopic compound such as calcium chloride, silica gel or sulphuric acid and observing the increase in weight. The method is slow and somewhat cumbersome. However, it gives a direct measure of the total moisture content of an airstream and, with care, will give results of high accuracy.

14.3.2 Electrical methods (electronic psychrometers or humidity meters)

The changes in the electrical properties (resistivity, dielectric constants) of some compounds in the presence of water vapour are used in a variety of instruments to give a rapid indication of humidity. Such equipment needs to be compensated against the effect of variations in ambient temperature. Modern devices of this type may employ semiconductors and electronic microprocessors to improve the versatility, reliability and stability of the instrument. Nevertheless, for accurate work, it is advisable to check the calibrations of electronic psychrometers against a more direct method.

14.3.3 Hair hygrometers

Many organic compounds such as bone, hair or other fibrous materials exhibit changes in their volume and elasticity when exposed to water vapour. An example is the opening and closing of fir cones as the weather changes. Human hair is particularly reactive.

Most inexpensive humidity meters sold for domestic display incorporate strands of hair maintained under spring tension. Variations in humidity cause small changes in the length of the hair. These are amplified mechanically through a lever arrangement. Recording instruments of this type are used to move a pen across a revolving paper chart.

14.3.4 Dew point hygrometers

If an air–vapour mixture is cooled at constant barometric pressure then the actual vapour pressure remains unchanged (see equation (14.4)). However, the falling temperature will cause the corresponding saturation vapour pressure to decrease (equation (14.10)). At some point, the saturation vapour pressure will become equal to the actual vapour pressure. According to classical theory, condensation will then commence.

The temperature at which saturation conditions are attained is known as the dew point temperature. An inversion of equation (14.10) gives

$$t(\text{dew point}) = \frac{237.3 \ln(e/610.6)}{17.27 - \ln(e/610.6)} \quad {}^\circ\text{C} \qquad (14.27)$$

In a dew point hygrometer, a sample of air is drawn over a mirrored surface. The rear of the mirror is cooled, often by evaporation of a volatile liquid. A thermometer attached to the mirror gives the temperature of its surface. This is read at the moment when misting first appears on the mirror to find the dew point temperature. While dew point apparatus gives a good demonstration of the effect of condensation by cooling, the visual detection of the initial film of condensate on the mirror is somewhat subjective. Such instruments are used little in practice.

14.3.5 Wet and dry bulb hygrometers (psychrometers)

These are the most widely used types of hygrometers in subsurface ventilation engineering. They give reliable results when employed by competent personnel and are simple to use. Because of their widespread employment, the theory of the wet bulb thermometer is developed in section 14.4.

A wet and dry bulb hygrometer is simply a pair of balanced thermometers, one of which has its bulb shrouded in a water-saturated muslin jacket. Air passing over the two bulbs will cause the dry bulb thermometer to register the ordinary temperature of the air. However, the cooling effect of evaporation will result in the wet bulb thermometer registering a lower temperature. A knowledge of the wet and dry bulb temperatures, together with the barometric pressure, allow all other psychrometric parameters to be calculated.

Instruments vary in (a) the precision of the thermometers, (b) the manner in which water is supplied to the wet bulb and (c) the means of providing the required airflow over the thermometer bulbs. The term 'psychrometer', rather than 'hygrometer', tends to be used for the more accurate instruments.

The accuracy of the thermometers depends on the quality of their manufacture and calibration, and also on their length. The better psychrometers can be read to the nearest 0.1 °C. Variations considerably greater than this may occur over the cross-section of an airway, particularly near shaft stations. In order to maintain the instrument at a portable size, some models achieve accuracy at the expense of range. The approximate extremes of temperature should be known prior to conducting an important psychrometric survey, and an appropriate instrument selected. Replacement thermometers should be purchased as balanced pairs. The difference between the wet and dry bulb temperatures is more important than the absolute temperature. Significant errors in both the wet and dry bulb temperature readings may occur by radiation from surrounding surfaces or between the bulbs. This can be reduced by increasing the air velocity over the thermometer bulbs. However, for precise results a polished radiation shield should be provided separately around each of the bulbs.

Many hygrometers are fitted with a small reservoir of water into which dips one end of the muslin wick. Capillary action draws water through the muslin to feed the wet bulb. In such devices, it is important that the water reaching the wet bulb is already at wet bulb temperature. The water within the reservoir will be at dry bulb temperature and, if supplied to the wet bulb at too liberal a rate, will give a falsely elevated wet bulb temperature. Conversely, if the water supply is insufficient then drying out of the wick will occur and, again, the wet bulb thermometer will read too high. Because of these difficulties, the more precise psychrometers have no integrated water reservoir and must have their wet bulbs wetted manually with distilled water.

If a wet bulb thermometer is located in a wind tunnel and air of fixed psychrometric condition passed over it at increasing velocity, the indicated wet bulb temperature will decrease initially, then level off. The reason for this phenomenon is that the envelope of saturated air surrounding the wet bulb must be removed efficiently to maintain the evaporative process. The steady-state reading attained is essentially identical to the true thermodynamic wet bulb temperature attained during an adiabatic saturation process (see section 14.5.2). The minimum air velocity which gives the true wet bulb temperature is a function of the size and shape of the wet bulb, and its orientation with respect to the direction of the airflow. For most commercially available instruments, an air velocity of at least 3 m/s over the bulbs is suggested.

The static hygrometer (sometimes known as the Mason hygrometer) has no intrinsic means of creating the required air velocity and relies on being hung in a location where the thermometer bulbs are exposed to an external air velocity. The 'whirling' or 'sling' psychrometer has its thermometers and water reservoir mounted on a frame which may be rotated manually about its handle. Whirling the instrument at about 200 rpm will give the required air velocity over the bulbs. After 30 s of rotation, the whirling should be terminated (but not by clamping one's hands around the bulbs), the instrument held such that the observer is not breathing on it, and the wet bulb temperature read immediately. The process of whirling should be repeated until the reading becomes constant.

Aspirated psychrometers have small fans driven by clockwork or batteries. Air is drawn through the radiation shield surrounding each bulb. Again, the wet bulb temperature should be observed until it becomes constant. While the whirling hygrometers are, by far, the most widely used instruments for routine measurements, aspirated psychrometers are recommended for important surveys.

14.4 THEORY OF THE WET BULB THERMOMETER

The wet bulb temperature is a most important parameter in hot climatic conditions for two separate but interrelated purposes. The first lies in its vital importance in evaluating the ability of the air to remove metabolic heat from personnel. The second is the use of the wet bulb temperature in quantifying the humidity of the air.

14.4.1 Heat balance on a wet bulb

Figure 14.1 illustrates $1 + X$ kg of moist unsaturated air approaching and flowing closely over the surface of a wet bulb. On leaving the wet surface, the 1 kg of air remains the same but the mass of associated water vapour has increased from X kg to X_s kg and the air has become saturated. Thus a mass of $X_s - X$ kg of water has evaporated from the wet surface for each kilogram of 'dry' air passing. The reason for employing 1 kg of dry air as the basis of measurement is, again, apparent—this remains constant, while the mass of water vapour varies.

As the molecules of water leave the wet surface they take energy with them. This 'latent heat' energy is transferred to the air leaving the wet bulb at a reduced energy level. This is reflected by a drop in temperature (the evaporative cooling effect) and the wet bulb thermometer gives a reading depressed below that of the dry bulb. The greater the rate of evaporation, the greater the wet bulb depression.

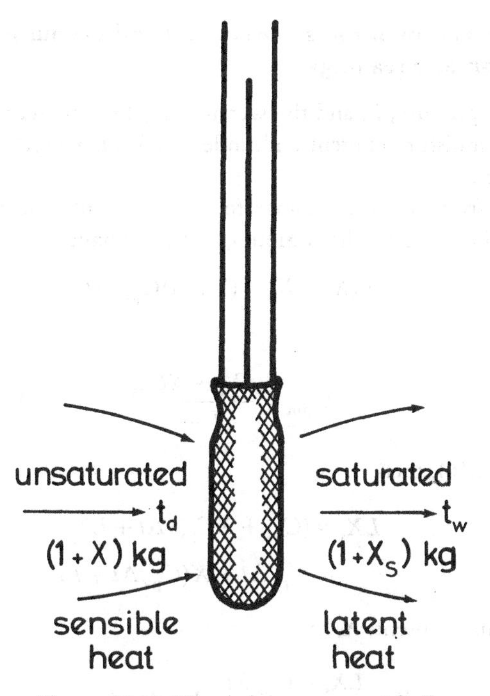

Figure 14.1 Heat balance on a wet bulb.

As there is now a difference between the air temperature and the cooler wet surface, a transfer of sensible heat will occur by convection from the air to the wet bulb. A dynamic equilibrium will be established at which the heat loss from the wet bulb by evaporation is balanced by the convective sensible heat gain:

$$\text{latent heat loss from wet bulb} = \text{sensible heat gain by wet bulb} \quad (14.28)$$

Each of these terms can be quantified. Latent heat transfer from the wet bulb to the air

$$q = L(X_s - X) \quad \text{J/(kg dry air)} \quad (14.29)$$

where L = latent heat of evaporation (J/(kg evaporated)) at wet bulb temperature.

The air–vapour mixture approaches the wet surface at dry bulb temperature, t_d, and leaves at wet bulb temperature, t_w. The sensible heat transfer from the air to the wet bulb is then

$$q = \text{mass} \times \text{specific heat} \times \text{change in temperature}$$

$$= (1 + X) \times C_{pm} \times (t_d - t_w) \quad \text{J/(kg dry air)} \quad (14.30)$$

At equilibrium, latent heat transfer equals sensible heat transfer. Equations (14.29) and (14.30) quantify the heat balance that exists on the wet bulb:

$$L(X_s - X) = (1 + X)C_{pm}(t_d - t_w) \quad \text{J/(kg dry air)} \quad (14.31)$$

14.4.2 Determination of moisture content and vapour pressure from psychrometer readings

If the barometric pressure, P, and the wet and dry bulb temperatures, t_w and t_d, are known then the moisture content and, indeed, all other psychrometric parameters can be determined.

Commencing from the heat balance for the wet bulb, and using the shorthand notation Δt for the wet bulb depression, $t_d - t_w$, we have

$$L(X_s - X) = (1 + X)C_{pm}\,\Delta t$$

However,

$$C_{pm} = \frac{C_{pa} + XC_{pv}}{1 + X}$$

from equation (14.16); then

$$LX_s = (C_{pa} + XC_{pv})\,\Delta t + LX$$
$$= C_{pa}\,\Delta t + X(C_{pv}\,\Delta t + L)$$

giving the moisture content, X, as

$$X = \frac{LX_s - C_{pa}\,\Delta t}{C_{pv}\,\Delta t + L} \quad \text{kg/(kg dry air)} \tag{14.32}$$

All variables on the right-hand side of this equation are known, or can be calculated easily for the given P, t_d and t_w: hence, the moisture content, X, is defined.

The actual vapour pressure then follows from a transposition of equation (14.4):

$$e = \frac{PX}{0.622 + X} \quad \text{Pa} \tag{14.33}$$

Example Determine the moisture content and vapour pressure for a barometric pressure of 100 kPa and wet and dry bulb temperatures of 20 and 30 °C respectively.

Solution At the wet bulb temperature, the variables required by equation (14.32) are calculated as follows:

$$L = (2502.5 - 2.386 \times 20)1000 \quad \text{(from equation (14.6))}$$
$$= 2454.78 \times 10^3 \text{ J/kg or } 2454.78 \text{ kJ/kg}$$

$$e_{sw} = 610.6 \exp\left(\frac{17.27 \times 20}{237.3 + 20}\right) \quad \text{(equation (14.10))}$$
$$= 2.3375 \text{ kPa}$$

$$X_s = 0.622 \times \frac{2.3375}{100 - 2.3375} \quad \text{(equation (14.4))}$$

$$= 0.014\,887 \text{ kg/(kg dry air)}$$

Then equation (14.32) gives

$$X = \frac{2454.78 \times 10^3 \times 0.014\,887 - 1005(30 - 20)}{1884(30 - 20) + 2454.78 \times 10^3}$$

$$= \frac{36\,544 - 10\,050}{18\,840 + 2\,454\,780}$$

$$= 0.010\,71 \text{ kg/(kg dry air)}$$

The actual vapour pressure is then given by equation (14.33)):

$$e = \frac{100 \times 0.010\,71}{0.622 + 0.010\,71} = 1.6927 \text{ kPa}$$

This example illustrates that the term involving C_{pv} (1884 J/(kg K)) in equation (14.32) is very small compared with L. A more traditional equation for actual vapour pressure follows from ignoring the smaller term. Then equation (14.32) simplifies to

$$X = X_s - \frac{C_{pa}\,\Delta t}{L}$$

Substituting for X and X_s from equation (14.4) gives

$$0.622\frac{e}{P - e} = 0.622\frac{e_{sw}}{P - e_{sw}} - \frac{C_{pa}\,\Delta t}{L}$$

or

$$e = e_{sw}\frac{P - e}{P - e_{sw}} - \frac{C_{pa}\,\Delta t}{0.622\,L}(P - e)$$

If we now assume that e and e_{sw} are small compared with P, the equation simplifies further to

$$e = e_{sw} - \frac{C_{pa}}{0.622\,L}P\,\Delta t \quad \text{Pa} \tag{14.34}$$

or

$$e = e_{sw} - AP\,\Delta t \quad \text{Pa} \tag{14.35}$$

The parameter A is known as the psychrometric 'constant'. However, because of the simplifications made in the derivation, it is not a true constant; neither is it given precisely as $C_{pa}/0.622L$ but varies non-linearly with temperature and pressure between 0.000 58 and 0.000 648 °C^{-1} over the atomspheric range. This variation of the psychrometric 'constant' is a weakness in the relationship. However, a value of 0.000 644 °C gives acceptable results for most practical purposes.

Example Determine the actual vapour pressure from equation (14.35) for $P = 100$ kPa, $t_w = 20\,°C$ and $t_d = 30\,°C$.

Solution These are the same conditions as specified in the previous example. This gave

$$e_{sw} = 2.3375 \text{ kPa}$$

Then equation (14.35) gives

$$e = 2.3375 - 0.000\,644 \times 100 \times (30 - 20)$$
$$= 1.6935 \text{ kPa}$$

(a difference of less than 0.05% from the earlier solution).

14.5 FURTHER PSYCHROMETRIC RELATIONSHIPS

14.5.1 Enthalpy of moist air

In Chapter 3, the steady flow energy equation was derived as

$$\frac{u_1^2 - u_2^2}{2} + (Z_1 - Z_2)g + W_{in} = \int_2^1 V\,dP + F_{12} = H_2 - H_1 - q_{in} \quad \text{J/kg} \qquad (14.36)$$

where u = air velocity (m/s), Z = height above datum, (m), g = gravitational acceleration (m/s^2), W_{in} = mechanical (fan) work input (J/kg), V = specific volume, (m^3/kg), F = frictional conversion of mechanical to heat energy (J/kg), H = enthalpy, (J/kg) and q_{in} = heat input from external sources (J/kg).

In its application to mine ventilation thermodynamics (Chapter 8), the change in enthalpy between end stations 1 and 2 was earlier assumed to be

$$H_2 - H_1 = C_{pa}(t_2 - t_1) \quad \text{J/kg} \qquad (14.37)$$

where t_1 and t_2 were the dry bulb temperatures. This assumption was based on the premise that the airway was dry—neither evaporation nor condensation was considered. For moist, but unsaturated, air, the specific heat term may be replaced by C_{pm} for the actual air–vapour mixture. However, if evaporation or condensation does take place then the exchange of latent heat will have a large additional effect on the dry bulb temperature. Equation (14.37) no longer applies. As this is the situation in most underground airways, we must seek a method of evaluating enthalpy that takes the moisture content of the air into consideration.

Let us carry out another imaginary experiment. Suppose we have 1 kg of dry air at 0 °C and, in a separate container, a small amount, X kg, of liquid water. Now, let us heat the air until it reaches a dry bulb temperature of t_d. Similarly, we add heat to the water until it vaporizes and the vapour also reaches a temperature of t_d. Finally, we mix the two components to obtain $1 + X$ kg of moist unsaturated air at a dry bulb temperature of t_d and some lower value of wet bulb temperature, t_w.

The total amount of heat that we have added to the air and water during this experiment represents the increase in enthalpy of the system over its starting condition at 0 °C. If we choose 0 °C as our enthalpy datum, that same added heat represents the enthalpy level of the final $1 + X$ kg of mixture.

In order to quantify the added heat, consider the air and water separately.

1. The heat required to raise the temperature of 1 kg of dry air from 0 °C to t_d is simply mass × specific heat × change in temperature:

$$q_{air} = 1 \times C_{pa} \times t_d \quad \text{J}$$

2. We can convert the water at 0 °C to vapour at t_d in any of three ways, depending on the pressure that we maintain in the vessel.
   (a) We could consider evaporating the X kg at water at 0 °C then raising the temperature of the vapour to t_d:

$$q_{water} = L_{(0)}X + XC_{pv}t_d \quad \text{J}$$

   where $L_{(0)}$ = Latent heat of evaporation at 0 °C.
   (b) Alternatively, we might raise the temperature of the liquid water to t_d and then evaporate at that temperature:

$$q_{water} = XC_w t_d + L_{(t_d)}X \quad \text{J}$$

   where C_w = specific heat of liquid water (4187 J/(kg K)).
   (c) We could raise the temperature of the liquid to any intermediate value, t, evaporate at that temperature, then continue to superheat the vapour until it reaches the required temperature, t_d:

$$q_{water} = XC_w t + L_{(t)}X + XC_{pv}(t_d - t) \quad \text{J} \tag{14.38}$$

In fact, as the two end conditions are precisely defined, it does not matter which method is used. They all give the same result provided that account is taken of the variations of latent heat and specific heat with respect to temperature.

For a reason that will become apparent later, let us choose method (c) with the intermediate value of temperature chosen to be t_w, the eventual wet bulb temperature of the air–vapour mixture.

We have now found the total amount of heat added to the system and, hence, its enthalpy relative to a 0 °C datum:

$$\begin{aligned} H &= q_{air} + q_{water} \\ &= C_{pa}t_d + X[C_w t_w + L + C_{pv}(t_d - t_w)] \quad \text{J/(kg dry air)} \end{aligned} \tag{14.39}$$

Here, the symbol L has reverted to its earlier meaning of latent heat of evaporation at wet bulb temperature. All of the parameters on the right-hand side of the equation can be measured, are constant or can be calculated from equations (14.6) and (14.32). Hence, the enthalpy of an air–vapour mixture can be determined from psychrometric measurements of air pressure and wet and dry bulb temperatures. If such values of enthalpy are employed in the steady-flow energy equation (14.36) then that equation may continue to be used for airways that involve evaporation or condensation processes.

14.5.2 The adiabatic saturation process

Consider a long level airway with no heat additions from any source and free-standing water covering the floor (Fig. 14.2). Unsaturated air enters at one end and moves sufficiently slowly to allow full saturation to occur. The air exits at saturation conditions.

The surface of the liquid water reacts in the same way as that of a wet bulb thermometer (section 14.4.1) and, indeed, will be at wet bulb temperature. At inlet, the dry bulb temperature will be at a higher value. Hence, sensible heat transfer will occur from the air to the water resulting in a reduction in dry bulb temperature. Simultaneously, heat and mass transfer from the water to the air will take place as water molecules escape from the liquid surface—a latent heat gain by the air. These exchanges will continue until saturation when wet bulb, dry bulb and water temperature all become equal (thermodynamic wet bulb temperature).

This is known as an **adiabatic saturation process**. The sensible heat lost by the air is balanced by the latent heat gained by the air and the process involves no net addition or loss of heat. This latter statement is true for the combination of air and water, but not quite true for the air alone. Mass has been added to the airstream in the form of water molecules, and those molecules already contained sensible heat before they were evaporated. Hence, the enthalpy of the air does not quite remain

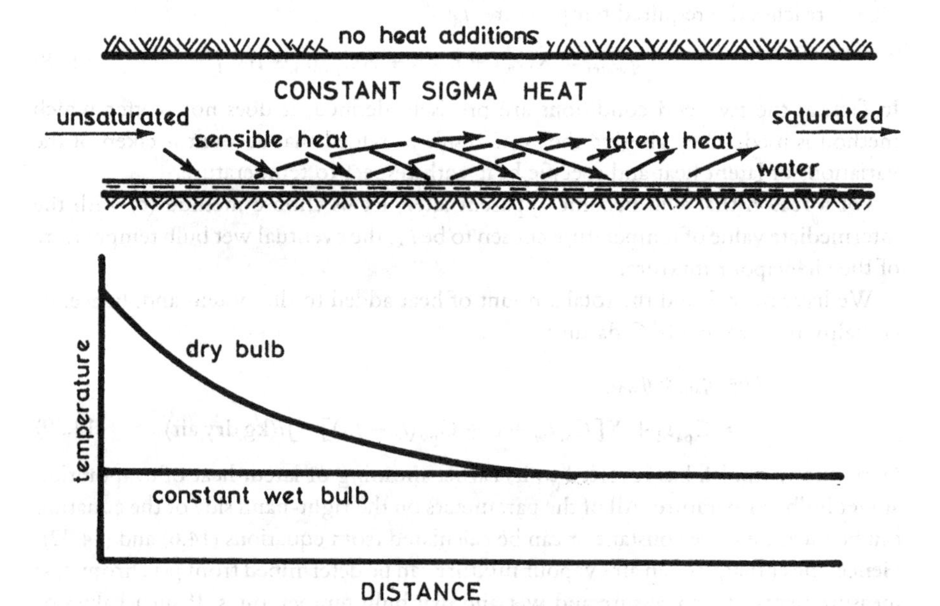

Figure 14.2 An adiabatic saturation process.

constant. However, if we were to move along the airway taking psychrometric readings, calculating enthalpy from equation (14.39) but omitting the term for the sensible heat of liquid water, $XC_w t_w$, then the result would be a property value that remained truly constant throughout the adiabatic saturation process. The significance of this property seems first to have been recognized by Carrier and was named sigma heat, S, to distinguish it from its near neighbour, enthalpy.

14.5.3 Sigma heat, *S*

Sigma heat is much more than simply an interesting property of an adiabatic saturation process. Indeed, it features in the majority of analyses concerning subsurface climatic changes.

Ignoring the sensible heat of the liquid water in equation (14.39) gives

$$S = H - XC_w t_w \quad \text{J/(kg dry air)} \tag{14.40}$$

and

$$S = C_{pa} t_d + X[L + C_{pv}(t_d - t_w)] \quad \text{J/(kg dry air)} \tag{14.41}$$

However, from equation (14.32), and replacing Δt with $t_d - t_w$,

$$X[L + C_{pv}(t_d - t_w)] = LX_s - C_{pa}(t_d - t_w)$$

Notice that the left-hand side of this relationship appears in equation (14.41). Substituting for $X[L + C_{pv}(t_d - t_w)]$ in that equation gives

$$S = C_{pa} t_d + LX_s - C_{pa}(t_d - t_w)$$

or

$$S = LX_s + C_{pa} t_w \quad \text{J/(kg dry air)} \tag{14.42}$$

The t_d term cancels to produce a really neat equation for sigma heat. Now we are about to discover a phenomenon of major significance in psychrometric processes. Recalling that X_s is the saturation moisture content dependent only on wet bulb temperature and pressure (equations (14.4) and (14.10)), and that L is the latent heat of evaporation, also at wet bulb temperature, it follows that **sigma heat, *S*, is a function of wet bulb temperature only, for any given barometric pressure**. Furthermore, as sigma heat remains constant during an adiabatic saturation process so must also the wet bulb temperature remain constant. This is why the wet bulb temperature appears as a horizontal line in Fig. 14.2.

Using the concept of sigma heat, thermal additions or losses from a airstream can readily be quantified from psychrometric observations. Furthermore, the behaviour of the wet bulb temperature, a directly measurable parameter, is an immediate indication of heat transfer from the strata, machines, coolers, potential energy or any other source. Contrasting this with dry bulb temperature, which varies during evaporation or condensation, illustrates the fundamental importance of the wet bulb temperature in psychrometric processes.

14.6 SUMMARY OF PSYCHROMETRIC EQUATIONS

This is a convenient point to make a reference list of the more important equations that have been derived in the previous sections, and to re-order them in the sequence that they are required for most psychrometric calculations where P, t_w, t_d are the measured variables. All temperatures are expressed in degrees Centigrade and pressures in pascals.

$$e_{sw} = 610.6 \exp\left(\frac{17.27 t_w}{237.3 + t_w}\right) \quad \text{Pa} \qquad (14.43)$$

$$X_s = 0.622 \frac{e_{sw}}{P - e_{sw}} \quad \text{kg/(kg dry air)} \qquad (14.44)$$

$$L_w = (2502.5 - 2.386 t_w)1000 \quad \text{J/kg} \qquad (14.45)$$

$$S = L_w X_s + 1005 t_w \quad \text{J/(kg dry air)} \qquad (14.46)$$

$$X = \frac{S - 1005 t_d}{L_w + 1884(t_d - t_w)} \quad \text{kg/(kg dry air)} \qquad (14.47)$$

or

$$X = \frac{L_w X_s - 1005(t_d - t_w)}{L_w + 1884(t_d - t_w)} \quad \text{kg/(kg dry air)} \qquad (14.48)$$

$$e = \frac{PX}{(0.622 + X)} \quad \text{Pa} \qquad (14.49)$$

$$\rho_m(\text{apparent}) = \frac{P - e}{287.04(t_d + 273.15)} \quad \text{kg dry air/m}^3 \qquad (14.50)$$

$$\rho_m(\text{actual}) = \frac{P - 0.378e}{287.04(t_d + 273.15)} \quad \text{kg moist air/m}^3 \qquad (14.51)$$

$$H = S + (4187 \times t_w \times X) \quad \text{J/(kg dry air)} \qquad (14.52)$$

$$\text{rh} = \frac{e}{e_{sd}} \times 100\% \qquad (14.53)$$

(equation (14.44) is used for e_{sd} with t_w replaced by t_d). Note that in this list of important relationships, the need for the old troublesome psychrometric 'constant' (equation (14.35)) has been entirely eliminated.

Example At entry to a level continuous underground airway of constant cross-section, psychrometer readings give $P_1 = 110.130$ kPa, $t_{w1} = 23$ °C and $t_{d1} = 28$ °C. The corresponding readings at exit are $P_2 = 109.850$ kPa, $t_{w2} = 26$ °C and $t_{d2} = 32$ °C. If the volume flow of air at inlet is 25 m³/s, calculate the heat and moisture added to the airstream during its passage through the airway.

Solution

Equation	*Parameter*	*Inlet*	*Outlet*	*Units*
Measured	P	110 130	109 850	Pa
Measured	t_w	23	26	°C
Measured	t_d	28	32	°C
(14.43)	e_{sw}	2808.5	3360.3	Pa
(14.44)	X_s	0.016 277	0.019 628	kg/(kg dry air)
(14.45)	L_w	2447.6×10^3	2440.5×10^3	J/(kg water)
(14.46)	S	62.955×10^3	74.032×10^3	J/(kg dry air)
(14.47) or (14.48)	X	0.014 170	0.017 078	kg/(kg dry air)

At inlet, equation (14.49) gives the actual vapour pressure

$$e_1 = \frac{110\,130 \times 0.014\,17}{0.622 + 0.014\,17} = 2453.0\,\text{Pa}$$

and apparent density (equation (14.50))

$$\rho_m(\text{apparent}) = \frac{110\,130 - 2453}{287.04(28 + 273.15)}$$

$$= 1.2457\,\text{kg dry air/m}^3$$

Mass flow of dry air

$$M = Q\rho_m(\text{apparent})$$

$$= 25 \times 1.2457 \frac{\text{m}^3}{\text{s}} \frac{\text{kg dry air}}{\text{m}^3}$$

$$= 31.141\,\text{kg dry air/s}$$

Increase in sigma heat along airway

$$S_2 - S_1 = 74.032 - 62.955$$

$$= 11.077\,\text{kJ/(kg dry air)}$$

Heat added

$$q_{in} = M(S_2 - S_1)$$

$$= 31.141 \times 11.077 \frac{\text{kg dry air}}{\text{s}} \frac{\text{kJ}}{\text{kg dry air}}$$

$$= 344.9\,\text{kJ/s or kW}$$

Increase in moisture content along airway

$$X_2 - X_1 = 0.017\,078 - 0.014\,17$$
$$= 0.002\,908 \text{ kg/(kg dry air)}$$

$$\text{rate of evaporation} = M(X_2 - X_1)$$
$$= 31.141 \times 0.002\,908 = 0.0906 \text{ kg/s or l/s}$$

14.7 DEVIATIONS FROM CLASSICAL THEORY

14.7.1 Fogged air

The classical theory of condensation makes the premise that condensation cannot commence until saturation conditions are reached. This is not quite in agreement with observable phenomena. The formation of fog is likely to commence before 100% relative humidity is reached although the process of condensation accelerates rapidly as saturation is approached. Smoke fogs (smogs) can occur over cities at relative humidities of less than 90%.

The physical mechanism of condensation is complex. Condensation can occur only in the presence of hygroscopic and microscopic nuclei on which the process can commence. It has been estimated (Brunt) that there are between 2000 and 50 000 hygroscopic nuclei in each cubic centimetre of the atmosphere, having radii of 10^{-6} to 10^{-5} cm. The most productive sources of these hygroscopic particles are the oceans and, in localized areas, sulphurous smokes from burning hydrocarbon fuels.

The natural formation of fog would appear to be a continuous process of condensation commencing at comparatively low relative humidities when the more strongly hygroscopic nuclei begin to attract water. This can often result in a noticeable haze in the atmosphere. As the relative humidity approaches 100%, the rate of condensation on all hygroscopic nuclei increases; the droplets of water so formed increase rapidly in size and the haze develops into fog.

Fogging in subsurface ventilation systems occurs in two situations. First, when the strata are cooler than the dew point temperature of the incoming air and, secondly, as a result of decompressive cooling of humid return air. Hence, fogging in ascending return airways and, especially, upcast shafts is not uncommon.

The formation of fogs in underground structures is undesirable for a number of reasons. The reduction in visibility may produce a safety hazard, particularly where moving vehicles are involved. High humidity can result in physical and chemical reactions between the airborne water and hygroscopic minerals within the strata, and may produce falls of roof and spalling of the sides of airways. Such problems can reach serious proportions in certain shales and evaporite mines. Another difficulty that may arise in salt mines is absorbance of water vapour by the mined rock. If the air is humid then the material may become 'sticky' and difficult to handle. The problem may be alleviated by transporting the ore in the return airways.

One further problem of fogging in upcast shafts is that it creates very wet conditions throughout the shaft, headgear and/or exhaust fans. If the air velocity within the shaft lies within the range 7 to 12 m/s then droplets will remain in suspension, encouraging water blankets to form. The resulting increase in shaft resistance causes a fluctuating load to be imposed on main fans, even to the extent of stalling them. The sudden reduction in airflow results in the water cascading to the bottom of the shaft and the whole process will be repeated in a cyclic manner. In extreme cases this may lead to failure of the fan blades.

If it becomes necessary to dehumidify the air because of any of these problems then there are two possibilities. The less expensive option, where it can be used, is to divert the air through old stopes or workings that contain water-absorbant minerals in the broken strata. Alternatively, the air temperature may be reduced to below dew point in air coolers and the condensed water removed from the system. All, or part, of the reject heat produced by the refrigeration plant may be returned to the downstream airflow to control its temperature. Chemical dehumidification is seldom practicable in mining circumstances because of the large volumes of air involved.

14.7.2 Imperfect gas behaviour

The equations given for saturation vapour pressure (equations (14.10) and (14.11)) took deviation from perfect gas behaviour into account. However, the other relationships derived in this chapter assume that dry air, water vapour, and air–vapour mixtures all obey the perfect gas laws. These assume that the molecules of the air are perfectly elastic spheres of zero volume and that they exhibit no attractive or repulsive forces.

For the ranges of pressures and temperatures encountered in atmospheric psychrometrics, assumption of the perfect gas laws gives results of acceptable accuracy. Where additional precision is required, Hemp (1982) has suggested the following two corrections based on the earlier work of Goff and Gratch.

Equation (14.3) for moisture content

$$X = \frac{R_a}{R_v}\frac{e}{P-e}$$

becomes

$$X = \frac{R_a}{R_v}\frac{1.0048e}{P-1.0048e} \quad \text{kg/(kg dry air)} \tag{14.54}$$

and gas constants, R, are corrected for pressure and temperature:

$$R(\text{corrected}) = R\{1 - [(5.307 \times 10^{-9}P + 9.49 \times 10^{-6}) - (8.115 \times 10^{-11}P + 2.794 \times 10^{-6})t_d]\} \tag{14.55}$$

where P is in Pa and t_d in °C. This gives a maximum change in R of 0.06% over the ranges 80 to 120 kPa and 0 to 60 °C. At the mid-range value of 100 kPa and 30 °C, the difference is 0.02%.

14.8 PSYCHROMETRIC CHARTS

In Chapters 3 and 8, we saw how useful it was to plot the behaviour of one thermodynamic property against another during any given process. A powerful equivalent in the study of moisture in air is the psychrometric chart. This is a graphical representation of the majority of the psychrometric parameters for a fixed barometric pressure. Figure 14.3 gives an example of a 100 kPa chart. This is one of a series of such charts ranging from 80 to 130 kPa at intervals of 2.5 kPa. As pressure is the least sensitive of the independent variables, interpolation between charts is unnecessary for most practicable accuracies. More sophisticated types of charts have been produced that allow for pressure variations. However, these seem not to have found widespread use.

Returning to Fig. 14.3, it is clear that a knowledge of any two of the variables will fix the location on the chart, describing the climatic condition of the air, and enabling other psychrometric variables to be read directly or determined with a minimum of calculation. The usual situation is one in which the wet and dry bulb temperatures are known. The programs that are available for psychrometric calculations on personal computers and programmable calculators have reduced the need for psychrometric charts somewhat besides giving more precise results. However, those charts provide a most powerful visual representation of process lines that represent changes in the psychrometric condition of airstream. Engineers concerned with climatic variations should be adept at using psychrometric charts.

Figure 14.4 illustrates the ease with which processes can be followed. Movement along a constant wet bulb line (i.e. constant sigma heat) indicates adiabatic conditions. Any movement to a higher or lower wet bulb temperature indicates that the air has been heated or cooled respectively. A path to a higher moisture content (labelled as apparent specific humidity on Fig. 14.3) indicates evaporation while movement along a constant moisture line denotes sensible heating or cooling. Similarly, pure latent heating or cooling is indicated by a constant dry bulb temperature.

Example Air passes through a cooler at a mean pressure of 100 kPa. It enters at wet and dry bulb temperatures of 30 and 40 °C respectively, and exits at 25 °C saturated. If the volume flow at inlet, Q, is 20 m^3/s, use the psychrometric chart to determine the rates of heat exchange and production of condensate.

Solution On the 100 kPa psychrometric chart, plot the two points (30°C, 40 °C) and (25 °C). The latter lies on the saturation line (100% relative humidity). It is immediately obvious that both the moisture content (specific humidity) and sigma heat have decreased. The following values are read from the chart. The corresponding values calculated from the equations in section 14.6 are given in parentheses for comparison.

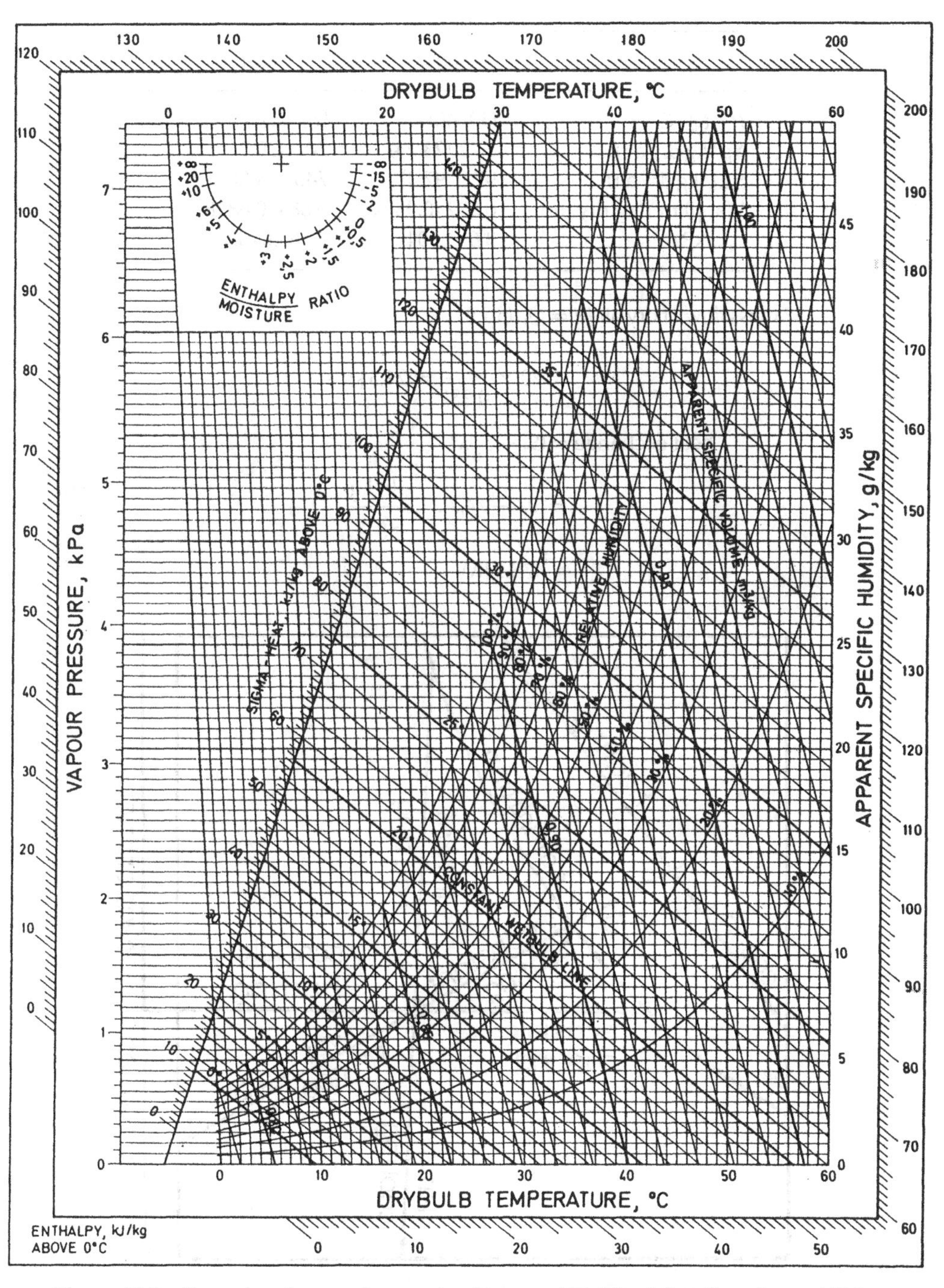

Figure 14.3 Example of a psychrometric chart, at 100 kPa (after Barenbrug, 1974). (*Reproduced by courtesy of the Chamber of Mines of South Africa.*)

	Inlet	*Outlet*
(t_w, t_d) °C	(30, 40)	(25, 25)
Xg/(kg dry air)	23.2 (23.24)	20.3 (20.34)
S kJ/(kg dry air)	97.0 (97.1)	74.7 (74.8)
V(app) m^3/(kg dry air)	0.932 (0.9324)	0.884 (0.8838)
e_{act} kPa	3.6 (3.601)	3.16 (3.167)

mass flow of dry air $M = Q\rho(\text{app})$

$$= 20 \times \frac{1}{0.932} = 21.5 \text{ kg/s}$$

$$\text{heat exchange} = (S_1 - S_2)M$$

$$= (97 - 74.7)21.5 \frac{\text{kJ}}{\text{kg}} \frac{\text{kg}}{\text{s}}$$

$$= 479.5 \text{ kW}$$

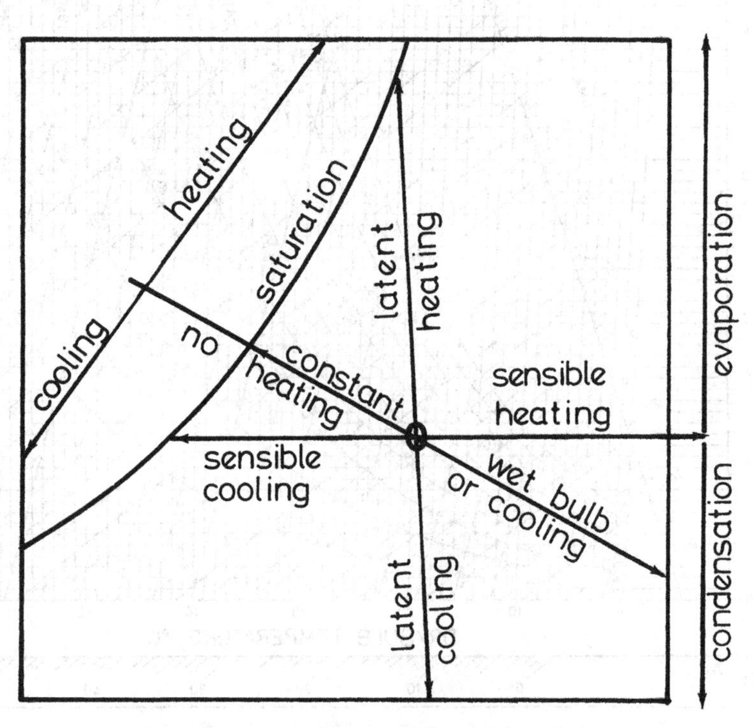

Figure 14.4 Process lines on a psychrometric chart.

rate of condensation $= (X_1 - X_2)M$

$= (0.0232 - 0.0203)21.5$

$= 0.0623$ kg water/s or $0.0623 \times 60 = 3.741$ l/min.

A14 APPENDIX: DERIVATION OF THE CLAUSIUS–CLAPEYRON EQUATION

When the enthalpy, or overall energy, of a system is increased by doing mechanical work on it, then this work represents available energy all of which can theoretically be utilized. On the other hand, if the enthalpy increase is brought about by the addition of heat then the second law of thermodynamics insists that only a part of this heat is available energy. The unavailable fraction is utilized in increasing the internal energy of the system. Thus if we have 1 kg of liquid water of volume V, internal energy U and pressure e_s then if we add an increment of heat energy the corresponding increase in enthalpy will be

$$dH = dU + e_s\,dV \tag{14A.1}$$

Now a change of phase (evaporation from liquid to vapour) occurs at constant temperature and constant pressure. During such a process the enthalpy change is also related to the increase in entropy of the system

$$dH = T\,ds \tag{14A.2}$$

where s = entropy. Hence equations (14A.1) and (14A.2) give

$$dH = dU + e_s\,dV = T\,ds \tag{14A.3}$$

If the 1 kg of water is completely evaporated, then the total heat added to accomplish this is the latent heat of evaporation, L.

Equation (14A.3) can be integrated over the change of phase from water (subscript L) to vapour (subscript s):

$$L = H_s - H_L = U_s - U_L + e_s(V_s - V_L) = T(s_s - s_L) \tag{14A.4}$$

or

$$U_s + e_sV_s - Ts_s = U_L + e_sV_L - Ts_L \tag{14A.5}$$

This relationship defines a function of state, Φ, known as thermodynamic potential which remains constant during any isobaric–isothermal change of phase:

$$\Phi = U + e_sV - Ts \tag{14A.6}$$

Now consider the situation at a slightly different temperature $T + dT$ and pressure $e_s + de_s$. The thermodynamic potential that will apply under these conditions will be $\Phi + d\Phi$. Differentiating equation (14A.6) to find $d\Phi$ gives

$$d\Phi = dU + e_s\,dV + V\,de_s - T\,ds - s\,dT \tag{14A.7}$$

but from equation (14A.3)

$$dU + e_s \, dV - T \, ds = 0$$

Hence

$$d\Phi = V \, de_s - s \, dT \tag{14A.8}$$

As Φ is constant during the evaporation at conditions T and e_s, and $\Phi + d\Phi$ remains constant during evaporation at conditions $T + dT$ and $e_s + de_s$, it must follow that $d\Phi$ remains constant as T approaches $T + dT$ and e_s approaches $e_s + de_s$. Therefore from equation (14A.8)

$$V_L \, de_s - s_L \, dT = V_s \, de_s - s_s \, dT \tag{14A.9}$$

or

$$\frac{de_s}{dT} = \frac{s_L - s_s}{V_L - V_s} \tag{14A.10}$$

This equation was first derived by Clapeyron in 1832.

The volume of liquid water, V_L, is negligible compared with that of the vapour, V_s (at normal atmospheric conditions V_s/V_L is of the order of 2500). Furthermore, from equation (14A.4)

$$-(s_L - s_s) = \frac{L}{T}$$

and equation (14A.10) becomes

$$\frac{de_s}{dT} = \frac{L}{T} \frac{1}{V_s} \tag{14A.11}$$

but from the general gas law for 1 kg of water vapour

$$V_s = \frac{R_v T}{e_s}$$

Hence

$$\frac{1}{e_s} \frac{de_s}{dT} = \frac{L}{R_v T^2}$$

REFERENCES

Barenbrug, A. W. T. (1974) *Psychrometry and Psychrometric Charts*, 3rd edition, Chamber of Mines of South Africa.

Brunt, D. (*c.* 1980) *Physical and Dynamical Meteorology*, Cambridge Press.

Hemp, R. (1982) Psychrometry. *Environmental Engineering in South African Mines*, Chapter 18, 435–63, Mine Ventilation Society of South Africa.

FURTHER READING

Goff, J. A. and Gratch, S. (1945) Thermodynamic properties of moist air. *ASHVE Trans.* **51**, 125–58.

Van der Walt, N. T. and Hemp, R. (1982) Thermometry and temperature measurements. *Environmental Engineering in South African Mines*, Chapter 17, 413–33, Mine Ventilation Society of South Africa.

Whillier, A. (1971) Psychrometric charts for all barometric pressures. *J. Mine Vent. Soc. S. Afr.* **24**, 138–43.

15

Heat flow into subsurface openings

15.1 INTRODUCTION

Heat is emitted into subsurface ventilation systems from a variety of sources. In the majority of the world's coal mines, the airstream itself is sufficient to remove the heat that is produced. In deep metal mines, heat is usually the dominant environmental problem and may necessitate the use of large-scale refrigeration plant. Conversely, in cold climates, the intake air may require artificial heating in order to create conditions that are tolerable for both personnel and equipment.

In section 9.3.4, quantification of the heat emitted into a mine or section of a mine was required in order to assess the airflow needed to remove that heat. Hence, a sensible place to commence the study of heat flow into mine openings is to classify, analyse and attempt to quantify the various sources of heat. The three major heat sources in mines are the conversion of potential energy to thermal energy as air falls through downcasting shafts or slopes (autocompression), machinery, and geothermal heat from the strata. The latter is, by far, the most complex to analyse in a quantitative manner. We shall deal with this separately in section 15.2, then quantify other sources of heat in section 15.3.

15.2 STRATA HEAT

15.2.1 Methods of determining strata heat load

In reviewing the literature for means of determining the amount of heat that will be emitted from the strata, the ventilation engineer is faced with a bewildering array of methods varying from the completely empirical, through analytical and numerical, to computer simulation techniques. The basic difficulty is the large number of variables, often interacting with each other, that control the flow of strata heat into mine airways. These include

1. the length and geometry of the opening,

2. depth below surface and inclination of the airway,
3. wetness of the surfaces,
4. roughness of the surfaces,
5. rate of mineral production or rock breaking,
6. time elapsed since the airway was driven,
7. volume flow of air,
8. barometric pressure, and wet and dry bulb temperatures at the inlet,
9. virgin (natural) rock temperature,
10. distance of the workings from downcast shafts or slopes,
11. geothermic step or geothermic gradient,
12. thermal properties of the rock, and
13. other sources of heating or cooling such as machines and cooling plant.

With such a variety of parameters, it is hardly surprising that traditional methods of predicting strata heat loads have been empirical. Perhaps the simplest and most common of these has been to quote strata heat flux in terms of heat load per unit rate of mineral production; for example, kW per tonne per day. As the rate of production is only one of the several variables listed above, it is obvious that this technique may lead to gross errors if it is applied where the value of any one of those variables is significantly different from the original sets of measurements used used to establish the (kW/tonne)/day value.

The more sophisticated empirical techniques extend their range of application by incorporating estimated corrections for depth, distance, age, inlet conditions or, indeed, any of the listed variables considered to be of local importance.

The purely analytical methods of quantifying heat flow from the strata are somewhat limited for direct practical application because of the complexity of the equations that describe three-dimensional, time-transient heat conduction. Indeed, they can be downright frightening. However, the theory that has evolved from analytical investigations has provided the basis for numerical modelling which, in turn, has resulted in the development of pragmatic computer simulation packages for the detailed prediction of variations in the mine climate.

A hybrid method has grown out of experience in running climatic simulation packages. It is often the case that, for particular conditions, some of the input variables have a very limited effect on the results. By ignoring those weaker parameters it is then sometimes possible to develop simple equations that give an approximation of the heat flow.

In view of these alternative methodologies, what is the mine environmental engineer to do when faced with the practical problems of system design? Experience gained from major planning projects has indicated the following recommended guidelines.

1. If the objective is to plan the further development of an existing mine, or if there are neighbouring mines working similar deposits at equivalent depths and with the same methods of working, then the **empirical approach** should be adopted for the overall strata heat load on the whole mine or major sections of

the mine. This presupposes the existence of data that allow acceptable empirical relationships to be established and verified.

Employing past and relevant experience in this way provides a valuable and fairly simple means of arriving at an approximate heat load which, when combined with the methodology of section 9.3.4, will give an indication whether the heat can be removed by the airflow alone, or if refrigeration is required.

However, let the user beware. If the proposed mine project deviates in any siginificant manner (check the list of variables at the beginning of this section) from the conditions in which the empirical data were compiled, then the results may be misleading. In particular, great caution should be exercised in employing empirical relationships established in other geographical regions. A phrase commonly heard at mine ventilation symposia is 'what works there, doesn't work here'.

2. The **hybrid** equations are very useful for rapid approximations of heat flow into specified types of openings. Dr. Austin Whillier of the Chamber of Mines of South Africa produced many hybrid equations for easy manual application, including the following.

(a) *Radial heat flow into established tunnels.*

$$q = 3.35\, Lk^{0.854}\,(\mathrm{VRT} - \theta_d) \quad \mathrm{W} \tag{15.1}$$

where q = heat flux from strata (W), L = length of tunnel (m), k = thermal conductivity of rock (W/m °C), VRT = virgin (natural) rock temperature (°C) and, θ_d = mean dry bulb temperature (°C). Throughout this section we shall use the symbol k for thermal conductivity and θ for temperature.

(b) *Advancing end of a heading.*

$$q = 6\,k\,(L + 4\,\mathrm{DFA})\,(\mathrm{VRT} - \theta_d) \quad \mathrm{W} \tag{15.2}$$

where L = length of the advancing end of the heading (m) (this should be not greater than the length advanced in the last month; equation (15.1) may be used for the older sections of the heading, and DFA = daily face advance (m).

(c) *One-dimensional heat flow towards planar surfaces.* Assuming good convective or evaporative cooling of the surfaces,

$$q = A\,(k\rho C)^{0.5}\,(\mathrm{VRT} - \theta_d)/t^{0.5} \quad \mathrm{W} \tag{15.3}$$

where A = area of surface (m), ρ = rock density (kg/m^3), C = specific heat of rock (J/(kg °C)) and t = time since the surface was exposed (s).

A weakness of these equations is that they each contain the mean dry bulb temperature, θ_d. As this is initially unknown, it must be estimated. If, when the resulting value of q is used to determine the temperature rise, the initial estimate of θ_d is found to have been significantly in error, then the process may be repeated. However, as the hybrid equations promise nothing more than rough approximations there is little point in progressing beyond a single iteration.

3. For accurate and detailed planning, a mine climate simulation package should be employed. These are computer programs that have been developed to take all of the relevant variables into account. They may be used both for single airways or combined into a total underground layout. Climate simulation programs go beyond the calculation of heat load by predicting the effects of that heat on the psychrometric conditions in the mine. The principles of a climate simulation program are given in Chapter 16.

   For major projects, estimates of heat loads may be based on empirical and hybrid methodologies for initial conceptual planning, progressing to simulation techniques for detailed analysis.

15.2.2 Qualitative observations

Before embarking upon a quantitative analysis of strata heat flow, it will be useful to introduce some of the observable phenomena from a purely qualitative viewpoint.

First, when cool air passes through a level airway, its temperature usually increases. This is caused by natural geothermal heat being conducted through the rock towards the airway, then passing through the boundary layers that exist in the air close to the rock surface. In working areas, the newly exposed rock surfaces are often perceptibly warmer than the air. However, those surfaces will cool with time until they may be only a fraction of a degree Centigrade higher than the temperature of the air.

If the airway is wet then the increase in dry bulb temperature is less noticeable. Indeed, that temperature may even fall. This is a result of the cooling effect of evaporation. Heat may still emanate from the strata but all, or much, of it is utilized in exciting water molecules to the extent that they leave the liquid phase and form water vapour. The heat content of the air–vapour mixture then rises because of the internal energy of the added water vapour.

Another observation that can be made in practice is that although the air temperature in main intake arteries rises and falls in sympathy with the surface climate, the temperatures in main returns remain remarkably constant throughout the year. This is because cool air will encourage heat to flow from the rock. However, as the temperature of the air approaches the natural temperature of the rock such heat transfer will diminish. It can, of course, work in reverse. For example, the temperature of the air leaving an intensively mechanized working area may be greater than the local strata temperature. In that case, heat will pass from the air to the rock. The air will cool and, again, approach equilibrium when its temperature equals that of the strata. A mine is an excellent thermostat.

The envelope of rock immediately surrounding a newly driven airway will cool fairly rapidly at first. There will, accordingly, be a relatively high rate of initial heat release into the air. This will decline with time. A well-established return airway may have reached near thermal equilibirum with the surrounding strata. However, the linings and envelope of rock around downcast shafts or main intakes will emit heat during the night when the incoming air is cool and, conversely, absorb heat during the day if the air temperature becomes greater than that of the surrounding envelope of rock. This cyclic phenomenon, sometimes known as the 'thermal flywheel' (Stroh, 1979) is superimposed upon longer-term cooling of the larger mass of rock around the opening.

The boundary layers that exist within the airflow close to the rock surface act as insulating layers and, hence, tend to inhibit heat transfer between the rock and the main airstream. It follows that any thinning or disturbance of those boundary layers will increase the rate at which heat transfer takes place. This can occur either through a rise in air velocity or because of a greater degree of roughness on the surface (section 2.3.6).

Finally, although there may be significant rises in the air temperature along intake airways, the most noticeable increase usually occurs in the mineral winning areas. This is because, first, the newly exposed and warm surfaces of both the solid and broken rock give up their heat readily and, secondly, because of the mechanized equipment that may be concentrated in stopes or working faces.

Having introduced these concepts in a purely subjective manner, we now have a better intuitive understanding from which to progress into a quantified and analytical approach.

15.2.3 Fourier's law of heat conduction

When a steady heat flux, q, passes through a slab of homogeneous material, the temperature will fall from θ_1 at entry to θ_2 at exit (Fig. 15.1). Planes of constant temperature, or isotherms, will exist within the material. Figure 15.1 shows two isotherms at a short distance, dx, apart and with temperatures θ and $\theta + d\theta$.

The heat flux is proportional to both the orthogonal area, A, through which the heat travels and the temperature difference, $d\theta$, between isotherms. It is also inversely proportional to the distance, dx, between those isotherms. Hence,

$$q \propto -A\frac{d\theta}{dx}$$

where q = heat flux (W), A = area through which q passes (m^2), θ = temperature (°C) and x = distance (m). The negative sign is necessary since θ reduces in the direction of heat flow. To convert this relationship into an equation, a constant of proportionality, k, is introduced, giving

$$q = -kA\frac{d\theta}{dx} \quad \text{W} \quad \text{(Fourier's law)} \tag{15.4}$$

k is termed the thermal conductivity of the material and has units of W/(m °C).

To be precise, k is a slowly changing function of temperature. It may also vary with the mechanical stress applied to the material. In the strata around mine openings, the **effective** thermal conductivity of the strata can be significantly different from those given by samples of the rock when measured in a laboratory test (section 15.2.10).

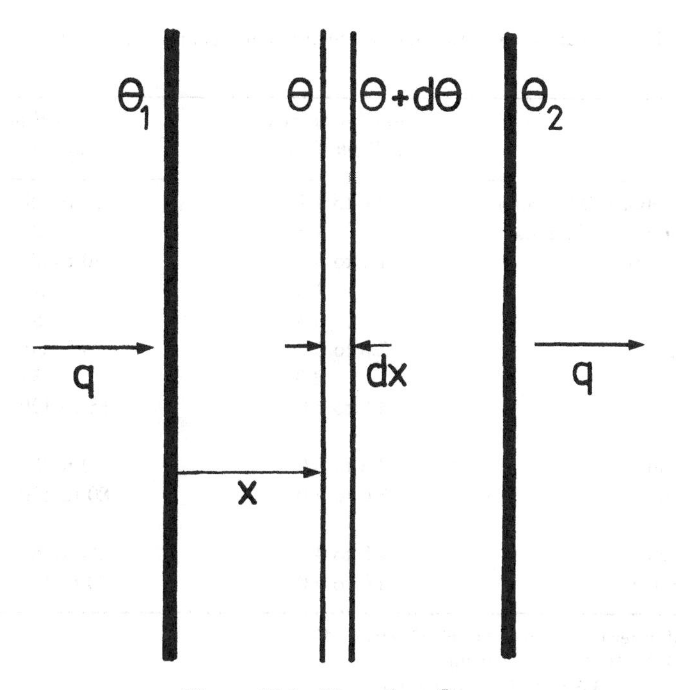

Figure 15.1 Linear heat flow.

The reasons for such differences include natural or induced fractures in the strata, variations in mineralogy that may be direction dependent, movements of groundwater, radioactive decay and local geothermal anomalies.

15.2.4 Geothermic gradient, geothermal step and thermal conductivity

The crustal plates upon which the continents drift over geological time are relatively thin compared with the diameter of the earth. Furthermore, it is only in the upper skin of those plates that mining takes place at the present time. The geothermal flow of heat emanating from the earth's core and passing through that skin has an average value of 0.05 to 0.06 W/m^2. It can, of course, be much higher in regions of anomalous geothermal activity.

In Fourier's law, equation (15.4), if we use the value of 0.06 W for each square metre of land surface to give the variation of temperature, θ, with respect to depth $D(=-x)$, then

$$\frac{d\theta}{dD} = \frac{0.06}{k} \quad ^\circ\text{C/m} \qquad (15.5)$$

The increase of strata temperature with respect to depth is known as the **geothermic gradient**. In practical utilization, it is often inverted to give integer values and is

Table 15.1 Typical values of thermal conductivity and geothermal step for a range of rock types

Rock type	*Thermal conductivity (W/(m °C))*	*Geothermal step (m/°C)*
Copper orebody (Montana)	0.8 to 1.1	13 to 18
Copper orebody (Arizona)	1.3	22
Carboniferous	1.2 to 3.0	20 to 50
Clays	1.8	30
Limestone	3.3	55
Sandstone	2.0 to 3.6	30 to 60
Dolerite	2.0	33
Quartzite	4.0 to 7.0	65 to 120
Potash		
Low grade	3.5 to 5.0	60 to 80
High grade	5.0 to 7.0	80 to 120
Halite		
Low grade	1.5 to 4.0	25 to 70
High grade	4.0 to 6.0	70 to 100

For air and water in the range 0 to 60 °C (Hemp, 1985)

$k_a = 2.2348 \times 10^{-4}\ T^{0.8353}$ W/(m °C)

$k_w = 0.2083 + 1.335 \times 10^{-3}\ T$ W/(m °C)

where T = absolute temperature (K).

then referred to as the

$$\text{geothermal step} = \mathrm{d}D/\mathrm{d}\theta \quad \text{m/°C} \tag{15.6}$$

It is clear from equations (15.5) and (15.6) that as we progress downwards through a succession of strata, the geothermal step will vary according to the thermal conductivity of the local material.

Table 15.1 has been assembled from several sources as a guide to the thermal conductivities and corresponding geothermal steps that may be expected for a range of rock types. However, it should be remembered that these parameters are subject to significant local variations. Site specific values should be obtained, preferably from *in situ* tests, for any important planning work.

15.2.5 An analysis of three-dimensional radial heat conduction

During the practical application of empirical, hybrid or simulation techniques of assessing strata heat loads (section 15.2.1) the mine environmental engineer need give little conscious thought to the theory of heat conduction or the derivations given in this chapter. These analyses are included here because they provide the essential background to the development of numerical models and, hence, mine climate simulation packages.

Within the envelope of rock that surrounds an underground airway, the temperature

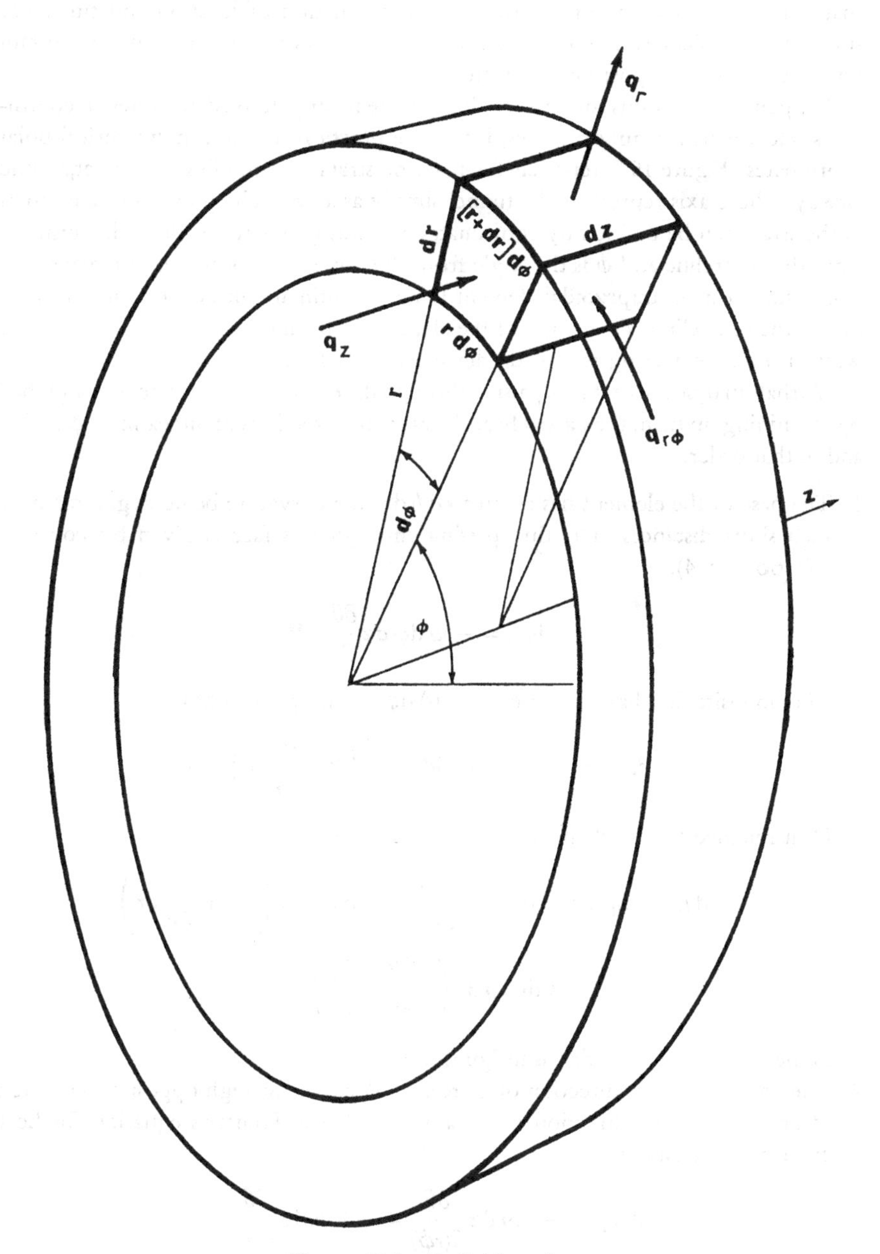

Figure 15.2 Radial heat flow.

varies with both position and time. Our first task is to derive a general relationship that defines the temperature of the rock as a function of location and time. For simplicity, we shall assume that the thermal properties of the rock remain constant with respect to both position and time.

The general equations for heat conduction are readily derived in Cartesian coordinates. However, for mine airways, it is more practical to work in cylindrical polar coordinates. Figure 15.2 represents a mass of strata surrounding an underground airway. The z axis represents the tunnel simply as a line. The position of any point in the strata can be defined by coordinates z, r and ϕ where r is the radial distance from the centreline and ϕ is the angle from the horizontal measured in radians.

Consider a small trapezoidal element lying in a thin annulus of rock at a distance r from the centreline. The length of the element is dz, and its height is dr. The inner width is $r\,d\phi$, increasing to $(r + dr)\,d\phi$ at its outer limit.

We shall analyse the heat flux passing through the element. This can be accomplished by examining, in turn, the partial heat flows in the directions of increasing r, ϕ and z, and in that order.

1. The base of the element has an area $r\,d\phi\,dz$, the curvature being negligible over such short distances. The flux passing through this face is given by Fourier's equation (15.4):

$$dq_{1,r} = -kr\,d\phi\,dz\frac{\partial\theta}{\partial r} \quad \text{W}$$

The opposite face has an area of $(r + dr)\,d\phi\,dz$ and passes a heat flow of

$$dq_{2,r} = -k(r + dr)\,d\phi\,dz\frac{\partial}{\partial r}\left(\theta + \frac{\partial\theta}{\partial r}dr\right) \quad \text{W}$$

Heat retained by the element in the r direction is

$$dq_{1,r} - dq_{2,r} = kr\,d\phi\,dz\frac{\partial^2\theta\,dr}{\partial r^2} + k\,dr\,d\phi\,dz\left(\frac{\partial\theta}{\partial r} + \frac{\partial^2\theta}{\partial r^2}dr\right)$$

$$= k\,dr\,d\phi\,dz\left(\frac{r\partial^2\theta}{\partial r^2} + \frac{\partial\theta}{\partial r}\right) \qquad (15.7)$$

as the term involving $(dr)^2$ is insignificant.

2. The heat flux in the direction of increasing ϕ passes through opposite faces, each of area $dr\,dz$. The direction is actually $r\phi$. Hence, Fourier's equation for heat entering the element is

$$dq_{1,\phi} = -k\,dr\,dz\frac{\partial\theta}{\partial(r\phi)} = -k\,dr\,dz\frac{\partial\theta}{r\partial\phi} \quad \text{W}$$

while the heat leaving the element is

$$dq_{2,\phi} = -k\,dr\,dz\frac{\partial}{r\partial\phi}\left(\theta + \frac{\partial\theta}{r\partial\phi}r\,d\phi\right) = -k\,dr\,dz\left(\frac{\partial\theta}{r\partial\phi} + \frac{\partial^2\theta}{r\partial\phi^2}d\phi\right) \quad \text{W}$$

Heat gain in the $r\phi$ direction is

$$dq_{1,\phi} - dq_{2,\phi} = k\,dr\,dz\,d\phi \frac{\partial^2\theta}{r\partial\phi^2} \quad \text{W} \tag{15.8}$$

3. Repeating the exercise for the z direction, the opposite faces are trapeziums of height dr and opposite edges of length $r\,d\phi$ and $(r + dr)\,d\phi$, the curvature of the lines being insignificant over such small distances. The latter have an average length of

$$\frac{r\,d\phi + (r + dr)\,d\phi}{2} \quad \text{m}$$

The product of the differentials $dr\,d\phi$ is negligible compared with $r\,d\phi$, giving the width to be $r\,d\phi$ and face area of $r\,d\phi\,dr$.

The heat entering the element in the z direction is

$$dq_{1,z} = -kr\,d\phi\,dr \frac{\partial\theta}{\partial z} \quad \text{W}$$

and leaving the opposite face,

$$dq_{2,z} = -kr\,d\phi\,dr \frac{\partial}{\partial z}\left(\theta + \frac{\partial\theta}{\partial z}dz\right) \quad \text{W}$$

Then rate of heat accumulation in the z direction is

$$dq_{1,z} - dq_{2,z} = kr\,d\phi\,dr\,dz \frac{\partial^2\theta}{\partial z^2} \quad \text{W} \tag{15.9}$$

The total rate of heat gain by the element, dq, is given by summing equations (15.7), (15.8) and (15.9):

$$dq = k\,dr\,d\phi\,dz\left(r\frac{\partial^2\theta}{\partial r^2} + \frac{\partial\theta}{\partial r} + \frac{1}{r}\frac{\partial^2\theta}{r\partial\phi^2} + r\frac{\partial^2\theta}{\partial z^2}\right) \quad \text{W} \tag{15.10}$$

The heat gain by the element in time ∂t can also be expressed as

$$dq = mC\frac{\partial\theta}{\partial t} \quad \text{W}$$

where m = mass of element and C = specific heat of the material. However, m = volume × density (ρ) = $dr\,dz\,r\,d\phi\,\rho$. Then

$$dq = dr\,dz\,r\,d\phi\,\rho C\frac{\partial\theta}{\partial t} \quad \text{W} \tag{15.11}$$

Hence, from equations (15.10) and (15.11),

$$k\left(\frac{r\partial^2\theta}{\partial r^2} + \frac{\partial\theta}{\partial r} + \frac{1}{r}\frac{\partial^2\theta}{\partial\phi^2} + r\frac{\partial^2\theta}{\partial z^2}\right) = r\rho C\frac{\partial\theta}{\partial t} \quad \frac{\text{W}}{\text{m}^2} \tag{15.12}$$

or

$$\frac{k}{\rho C}\left(\frac{\partial^2\theta}{\partial r^2}+\frac{1}{r}\frac{\partial\theta}{\partial r}+\frac{1}{r^2}\frac{\partial^2\theta}{\partial\phi^2}+\frac{\partial^2\theta}{\partial z^2}\right)=\frac{\partial\theta}{\partial t}\quad ^\circ\text{C/s}$$

This fundamental relationship is the general three-dimensional equation for unsteady heat conduction expressed in cylindrical polar coordinates.

For most purposes in strata heat conduction towards airways, we can assume that

$$\frac{\partial\theta}{\partial z}=\frac{\partial^2\theta}{\partial z^2}=0$$

and that

$$\frac{\partial\theta}{\partial\phi}=\frac{\partial^2\theta}{\partial\phi^2}=0$$

These simplifications are based on the premise that the natural geothermic gradient is small compared with the radial variation in temperature around the incremental length of airway, and that the heat conduction is radial along the full length. Then

$$\frac{k}{\rho C}\left(\frac{\partial^2\theta}{\partial r^2}+\frac{1}{r}\frac{\partial\theta}{\partial r}\right)=\frac{\partial\theta}{\partial t}\quad ^\circ\text{C/s}$$

The term $k/\rho C$ is a constant for the material and is called the thermal diffusivity, α (m²/s), giving

$$\alpha\left(\frac{\partial^2\theta}{\partial r^2}+\frac{1}{r}\frac{\partial\theta}{\partial r}\right)=\frac{\partial\theta}{\partial t}\quad ^\circ\text{C/s}\qquad(15.13)$$

This is the form of the equation normally quoted for radial heat conduction and is the basis on which strata heat flow is determined in mine climate simulation programs. Figure 15.3 gives a graphical depiction of the time-transient heat conduction equation.

15.2.6 Solution of the radial heat conduction equation

Having derived equation (15.13) for the time–space variation in temperature around an underground airway, we must now attempt to transform it into a practical procedure to determine strata heat emission into the airway.

In order to facilitate further analysis and generality, it is convenient to express radial distances and time as dimensionless numbers. Dimensionless radius

$$r_d=\frac{r}{r_a}$$

where r_a = effective radius of the airway (= perimeter/2π). The actual shape of the cross-section has little effect on the influx of strata heat into airways. Dimensionless time

$$F_0=\frac{\alpha t}{r_a^2}$$

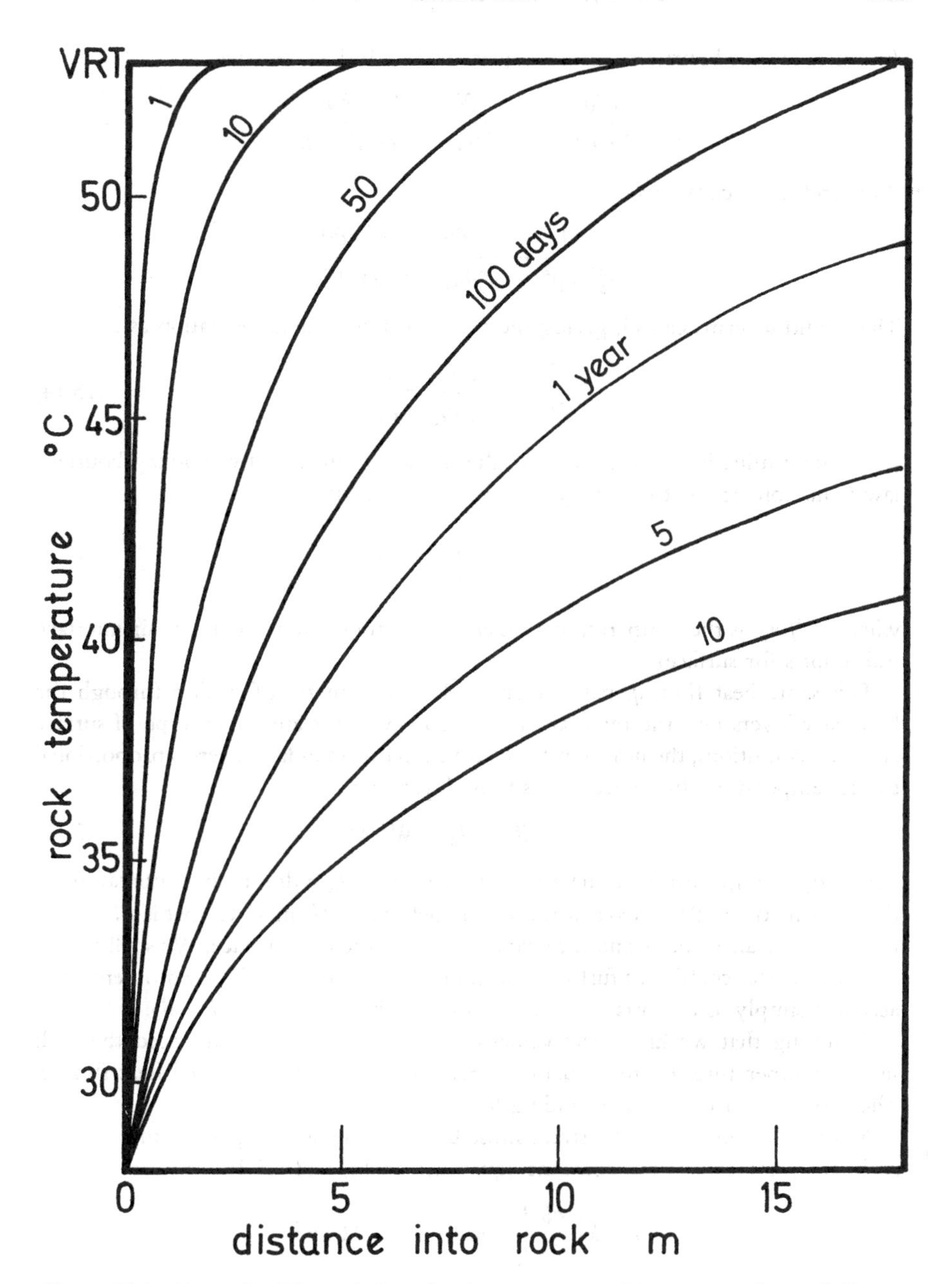

Figure 15.3 Example of the variation of rock temperature with respect to time and distance into the rock.

(this is called the Fourier number). Equation (15.13) then becomes

$$\frac{\partial^2\theta}{\partial(r_d r_a)^2}+\frac{1}{r_d r_a}\frac{\partial\theta}{\partial(r_d r_a)}=\frac{1}{\alpha}\frac{\partial\theta}{\partial(F_0 r_a^2/\alpha)}$$

As r_a and α are constants,

$$\frac{\partial^2\theta}{r_a^2\partial(r_d)^2}+\frac{1}{r_d r_a^2}\frac{\partial\theta}{\partial r_d}=\frac{1}{\alpha r_a^2/\alpha}\frac{\partial\theta}{\partial F_0}$$

The r_a^2 and α terms cancel, giving the radial heat conduction equation as

$$\frac{\partial^2\theta}{\partial(r_d)^2}+\frac{1}{r_d}\frac{\partial\theta}{\partial r_d}=\frac{\partial\theta}{\partial F_0} \qquad (15.14)$$

The determination of heat flux into the airway commences by applying Fourier's law (equation (15.4)) to each square metre of rock surface

$$q=k\left[\frac{\partial\theta}{\partial r}\right]_s \quad \text{W/m}^2 \qquad (15.15)$$

where $(\partial\theta/\partial r)_s$ is the temperature gradient in the rock but at the rock–air interface (subscript s for surface).

This same heat flux, q, passes from each square metre of surface through the boundary layers into the main airstream. However, for any given type of surface and flow conditions, the heat transferred through the boundary layers is proportional to the temperature difference across those layers, i.e.

$$q=h(\theta_s-\theta_d) \quad \text{W/m}^2 \qquad (15.16)$$

where θ_s = temperature of the rock surface (°C), θ_d = dry bulb temperature in the main airstream (°C), h = a heat transfer coefficient (W/m²)/°C) that is a function mainly of the air velocity and the characteristics of the rock surface. We shall discuss the heat transfer coefficient further in section 15.2.7. However, for the moment let us accept it simply as a constant of proportionality between q and $\theta_s-\theta_d$.

Assuming that we know the values of h and θ_d, then we only need the rock surface temperature, θ_s, for equation (15.16) to give us the required heat flux, q. The problem turns to one of finding θ_s.

As the heat flux from the strata must be the same as that passing through the rock–air interface, we can combine equations (15.15) and (15.16) to give

$$q=k\left[\frac{\partial\theta}{\partial r}\right]_s=h(\theta_s-\theta_d) \quad \text{W/m}^2 \qquad (15.17)$$

Again, for generally, the temperature gradient at the surface may be expressed in dimensionless form, G, where

$$G=\frac{r_a}{\text{VRT}-\theta_d}\left[\frac{\partial\theta}{\partial r}\right]_s \qquad (15.18)$$

Combining with equation (15.17) gives

$$G(\mathrm{VRT}-\theta_d)=\frac{hr_a}{k}(\theta_s-\theta_d) \quad ^\circ\mathrm{C} \tag{15.19}$$

The group hr_a/k is known as the **Biot number**, B, or dimensionless heat transfer coefficient. Then

$$G(\mathrm{VRT}-\theta_d)=B(\theta_s-\theta_d) \quad ^\circ\mathrm{C}$$

or

$$\theta_s=\frac{G}{B}(\mathrm{VRT}-\theta_d)+\theta_d \quad ^\circ\mathrm{C} \tag{15.20}$$

We have now found an expression for θ_s. However, it contains the dimensionless but yet unknown temperature gradient G. This may be obtained from a solution of the general radial heat conduction equation (15.13) using Laplace transforms. The mathematics of the solution process can be found in Carslaw and Jaeger (1956). The result involves a series of Bessel functions and is reproduced in section 15A.1 in the appendix at the end of this chapter. Unfortunately, the solution appears even more disconcerting for practical use then the original differential equation (15.13). However, it is now more amenable to numerical integration. The results are shown on Fig. 15.4 as a series of curves from which the dimensionless temperature gradient, G, can be read for given ranges of Fourier number, F_0, and Biot number, B.

An algorithm produced by Gibson (1976) allows G to be determined much more easily than the full numerical integration of the Carslaw and Jaeger solution. Gibson's algorithm (section 15A.2) is suitable for programming into a personal computer and gives an accuracy of within 2% over the majority of the ranges covered in Fig. 15.4.

Older solutions to the general equation (15.13) (Carrier, 1940; Goch and Patterson, 1940) assumed that the rock surface temperature was equal to the dry bulb temperature of the air. This ignored the insulating effect of the boundary layers or, put another way, inferred an infinite heat transfer coefficient and, hence, an infinite Biot number. The uppermost curve on Fig. 15.4 represents this bounding condition and gives the same results as the tables produced by Carrier, and Goch and Patterson.

Having established a value of G, equations (15.17) and (15.20) then combine to give the required heat flux

$$q=h\frac{G}{B}(\mathrm{VRT}-\theta_d) \quad \mathrm{W/m^2} \tag{15.21}$$

15.2.7 Heat transfer coefficient for airways

In section 15.2.6, we introduced the heat transfer coefficient, h, for the rock surface as the 'constant' of proportionality between heat flux across a boundary layer and the corresponding temperature difference (equation (15.16)):

$$q=h(\theta_s-\theta_d) \quad \mathrm{W/m^2}$$

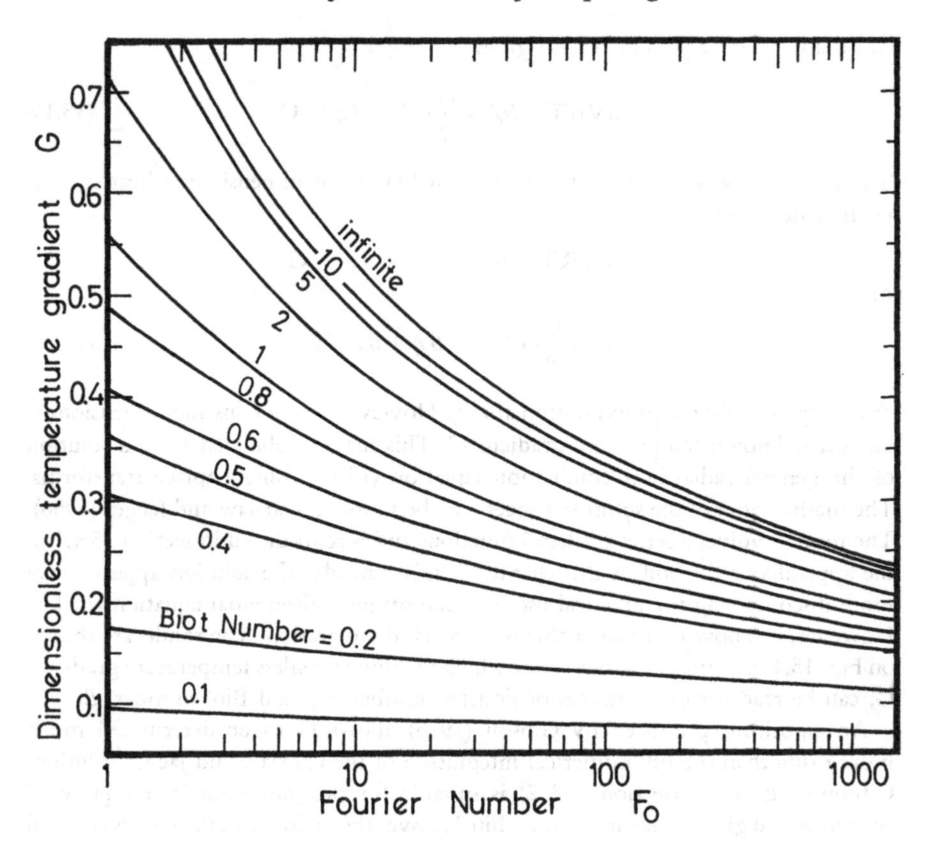

Figure 15.4 These curves enable the dimensionless temperature gradient in the rock at its surface to be determined for known Fourier and Biot numbers.

It is clear from this equation that for an established airway where the rock surface temperature, θ_s, is very close to the air dry bulb temperature, θ_d, the strata heat flux q will be small. The value of the heat transfer coefficient will then have little effect on climatic variations in the airway. Conversely, for newly exposed surfaces $\theta_s - \theta_d$ will be relatively high and the heat transfer coefficient will be a significant factor in controlling the flow of heat from the strata.

It can be seen from Fig. 15.4 that for Biot numbers ($B = hr_a/k$) of 10 or more, errors in the heat transfer coefficient and, hence, Biot number will have little effect on the dimensionless temperature gradient, G, which in turn controls heat flux into the airway. For typical sizes of modern mine airways, the Biot number is generally in excess of 5 where its accuracy (and, hence, the accuracy of the heat transfer coefficient) remains of limited consequence. It is, nevertheless, necessary to establish relationships that will allow us to evaluate the heat transfer coefficient.

The heat transfer coefficient is a constant only within the confines of equation (15.16). The overall heat transfer coefficient, h, for mine airways is made up of two parts, the convective heat transfer coefficient, h_c, and a component of the radiative heat transfer coefficient, h_r.

Convective heat transfer

In practice the convective heat transfer coefficient changes with those factors that cause variations in the thickness of the boundary layers, i.e. the air velocity and roughness on the surface that produce near-wall turbulence. Hence, the convective heat transfer coefficient depends upon the coefficient of friction, f (or Atkinson friction factor), and the Reynolds' number, Re. These parameters were introduced in section 2.3 as factors that influence airway resistance and losses of mechanical energy.

An increase in near-wall turbulence causes not only increased cross-flow momentum transfer close to the laminar sublayer but also helps to transport heat across the turbulent boundary layer into the mainstream. There is a close analogy between heat and momentum transfer across boundary layers. This has engaged the attention of researchers since the time of Osborne Reynolds and many equations relating heat transfer coefficients to fluid properties and flow regimes have been proposed for both hydraulically smooth and rough surfaces.

A relationship that has shown itself to agree with practical observations both underground (Mousset-Jones *et al.*, 1987; Danko *et al.*, 1988) and in scale models (Deen, 1988) is based on earlier work by Nunner (1956). For application in mine ventilation, it may be stated as

$$N_u = \frac{0.35 f\,\mathrm{Re}}{1 + 1.592(15.217 f\,\mathrm{Re}^{0.2} - 1)/\mathrm{Re}^{0.125}} \quad \text{dimensionless} \qquad (15.22)$$

The convective heat transfer coefficient, h_c, is then given as

$$h_c = \frac{N_u k_a}{d} \quad \mathrm{W/(m\,^\circ C)} \qquad (15.23)$$

where k_a = thermal conductivity of air ($2.2348 \times 10^{-4}\, T^{0.8353}\, \mathrm{W/(m\,^\circ C)}$, T being the absolute temperature in kelvins) and d = hydraulic mean diameter (m). The **Nusselt number** ($N_u = h_c d/k_a$) is simply a means of expressing the heat transfer coefficient as a dimensionless number in order to extend the generality of equation (15.22). It is analogous to the Biot number but is referred to the air and the size of the airway rather than the rock.

Here again, equations (15.22) and (15.23) may be programmed into a calculator or personal computer for a rapid determination of the heat transfer coefficient. For manual application, the Nusselt number may be read directly from Fig. 15.5 for any given values of f and Re. The background to equation (15.22) is given in section 15A.3.

A further phenomenon that has been observed experimentally and subjected to theoretical analyses is a tendency for the convective heat transfer coefficient to

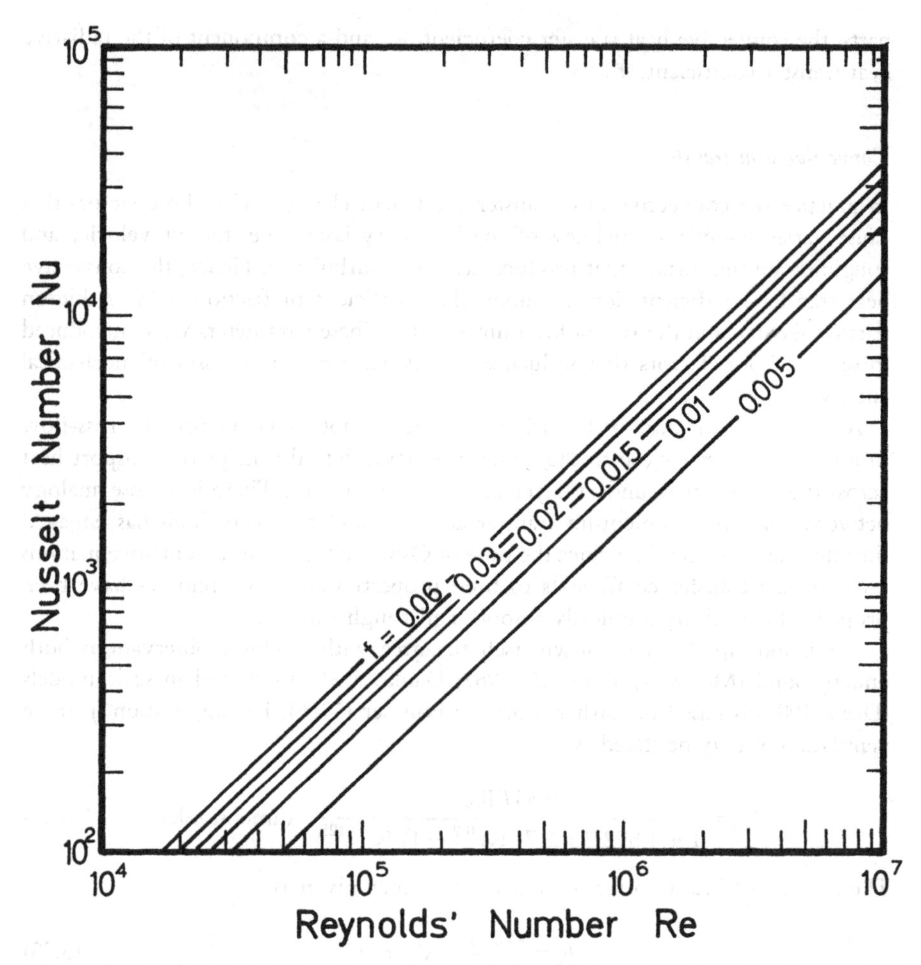

Figure 15.5 Variation of Nusselt number, N_u with coefficient of friction, f, and Reynolds' number, Re (based on air at 20° C).

decrease as the wall temperature rises for any given value of Reynolds' number. This is caused by a thickening of the temperature or thermal boundary layer and, hence, giving an enhanced insulating effect against heat transfer between the wall and the main airstream. However, the actual heat transfer will normally increase because of the greater value of $\theta_s - \theta_d$ in equation (15.16).

Radiative heat transfer

In addition to convective effects, heat may be lost from a rock surface by radiation. Oscillations in atomic energy levels at the surface produce electromagnetic radiation. This propagates through space until it reaches another solid surface where some of it is reflected, the remainder returning to thermal energy by exitation of atoms on the receiving surface. Additionally, some gases, including water vapour and carbon dioxide, will absorb a fraction of the thermal radiation causing a rise in temperature of the gas. Elemental gases such as oxygen and nitrogen, the major constituents of air, are not affected in this way.

According to the Stefan–Boltzmann equation, the net heat transfer by radiation from a hotter to a cooler surface is given as

$$q_r = 5.67 \times 10^{-8}(T_1^4 - T_2^4)\, A \times F_{ev} \quad \text{W} \tag{15.24}$$

where T_1 and T_2 are the absolute temperatures of the hotter and cooler surfaces respectively (K), A = the smaller of the two surfaces and

$$5.67 \times 10^{-8} \text{ is the Stefan–Boltzmann constant} \quad \text{W/(m}^2\text{°C)} \tag{15.25}$$

The parameter F_{ev} combines the thermal emissivity of the surfaces and the view factor which quantifies the degree to which the surfaces can 'see' each other (Whillier, 1982).

In subsurface environmental engineering, a more practical relationship for radiant heat transfer from airway surfaces is

$$q_r = h_r(\theta_s - \theta_d)F_{ev} \quad \text{W/m}^2 \tag{15.26}$$

where h_r = radiative heat transfer coefficient.

The emissivity of polished metal surfaces is fairly low. However, for rough natural surfaces it is usually more than 0.95. Furthermore, the view factor of each unit area of surface to water vapour in the passing airstream or to much of the nearby rock surface also approaches unity. Hence, we may approximate F_{ev} to 1.

The radiative heat transfer coefficient varies from 5 to 7 W/(m^2 °C) in the range of surface temperatures 10 to 40 °C. More precisely,

$$h_r = 4 \times 5.67 \times 10^{-8} \times T_{av}^3 \quad \text{W/(m}^2\text{°C)} \tag{15.27}$$

where T_{av} = average absolute temperature of the two surfaces (K). In practice, this may be taken as the dry bulb temperature of the air.

For a dry airway, the temperature of the rock surface will remain the same around any given perimeter. Hence, $T_1 = T_2$ in equation (15.24) and there will be no net transfer of radiant heat between surfaces. If part of the surface is wet, however, radiant heat will pass from the dry areas to the cooler wet surface. This will cause a slight increase in strata heat transfer from dry areas. However, the small rise in temperature of the wet surfaces will result in a diminished flow of strata heat to those surfaces. Although the two may not balance, the net effect is small. Radiative heat transfer between rock surfaces is usually ignored.

The amount of thermal radiation absorbed by water vapour in the air varies exponentially with the product of the vapour content and the distance travelled by the radiation through the air–vapour mixture. At a vapour content of 0.019 kg/(kg dry air), 22% of the radiant heat will be absorbed within 3 m and 47% within

30 m. The remaining radiation that is not absorbed will be received on other rock surfaces and, again, will have little impact on the mine climate.

The radiative heat transfer coefficient, h_r, is usually considerably lower than the convective heat transfer coefficient, h_c, in mine airways and may, indeed, be of the same order as the uncertainty in h_c. An estimate of the fraction a_b of thermal radiation absorbed by water vapour in the air may be obtained from

$$a_b = 0.104 \ln(147XL) \tag{15.28}$$

where ln = natural logarithm, X = water vapour content of air (kg/(kg dry air)) (section 14.2.2) and L = distance travelled by the radiation through the air (path length, m). This equation is based on curve fitting to empirical data.

The mean path of the electromagnetic waves propagating from a given point on a rock surface and before they strike another surface will depend upon the geometry of the mine opening. For an airway a typical mean path may be of the order of three times the hydraulic mean diameter.

The effective radiant heat transfer coefficient for thermal radiation between the rock and the air becomes

$$a_b h_r$$

The overall heat transfer coefficient is then

$$h = h_c + a_b h_r \quad \mathrm{W/(m^2\,°C)}$$

Analyses can also be carried out for the absorption of thermal radiation by carbon dioxide and dust particles. However, under normal conditions the corresponding rates of thermal absorption are very small.

Example The dry bulb temperature of the air in a 4 m by 3 m underground opening is 26°C. The corresponding moisture content is 0.015 kg/(kg dry air). Estimate the effective radiant heat transfer coefficient.

Solution Hydraulic mean diameter,

$$d_h = \frac{4 \times \text{area}}{\text{Perimeter}} = \frac{4 \times 12}{14}$$

$$= 3.43\ \text{m}$$

Assume the mean path length of the radiation to be

$$3d_h = 3 \times 3.43 = 10.3\ \text{m}$$

Fraction of radiation absorbed (from equation (15.28))

$$a_b = 0.104 \ln(147 \times 0.015 \times 10.3)$$

$$= 0.32$$

From equation (15.27)

$$h_r = 4 \times 5.67 \times 10^{-8}(273.15 + 26)^3$$
$$= 6.1\ \mathrm{W/(m^2\ °C)}$$

Then the effective radiant heat transfer coefficient is

$$a_b h_r = 0.32 \times 6.1 = 1.95\ \mathrm{W/(m^2\ °C)}$$

15.2.8 Summary of procedure for calculating heat flux at dry surfaces

Sections 15.2.5 to 15.2.7 have detailed the derivation of relationships that describe the radial flow of strata heat through the rock towards a mine opening and across a dry surface into the main airstream. As often occurs in engineering, the application of those relationships is straightforward compared with the theoretical analyses that have produced them.

Before moving on to consider heat exchange at wet surfaces, it is convenient to summarize the procedure for calculating the emission of strata heat across a dry surface, and to illustrate that procedure by a case study.

Calculation procedure for dry surface

1. Assemble the data:
   (a) airway dimensions (m)
   (b) coefficient of friction, f (= Atkinson friction factor/0.6)
   (c) age of airway, t (s)
   (d) airflow Q (m^3/s)
   (e) mean dry bulb temperature of air, θ_d(°C), wet bulb temperature, θ_w(°C), and barometric pressure, P (Pa)
   (f) rock thermal properties: thermal conductivity, k_r (W/(m °C)), density, ρ_r (kg/m^3), specific heat, C_r (J/(kg °C)), diffusivity, $\alpha_r = k_r/\rho_r C_r$ (m^2/s), and virgin rock temperature, VRT (°C)
2. Determine derived parameters:
   (a) cross-sectional, area, A (m^2)
   (b) perimeter, per (m)
   (c) hydraulic mean diameter, $d_h = 4A/\text{per}$ (m)
   (d) effective radius, $r_a = \text{per}/2\pi$ (m)
   (e) Reynolds' number, Re:
   for the purposes of this procedure, Re may be calculated from the approximation

$$\text{Re} = 268\,000 \frac{Q}{\text{per}}$$

   (f) moisture content of air, X (kg/(kg dry air)) (from section 14.6)
   (g) mean radiation path length, L (m)

3. Determine the Nusselt number, N_u, either from Fig. 15.5 or from equation (15.22), i.e.

$$N_u = \frac{0.35 f \,\mathrm{Re}}{1 + 1.592(15.217 f \,\mathrm{Re}^{0.2} - 1)/\mathrm{Re}^{0.125}}$$

4. Determine the overall heat transfer coefficient, h.
   (a) Convective heat transfer coefficient, h_c,

$$h_c = 0.026 \frac{N_u}{d_h} \quad \mathrm{W/(m^2\,°C)}$$

   where 0.026 W/(m °C) = thermal conductivity of air
   (b) Effective radiative heat transfer coefficient, $a_b h_r$,

$$h_r = 22.68 \times 10^{-8}(273.15 + \theta_d)^3 \quad \mathrm{W/(m^2\,°C)}$$

   Absorption fraction

$$a_b = 0.104 \ln(147 XL)$$

   Effective radiative heat transfer coefficient = $a_b h_r$.
   (c) Overall heat transfer coefficient,

$$h = h_c + a_b h_r \quad \mathrm{W/(m^2\,°C)}$$

5. Calculate Biot number, B:

$$B = \frac{h r_a}{k_r} \quad \text{(dimensionless)}$$

6. Calculate Fourier number, F_0:

$$F_0 = \frac{\alpha_r t}{r_a^2} \quad \text{(dimensionless)}$$

7. Determine dimensionless temperature gradient in the rock but at the surface, G, either from Fig. 15.4 or from Gibson's algorithm (section 15A.2).
8. Determine heat flux, q:

$$q = h \frac{G}{B} \quad (\mathrm{VRT} - \theta_d) \quad \mathrm{W/m^2}$$

9. Calculate heat emission into airway:

$$\frac{q \times \mathrm{per} \times \text{length of airway}}{1000} \quad \mathrm{kW}$$

Case study

This case study illustrates not only the calculation procedure but also typical magnitudes of the variables. The purpose of the exercise is to determine the strata heat that will flow into an incremental length of dry airway. The stages of calculation are numbered to follow the steps of the procedure given above.

1. Given data:
   (a) airway dimensions width = 3.5 m, height = 2.5 m, length = 20 m
   (b) Atkinson friction factor (at $\rho_a = 1.2\,\text{kg/m}^3$) = $0.014\,\text{kg/m}^3$, i.e. coefficient of friction,

$$f = 0.014/0.6$$
$$= 0.0233 \quad \text{(dimensionless)}$$

   (c) airway age = 3 months

$$t = \frac{365}{4} \times 24 \times 3600 = 7.884 \times 10^6\,\text{s}$$

   (d) airflow, $Q = 30\,\text{m}^3/\text{s}$
   (e) dry bulb temperature in airway, $\theta_d = 25\,°\text{C}$
   (f) from psychrometric data, the moisture content of the air has been determined to be $X = 0.01$ kg/(kg dry air)
   (g) rock thermal properties: conductivity, $k_r = 4.5\,\text{W/(m}\,°\text{C)}$, density, $\rho_r = 2200\,\text{kg/m}^3$, specific heat, $C_r = 950\,\text{J/(kg}\,°\text{C)}$, diffusivity,

$$\alpha_r = \frac{4.5}{2200 \times 950} = 2.153 \times 10^{-6}\,\text{m}^2/\text{s}$$

   virgin rock temperature, VRT = 42 °C

2. Further derived parameters:
   (a) cross-sectional area, $A = 3.5 \times 2.5 = 8.75\,\text{m}^2$
   (b) perimeter, per = 2(3.5 + 2.5) = 12 m
   (c) hydraulic mean diameter,

$$d_h = \frac{4A}{\text{per}}$$
$$= \frac{4 \times 8.75}{12} = 2.917\,\text{m}$$

   (d) effective radius,

$$r_a = \frac{\text{per}}{2\pi} = \frac{12}{2\pi} = 1.910\,\text{m}$$

   (e) Reynolds' number,

$$\text{Re} = 268\,000 \times \frac{30}{12} = 670\,000$$

   (f) mean radiation path length, L, is taken as $3d_h = 3 \times 2.917 = 8.751$ m

3. The Nusselt number, N_u, at Re = 670 000 and $f = 0.0223$ is estimated from

Fig. 13.5 to be 2400. Alternatively, it may be calculated as

$$N_u = \frac{0.35 \times 0.0233 \times 670\,000}{1 + 1.592(15.217 \times 0.0233 \times 670\,000^{0.2} - 1)/670\,000^{0.125}}$$
$$= 2433 \quad \text{dimensionless}$$

4. Overall heat transfer coefficient, h.
   (a) Convective heat transfer coefficient,

$$h_c = 0.026\,\frac{N_u}{d_h}$$
$$= 0.026 \times \frac{2433}{2.917} = 21.69\ \text{W/(m}^2\,{}^\circ\text{C)}$$

   (b) Effective radiative heat transfer coefficient, $a_b h_r$:

$$h_r = 22.68 \times 10^{-8}(273.15 + 25)^3$$
$$= 6.01\ \text{W/(m}^2\,{}^\circ\text{C)}$$

   Absorption fraction

$$a_b = 0.104 \ln (147 \times 0.01 \times 8.751)$$
$$= 0.266$$
$$a_b h_r = 0.266 \times 6.01 = 1.6$$

   (c) Overall heat transfer coefficient,

$$h = h_c + a_b h_r$$
$$= 21.69 + 1.6 = 23.29\ \text{W/(m}^2\,{}^\circ\text{C)}$$

   This result illustrates the limited effect of thermal radiation.

5. Biot number, B:

$$B = \frac{hr_a}{k_r} = \frac{23.29 \times 1.91}{4.5}$$
$$= 9.885 \quad \text{dimensionless}$$

6. Fourier number, F_0:

$$F_0 = \frac{\alpha_r t}{r_a^2}$$
$$= \frac{2.153 \times 10^{-6} \times 7.884 \times 10^6}{(1.91)^2}$$
$$= 4.653 \quad \text{dimensionless}$$

7. Dimensionless rock temperature at the surface, G, for $B = 9.885$ and $F_0 = 4.653$. This may be read from Fig. 15.4 or computed from Gibson's algorithm (section 15A.2) as 0.60. It is clear from the graphical method that errors in B (and, hence,

heat transfer coefficient) will have a very limited effect on G for the conditions cited in this case study.

8. Heat flux

$$q = h\frac{G}{B}(\text{VRT} - \Theta_d)$$

$$= 23.29 \times \frac{0.60}{9.885}(42 - 25)$$

$$= 24.0\ \text{W/m}^2$$

9. Heat emission into the 20 m length of airway

$$q \times \text{rock surface area} \quad \text{W}$$

or

$$\frac{24.0 \times 12 \times 20}{1000} = 5.76\ \text{kW}$$

15.2.9 Heat transfer at wet surfaces

The majority of underground openings have some degree of water evaporation or condensation occurring on rock surfaces even if no liquid water is visible. The cooling effect of evaporation or the heating effect of condensation will then result in a reduction or increase, respectively, of the wall surface temperature. This, in turn, will modify the strata heat flow toward that surface.

Figure 15.6 shows the heat balance that must exist for thermal equilibrium at the

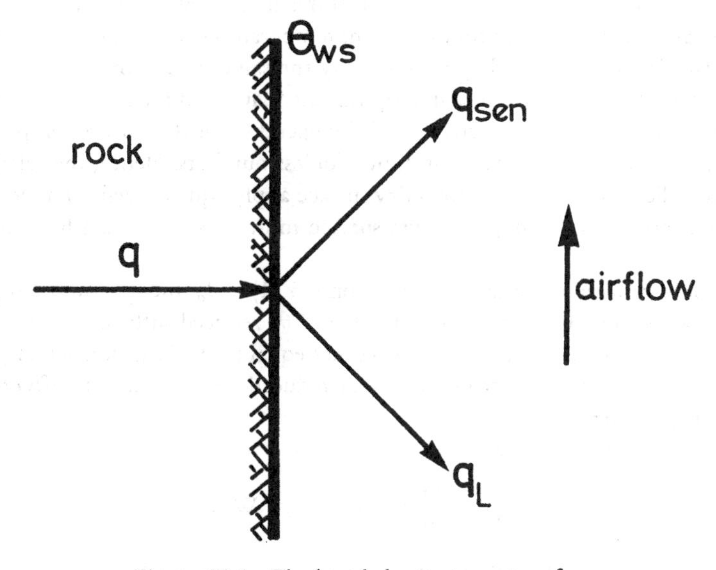

Figure 15.6 The heat balance at a wet surface.

rock–air interface.

$$q = q_L + q_{sen} \quad W/m^2 \tag{15.29}$$

where q = strata heat flux(W/m^2), q_L = latent heat transfer (W/m^2) and q_{sen} = sensible heat transfer (W/m^2). In practice, any of these heat flows may be positive or negative. Indeed, a common situation occurs when the wet surface temperature, θ_{ws}, is less than the dry bulb temperature of the air, θ_d. In this case, sensible heat will pass from the air to water on the rock surface; q_L must then accommodate the combined values of q and q_{sen}.

In order to prevent unnecessary repetition, we shall assume evaporation in the following analysis. The same logic applies for condensation except that q_L is negative.

Much of the theory pertaining to heat transfer at dry surfaces applies, also, to wet surfaces. In particular, let us re-examine equation (15.21) which was derived for a dry surface.

$$q = h_c \frac{G}{B}(\mathrm{VRT} - \theta_d) \quad W/m^2$$

(We shall ignore the small effect of radiative heat transfer. Hence, the overall heat transfer coefficient, h, becomes the convective heat transfer coefficient, h_c.)

Consider the functional dependence of each parameter in this equation in turn:

1. h_c depends on f, Re, k_a and d (equations (15.22) and (15.23)),
2. B depends on h_c, r_a and k_r (equation (15.19)),
3. G depends on the Fourier number $F_0 = \alpha_r t / r_a^2$ and the Biot number, B, and
4. VRT, the virgin rock temperature, is constant for any fixed location.

(The subscript r refers to the rock and a to the air or airway.) Notice that none of these parameters depends on time-transient temperatures within the rock or whether the surface is wet or dry. They have exactly the same values for both wet and dry surface conditions. It is particularly significant that the dimensionless temperature gradient in the rock at the surface, G, is independent of the surface temperature—another illustration of the value of dimensionless numbers. Hence, the relationships derived earlier for h_c, B and G for a dry surface apply equally well for a wet surface. Yet the evaporative cooling of a wet surface must cause the strata heat flux, q, to increase.

The only remaining variable in equation (15.21) is θ_d, the dry bulb temperature of the mainstream. We could select a value of θ_d reduced sufficiently such that the value of q given by equation (15.21) becomes equal to the heat flux actually passing through the wet surface. Let us name that reduced temperature the **effective dry bulb temperature, θ_{ef}.**

Then for the wet surface

$$q = h_c \frac{G}{B}(\mathrm{VRT} - \theta_{ef}) \quad W/m^2 \tag{15.30}$$

Provided that we replace θ_d by θ_{ef} then the equations for heat transfer at a dry surface become applicable also for a wet surface. The problem now is to find the value of θ_{ef}.

In order to evaluate q_L and q_{sen}, we shall require the wet surface temperature, θ_{ws}. The form of equation (15.20) gives a relationship between θ_{ws} and θ_{ef}.

$$\theta_{ws} = \frac{G}{B}(\mathrm{VRT} - \theta_{ef}) + \theta_{ef} \quad {}^\circ\mathrm{C} \tag{15.31}$$

This transposes to

$$\theta_{ef} = \left(\frac{G}{B}\mathrm{VRT} - \theta_{ws}\right)\frac{B}{G - B} \quad {}^\circ\mathrm{C} \tag{15.32}$$

Substituting for θ_{ef} in equation (15.30) and simplifying gives

$$q = h_c \frac{G}{B - G}(\mathrm{VRT} - \theta_{ws}) \quad \mathrm{W/m^2} \tag{15.33}$$

We now have an equation for the **strata heat flux** where everything is known except the wet surface temperature, θ_{ws}. Let us establish similar equations for q_L and q_{sen}, the other two terms in the heat balance of equation (15.29).

The **sensible heat** is given simply as (see equation (15.16))

$$q_{sen} = h_c(\theta_{ws} - \theta_d) \quad \mathrm{W/m^2} \tag{15.34}$$

The **latent heat** terms, q_L, is given by

$$q_L = 0.0007 h_c L_{ws} \frac{e_{ws} - e}{P} \quad \mathrm{W/m^2} \tag{15.35}$$

where L_{ws} = latent heat of evaporation at the wet surface temperature (J/kg), e_{ws} = saturated vapour pressure at the wet surface temperature (Pa), e = actual vapour pressure in the mainstream (Pa) and P = barometric pressure (Pa). The derivation of equation (15.35) is given in section 15A.4. Values of L_{ws}, e_{ws} and e may be determined from the relationships given in section 14.6. In particular L_{ws} and e_{ws} are functions of θ_{ws} only, while e depends on the wet and dry bulb temperatures of the mainstream, θ_w and θ_d, and the barometric pressure, P.

Equations (15.33), (15.34) and (15.35) allow q, q_{sen} and q_L to be determined on the basis of known mainstream conditions and an assumed wet surface temperature, θ_{ws}. If the assumed value of θ_{ws} were correct, then the heat balance

$$q = q_{sen} + q_L$$

from equation (15.29) would hold. However, an error in θ_{ws} will cause a corresponding error E in the heat balance

$$E = q - (q_{sen} + q_L) \tag{15.36}$$

The wet surface temperature, θ_{ws}, may now be corrected according to the value of E and the process repeated iteratively until E becomes negligible. All heat flows are then defined.

Example In order to illustrate the effect of water on heat flow across a rock surface, we shall use the same data given for the case study in section 15.2.8 but now assuming that the surface is fully wetted.

To recap on the data:

1. *Constants.* Fourier number,

$$F_0 = 4.653 \quad \text{dimensionless}$$

Convective heat transfer coefficient,

$$h_c = 21.69 \quad \text{W/(m}^2\,°\text{C)}$$

Biot number,

$$B = h_c \times \frac{r_a}{k_r} = 21.69 \times \frac{1.91}{4.5} = 9.207 \quad \text{dimensionless}$$

The value of B has changed slightly as we are now ignoring radiative effects, i.e. using h_c instead of the overall heat transfer coefficient, h. However, when we employ F_0 and the modified B on Fig. 15.4 or in Gibson's algorithm (section 15A.2), the value of dimensionless temperature gradient G is not significantly different from the 0.6 obtained previously.

2. *Psychrometric data.* The psychrometric relationships used here are listed in section 14.6. In addition to the dry bulb temperature of 25 °C, we shall also require the wet bulb temperature, θ_w, and the barometric pressure, P, of the mainstream. We shall take these to be 17.9 °C and 100 kPa respectively. Additionally, we need to choose a first estimate for the temperature of the wet surface. As this often lies between the wet and dry bulb temperatures of the main airstream, let us estimate θ_{ws} to be the mean of the two, i.e.

$$\theta_{ws} = (25.0 + 17.9)/2 = 21.45\,°\text{C}$$

(a) For the main airstream: saturated vapour pressure,

$$e_{sat,air} = 610.6 \exp\left(\frac{17.27 \times 17.9}{237.3 + 17.9}\right)$$

$$= 2050.4 \text{ Pa}$$

saturated moisture content,

$$X_{sat,air} = \frac{0.622 \times 2050.4}{100\,000 - 2050.4}$$

$$= 0.013\,02 \text{ kg/(kg dry air)}$$

latent heat of evaporation at wet bulb temperature,

$$L_{w,air} = (2502.5 - 2.386 \times 17.9)10^3$$

$$= 2459.79 \times 10^3 \text{ J/kg}$$

sigma heat,

$$S_{air} = (2459.79 \times 10^3 \times 0.013\,02) + (1005 \times 17.9)$$
$$= 50\,250.9 \text{ J/kg}$$

actual moisture content

$$X_{air} = \frac{50\,250.9 - 1005 \times 25}{2459.79 \times 10^3 + 1884(25 - 17.9)}$$
$$= 0.010\,09 \text{ kg/(kg dry air)}$$

actual vapour pressure,

$$e = \frac{100\,000 \times 0.010\,09}{0.622 + 0.010\,09}$$
$$= 1596 \text{ Pa}$$

(b) For the wet surface (at temperature, θ_{ws}): saturated vapour pressure,

$$e_{ws} = 610.6 \exp\left(\frac{17.27 \times 21.45}{237.3 + 21.45}\right)$$
$$= 2556 \text{ Pa}$$

latent heat of evaporation,

$$L_{ws} = (2502.5 - 2.386 \times 21.45)10^3$$
$$= 2451.3 \times 10^3 \text{ J/kg}$$

3. *Heat flows:* Strata heat,

$$q = \frac{h_c G}{B - G}(\text{VRT} - \theta_{ws})$$
$$= \frac{21.69 \times 0.6}{9.207 - 0.6}(42 - 21.45)$$
$$= 31.1 \text{ W/m}^2$$

Sensible heat,

$$q_{sen} = h_c(\theta_{ws} - \theta_d)$$
$$= 21.69\,(21.45 - 25)$$
$$= -77.0 \text{ W/m}^2$$

Latent heat,

$$q_L = 0.0007\, h_c\, L_{ws}\, (e_{ws} - e)/P$$
$$= 0.0007 \times 21.69 \times 2451.3 \times 10^3 \frac{2556 - 1596}{100\,000}$$
$$= 357.3 \quad \text{W/m}^2$$

4. *Correction of* θ_{ws}. The error in the heat balance is

$$E = q - (q_{sen} + q_L)$$
$$= 31.1 - (-77.0 + 357.3) = -249\ \text{W/m}^2$$

Corrected value

$$\theta_{ws} = 21.45 + (-249) \times 0.01$$
$$= 18.96\ ^\circ\text{C}$$

The correction function used here is $0.01E$. The constant 0.01 is chosen to give a reasonable rate of convergence. While mathematical techniques are available to find optimum correction functions for given iterative procedures, a more pragmatic approach (assuming computer availability) is simply to experiment with a set of typical data until satisfactory convergence is achieved.

The procedure from steps 2(b) to 4 is repeated iteratively until the absolute value of E becomes less than 0.01 W/m^2. This is clearly a task for a personal computer. Programming the equations and logic into an algorithm is a useful student exercise. The results of running such an algorithm on this particular example are shown on Table 15.2. Balance was achieved to within 0.01 W/m^2 after nine iterations. The first line of the table gives values that are in sensible agreement with the hand calculations given above. The slight differences are due to a coarser rounding-off during the manual calculations.

At balance, 36.1 W of strata heat arrive at each square metre of wet surface. Simultaneously, 148.1 W of sensible heat pass from the air to the wet surface (q_{sen} is negative). Hence, the sum of the two, 184.2 W/m^2, is emitted as latent heat to the air. The temperature of the wet surface is 18.17 °C.

The effect of water on the surface has now been quantified. The net heat emission of 36.1 W/m^2 compares with 24.0 W/m^2 for the corresponding dry

Table 15.2 Successive values computed during the iterative process for a wet surface

Iteration	*Wet surface temperature* θ_{ws}*(°C)*	*Strata heat* $q(W/m^2)$	*Latent heat* $q_L(W/m^2)$	*Sensible heat* $q_{sen}(W/m^2)$	*Error E* $(W/m)^2$
1	21.45	31.1	358.6	− 77.0	−250.49
2	18.95	34.9	222.7	−131.4	−56.43
3	18.38	35.8	194.5	−143.6	−15.11
4	18.23	36.0	187.1	−146.9	−4.19
5	18.19	36.1	185.0	−147.8	−1.17
6	18.18	36.1	184.4	−148.0	−0.33
7	18.17	36.1	184.3	−148.1	−0.09
8	18.17	36.1	184.2	−148.1	−0.03
9	18.17	36.1	184.2	−148.1	−0.01

surface of the earlier case study. For the 20 m length of airway at a perimeter of 12 m, this translates into a heat load of

$$\frac{36.1 \times 20 \times 12}{1000} = 8.66 \text{ kW}$$

compared with 5.76 kW when the surface was dry. However, whereas the 5.76 kW of heat from the formerly dry surface was sensible heat, the 8.66 kW emitted from the wet surface now takes the form of latent heat. This results in a significant increase in the moisture content of the air.

15.2.10 *In situ* measurement of rock thermal conductivity

The experimental determination of the thermal conductivity of solids depends essentially on Fourier's law (equation (15.4)). A source of heat is applied to one side of a sample. The heat flux across a known area and the corresponding temperature difference across the sample are measured leaving the thermal conductivity as the only unknown.

A number of laboratory test rigs have been developed for the determination of the thermal conductivities of samples. However, for subsurface environmental engineering, it is preferable to measure rock thermal conductivity *in situ*. The reason for this is the difference that may be found between an *in situ* value and the mean of numerous samples measured in the laboratory. If the rock is homogeneous, unfractured around mine openings, and if workings are above the water table then good correlation may be achieved between laboratory tests and *in situ* determinations. However, in the more usual situation of mining stresses having produced induced fractures in addition to existing natural fracture patterns, and particularly in the presence of groundwater migration, the effective thermal conductivity may be more than double that indicated by small and unfractured laboratory samples (Mousset-Jones and McPherson, 1984).

Water has a thermal conductivity of 0.62 W/(m °C) which is considerably lower than that of most rocks. Hence, the presence of still connate water in fissured or porous media will tend to inhibit heat flow. Unfortunately, water in the strata surrounding subsurface openings is seldom stationary and, dependent upon the rate of water migration, the transmission of heat by moving water may be far greater than that resulting from conduction through the rock (see, also, section 15.3.3.). This can result in considerable increases in the effective thermal conductivity of the strata.

The simplest method of determining rock thermal conductivity *in situ* is to use natural geothermal energy as the heat source. If possible, an airway should be found that is fairly well established, but sufficiently long to give a readily measured rise in wet bulb temperature. A thermocouple circuit to indicate the rise in wet bulb temperature over the selected length of airway will give better accuracy than mercury in glass thermometers. From a cross-section within the chosen length, four boreholes should be drilled radially outward to a depth of at least 10 m. These should each be fitted with strings of thermocouples or similar temperature transducers, positioned to indicate the variation in temperature along the hole. Precautions should be taken to ensure good thermal contact between the transducers and the side of the borehole, and to prevent convection currents of air or water within the hole.

Returning to Fig. 15.2, consider an elongated annulus of rock at radius r and length Y. The heat flux, q, passes radially through the annulus which has an orthogonal area $2\pi rY$ and thickness dr. Fourier's law (equation (15.4)) gives

$$q = -k_r 2\pi rY \frac{d\theta}{-dr}$$

(dr is negative as r is measured outward from the airway while q is usually directed toward the airway). Then

$$\frac{dr}{r} = \frac{2\pi k_r Y}{q} d\theta$$

If we assume steady-state conditions, q is constant for all values of r. We can then integrate to give

$$\ln(r) = \frac{2\pi k_r Y}{q}\theta + \text{constant} \qquad (15.37)$$

From the measurements taken in a borehole, θ can be plotted against $\ln(r)$ to give a straight line of slope

$$q/2\pi k_r Y \qquad (15.38)$$

Now, values of wet and dry bulb temperatures, together with barometric pressures measured at the two ends a and b of the airway, allow the corresponding values of sigma heat, S, to be established (section 14.6). Furthermore, anemometer traverses give the airflow, Q (section 6.2), and, hence, the mass flow of air, M:

$$M = Q\rho$$

where ρ = actual density of air (kg/m^2). Then

$$q = M(S_b - S_a) \quad \text{W} \qquad (15.39)$$

Substituting for q in the expression for the measured slope of the θ vs $\ln(r)$ graph allows k_r to be determined. An average value of rock thermal conductivity will be given by the data obtained from several boreholes.

Although simple in principle, this *in situ* method of measuring effective thermal conductivity of strata may involve several difficulties. First, the airflow in the airway and the psychrometric conditions at the inlet should, ideally, remain constant. Accuracy will be improved if all measurements are made electronically and data logged every few minutes for several days. The plots of θ against $\ln(r)$ are not always straight lines. A sudden change in slope along a given set of borehole data indicates that the borehole has intersected a change in rock type. A more gradual curvature suggests that the strata temperatures have not reached equilibrium. Deviations towards the mouth of the hole indicate non-steady conditions in the airway. In such situations, the slope of the line at the rock surface should be used in the determination of k_r. Differences in the slopes of the lines are often found between boreholes drilled in different directions from the same cross-section. Such deviations may indicate differences in rock type, or more usually, the effect of water distribution within the strata.

A major advantage of this method is that it gives the effective thermal conductivity of the complete envelope of rock around the chosen cross-section. Other techniques have been developed that produce an artificial source of heat within a borehole with instrumentation to monitor the resulting temperature field at other locations along the borehole (Danko *et al.*, 1987). Such equipment can produce results within a few hours for more limited representative volumes of rock.

15.3 OTHER SOURCES OF HEAT

15.3.1 Autocompression

When air or any other fluid flows downward, some of its potential energy is converted to enthalpy ($H = PV + U$, equation (3.19)) producing increases in pressure, internal energy and, hence, temperature. Actual examples are shown in Fig. 8.3.

The rise in temperature as air falls through a downcast shaft or other descentional airway is independent of any frictional effects (section 3.4.1). Ignoring the small change in kinetic energy, the steady flow energy equation (3.22) gives

$$H_2 - H_1 = (Z_1 - Z_2)g + q_{12} \quad \text{J/kg} \tag{15.40}$$

where H = enthalpy (J/kg), Z = height above datum (m), g = gravitational acceleration (m/s^2) and q = heat added from surroundings (J/kg) (subscripts 1 and 2 refer to the inlet and outlet ends of the airway respectively).

This equation shows that the increase in enthalpy is, in fact, due to two components: (a) the heat actually added, q_{12}, and (b) the conversion of potential energy $(Z_1 - Z_2)g$. The heat added, q_{12}, may be positive, negative or zero but the increase in temperature due to depth is definitive for any given dry shaft and moisture content of the air.

The effects of autocompression are virtually independent of airflow. In deep mines, the intake air leaving the bottoms of downcast shafts may already be at a temperature that necessitates air cooling. This is the inevitable result of autocompression.

The reverse effect, autodecompression, occurs in upcast shafts or ascentional airways. This is usually of less concern as upcast air will have no effect on conditions in the workings. However, the reduction in temperature due to autodecompression in upcast shafts can result in condensation and fogging (section 9.3.6). The mixture of air and water droplets may then reduce the life of the impellers of exhaust fans sited at the shaft top.

The increase in temperature due to depth is sometimes known as the **adiabatic lapse rate**. For dry airways, the change in enthalpy is given by equation (3.33):

$$H_2 - H_1 = C_{pm}(T_2 - T_1) \quad \text{J/kg}$$

where T = temperature (°C or K) and C_{pm} = specific heat of the actual (moist) air (we are reverting to the symbol T for temperature in this section in order that we may use it to denote absolute temperatures). Hence, equation (15.40) gives the adiabatic ($q_{12} = 0$) rise in dry bulb temperature to be

$$T_2 - T_1 = \frac{(Z_1 - Z_2)g}{C_{pm}} \quad °C \tag{15.41}$$

Substituting for C_{pm} from equation (14.16) gives

$$T_2 - T_1 = \frac{1 + X}{1005 + 1884\,X}(Z_1 - Z_2)g \quad °C \tag{15.42}$$

At a value of $g = 9.81\ m/s^2$, this gives an adiabatic dry bulb temperature lapse rate varying from 0.976 °C per 100 m depth for completely dry air to 0.952 °C per 100 m depth at a water vapour content of 0.03 kg/(kg dry air).

However, in the majority of cases, the shaft or airway is not completely dry and the rate of increase in dry bulb temperature is eroded by the cooling effect of evaporation. If the increase in vapour content of the air due to evaporation is ΔX kg/(kg dry air), then this will result in a conversion of sensible heat to latent heat, $L\,\Delta X$(J/kg), where L = latent heat of evaporation. Equation (15.41) then becomes

$$T_2 - T_1 = \frac{(Z_1 - Z_2)g - L\,\Delta X}{C_{pm}} \quad °C \tag{15.43}$$

Using mid-range values of $C_{pm} = 1010$ J/(kg K) and $L = 2453 \times 10^3$ J/kg gives the approximation

$$T_2 - T_1 = 0.009\,71\,(Z_1 - Z_2) - 2428.7\,\Delta X \quad °C \tag{15.44}$$

This equation is shown graphically in Fig. 15.7 and is useful only if the increase in vapour content is known. In practical design exercises, ΔX is normally unknown but an estimate may be made of the wetness of the shafts. Recourse must then be made to computer simulation techniques (Chapter 16). It must be emphasized that equations (15.41) to (15.44) refer to the dry bulb temperature only, and while the air remains unsaturated. Should saturation occur then the dry bulb temperature will follow the wet bulb temperature adiabatic lapse rate.

The behaviour of the wet bulb temperature due to adiabatic compression is more predictable than dry bulb temperature. The enthalpy (and sigma heat) of the air increases linearly with depth in the absence of any heat transfer with the surroundings. Furthermore, the increase in air pressure is also near linear with respect to depth. Hence, as wet bulb temperature varies only with sigma heat and pressure, the wet bulb adiabatic lapse rate in a shaft will be effectively constant irrespective of shaft wetness. Any further significant deviation in the wet bulb temperature can be due only to heat transfer with the walls or shaft fittings.

Unfortunately, calculation of the wet bulb adiabatic lapse rate is somewhat more involved than for the dry bulb.

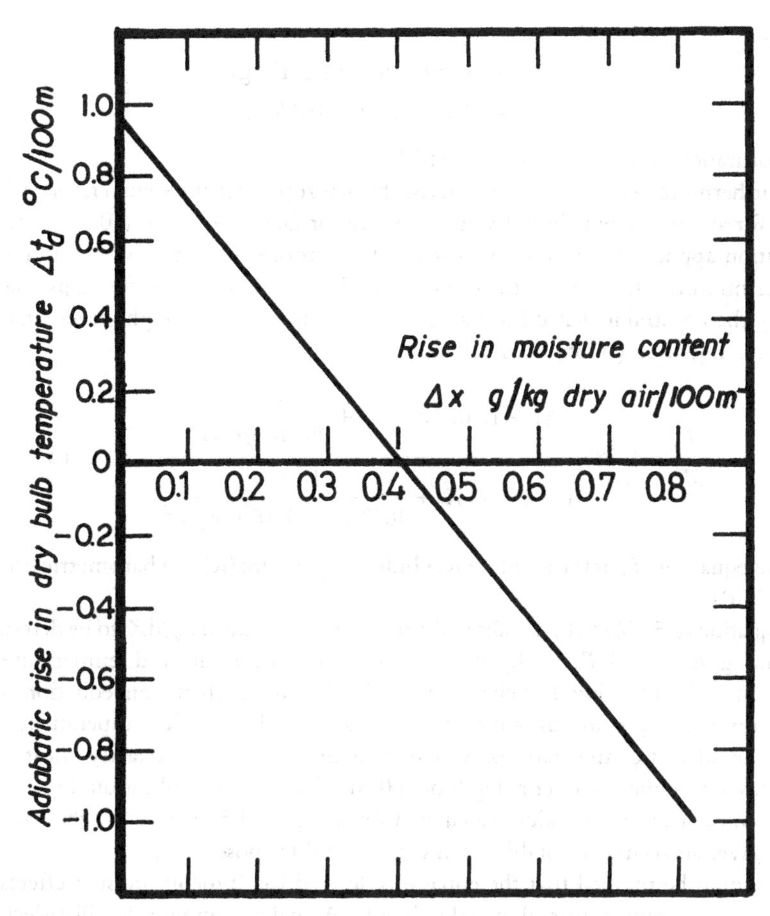

Figure 15.7 The dry bulb adiabatic lapse rate (variation of air temperature with depth) depends on the rate at which water is evaporated.

We can write the variation in wet bulb temperature, T_w, with respect to depth, Z, as

$$\frac{dT_w}{dZ} = \frac{dT_w}{dP}\frac{dP}{dZ} \quad °C \tag{15.45}$$

From the steady flow energy equation for isentropic conditions ($F = q = 0$) and negligible change in kinetic energy (equation (3.25)),

$$-g\,dZ = V dP$$

or

$$\frac{dP}{dZ} = \frac{-g}{V} \quad \frac{N}{m^3} \text{ or } \frac{Pa}{m} \tag{15.46}$$

where

$$V = \text{specific volume (m}^3\text{/kg)}$$
$$= 287.04\, T_w/(P - 0.378 e_s)$$

for saturation conditions (equation (14.19)).

Furthermore, we have already derived the isentropic pressure–temperature relationship for an air–vapour–liquid water mixture for fans as equation (10.56). The same equation applies to any other isentropic flow process involving air–vapour–liquid water mixtures. In order to track the adiabatic wet bulb temperature lapse rate, we apply the constraint that critical saturation is maintained throughout the shaft, i.e. $X = X_s$. Equation (10.56) then becomes

$$\frac{dT_w}{dP} = 0.286 \frac{(1 + 1.6078 X_s)\dfrac{T_w}{P} + \dfrac{L_w X_s}{287.04(P - e_s)}}{1 + 1.7921 X_s + \dfrac{L_w^2 P X_s}{463.81 \times 10^3 (P - e_s) T_w^2}} \quad \text{°C/Pa} \qquad (15.47)$$

In this equation, T_w is the absolute wet bulb temperature (K) and barometric pressure, P, is in Pa.

Equations (15.45) to (15.47) allow the wet bulb lapse rate, dT_w/dZ, to be determined for any given P and T_w, as X_s and e_s are functions of pressure and temperature only (section 14.6). Tracking the behaviour of dT_w/dZ down a shaft from equation (15.47) and commencing from known shaft top pressure and wet bulb temperature, shows that the adiabatic lapse rate for wet bulb temperature is not exactly constant but decreases by some 4% over a depth of 1000 m. For manual application, Fig. 15.8 has been constructed from values calculated for a depth of 500 m below the shaft top. This gives an accuracy suitable for most practical purposes.

It should be recalled that the equations derived for autocompression effects give the changes in temperature due to depth only. Actual measurements will reflect, also, heat transfer with the shaft walls or other sources. These are likely to be most noticeable in downcast shafts because of transients in the surface conditions.

15.3.2 Mechanized equipment

The operation of all mechanized equipment results in one, or both, of two effects; work is done against gravity and/or heat is produced. A conveyor transporting material up an incline, a shaft hoist and a pump are examples of equipment that work against gravity. Vehicles operating in level airways, rock-breaking machinery, transformers, lights and fans are all devices that convert an input power, via a useful effect, into heat.

With the exceptions of compressed air motors and devices such as liquid nitrogen engines, all other forms of power including electricity and chemical fuels produce thermal pollution that must be removed by the environmental control system.

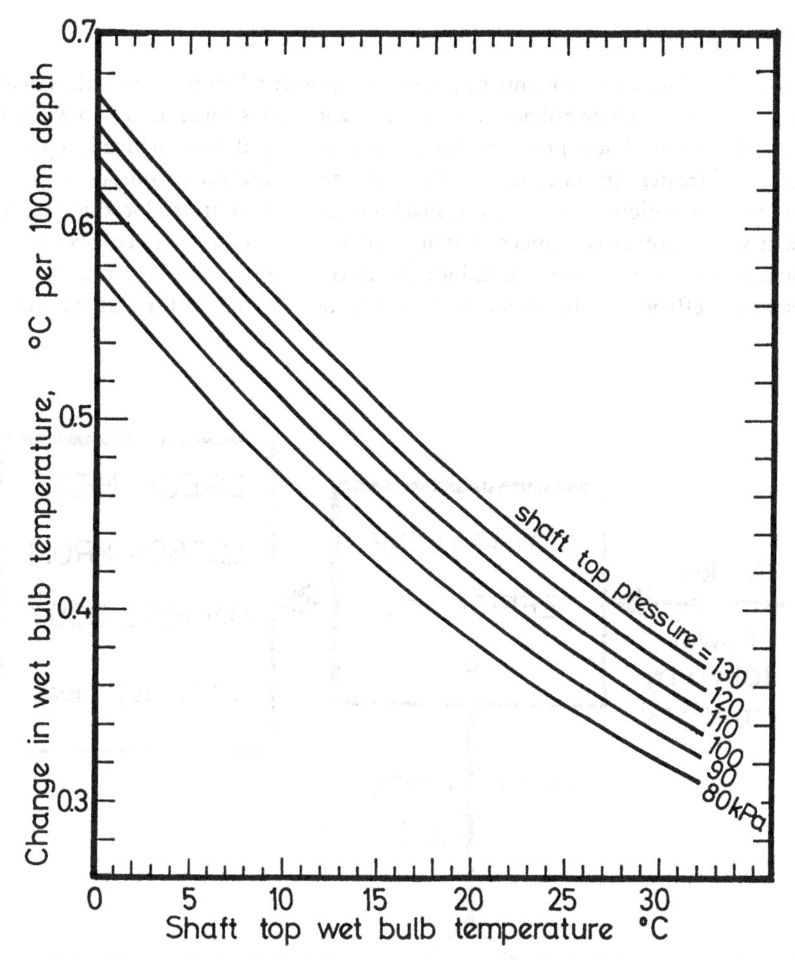

Figure 15.8 The wet bulb adiabatic lapse rate depends on the initial pressure and wet bulb temperature but is independent of evaporation within the shaft.

Increasing utilization and power of mechanization in mines and other subsurface facilities has resulted in such equipment joining geothermal effects and autocompression as a major source of heat. With machines consuming about 2 MW of electrical power on some highly mechanized longwall faces, a rising number of coal mines in the United Kingdom and Europe have had to resort to refrigeration equipment at depths below surface where it was not previously required. The calculation of equipment heat is straightforward compared with that for strata heat.

Electrical equipment

Figure 15.9 illustrates the manner in which the power taken by an electrical machine is utilized. The machine efficiency is relevant, within this context, in two ways. First, the total amount of heat produced can be reduced only if the machine is replaced by another of greater efficiency to give the same mechanical power output at a reduced power consumption. For any given machine, the total heat produced is simply the rate at which power is supplied, less any work done against gravity. Secondly, the efficiency of the machine determines the distribution of the heat produced. The greater the efficiency, the lower the heat produced at the motor and transmission,

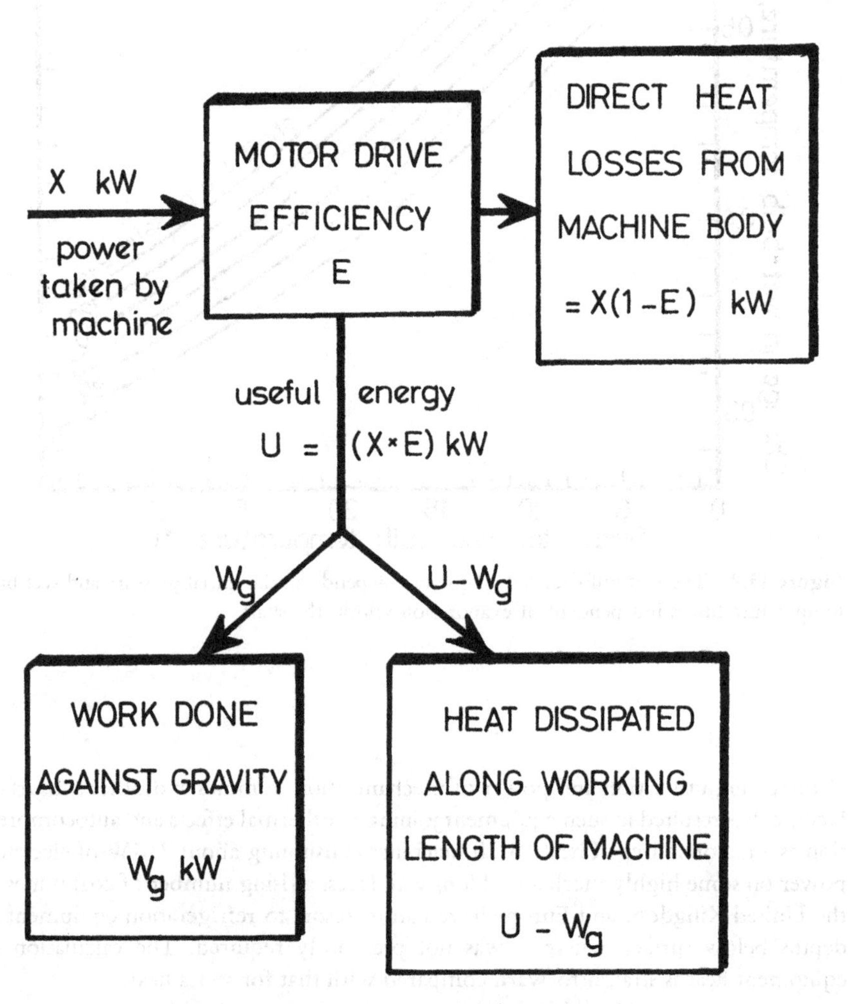

Figure 15.9 Heat produced by electrical machines.

and the greater is the heat produced at the pick-point, conveyor rollers, along the machine run or by any other frictional effects caused by the operation of the device.

Example A 2000 m long conveyor transports 500 t/h through a vertical lift of 200 m. If the conveyor motor consumes 1000 kW at a combined motor–transmission efficiency of 90%, calculate the heat emitted (1) at the gearhead and (2) along the length of the conveyor.

Solution Work done against gravity:

$$\text{mass flow} \times g \times \text{vertical lift} = \frac{500 \times 1000}{60 \times 60} \times 9.81 \times 200 \quad \frac{\text{kg}}{\text{s}}\frac{\text{m}}{\text{s}^2}\text{m}$$

$$= 272.5 \times 10^3 \text{ W or } 272.5 \text{ kW}$$

1. Heat generated at gearhead:

$$(100 - 90)\% \text{ of } 1000 \text{ kW} = 100 \text{ kW}$$

2. Heat generated along length of conveyor:

$$1000 - 272.5 - 100 = 627.5 \text{ kW}$$

Diesel equipment

The internal combustion engines of diesel equipment have an overall efficiency only about one-third of that achieved by electrical units. Hence, diesels will produce approximately three times as much heat as electrical equipment of the same mechanical work output. This can be demonstrated by taking a typical rate of fuel consumption to be 0.3 l per rated kW per hour. At a calorific value of 34 000 kJ/l for diesel fuel, the heat produced is

$$\frac{0.3}{60 \times 60} \frac{\text{litres}}{\text{kW output} \times \text{second}} \times 34\,000 \frac{\text{kJ heat}}{\text{litre}} \qquad (15.48)$$

$$= 2.83 \text{ kJ/s (or kWatts)}$$

of heat emitted for each kilowatt of mechanical output. This heat appears in three ways each of which may be of roughly the same magnitude. One-third appears as heat from the radiator and machine body, one-third as heat in the exhaust gases and the remaining third as useful shaft power which is then converted to heat by frictional processes as the machine performs its task.

As with other types of heat-emitting equipment there is little need, in most cases, to consider peak loads. It is sufficient to base design calculations on a average rate of machine utilization. The most accurate method of predicting the heat load is on the basis of average fuel consumption during a shift. However, in many mines, records of fuel consumption by individual machines or, even, in separate sections of the mine, do not seem to be maintained. The ventilation planner often must resort to the type of calculation shown above and using an estimated value for machine utilization. The latter is defined as the fraction of full load which if maintained continuously, would use the same amount of fuel as the actual intermittent load on the machine (section 16.2.3).

A difference between diesel and electrical equipment is that the former produces part of its heat output in the form of latent heat. Each litre of diesel fuel that is consumed produces approximately 1.1 l of water (liquid equivalent) in the exhaust gases (Kibble, 1978). This may be multiplied several times over by the evaporation of water from cooling systems and where water is employed in emission control systems. *In situ* tests have shown that the factor can vary from 3 to 10 l of water per litre of fuel (McPherson, 1986) depending on the design of the engine, exhaust treatment system and the proficiency of maintenance.

Example Two load–haul–dump vehicles consume 600 l of diesel fuel in an 8 h shift. Tests have shown that water vapour is produced at a rate of 5 l (liquid equivalent) per litre of fuel. If the combustion efficiency is 95% and the total calorific value of the fuel is 34 000 kJ/l, calculate the sensible and latent heat load on the stope ventilation system.

Solution Total amount of heat produced from burning 600 × 0.95 litres of fuel:

$$600 \times 0.95 \times 34\,000 = 19.38 \times 10^6 \text{ kJ in 8 h}$$

$$= \frac{19.38 \times 10^6}{8 \times 60 \times 60} = 673 \text{ kW}$$

(this is equivalent to the continuous running of diesel machinery of rated output = 673/2.83 = 238 kW).

Amount of water emitted as vapour:

$$600 \times 0.95 \times 5 = 2850 \text{ l} \quad \text{(liquid equivalent)}$$

i.e. 2850 kg of water.

Latent heat emitted in 8 h:

$$2450 \times 2850 = 6.982 \times 10^6 \text{ kJ}$$

where 2450 kJ/kg is an average value for the latent heat of evaporation of water, i.e.

$$\frac{6.982}{8 \times 3600} 242 \text{ kJ/s} \quad \text{or} \quad \text{kW}$$

Then,

$$\text{sensible heat produced} = \text{total heat} - \text{latent heat}$$
$$= 673 - 242$$
$$= 431 \text{ kW}$$

In summary, the diesels produce heat at an average rate of

431 kW sensible heat

242 kW latent heat

673 kW total heat

In addition to the particulate and gaseous pollutants emitted in diesel exhausts, the heat produced by these machines mitigates against their use in hot mines. Nevertheless, the flexibility and reliability of current diesel units will ensure their continued widespread employment underground.

Compressed air

When compressed air is used for drilling or any other purpose then there are two opposing effects that govern the heat load. First, the work output of the machine will result in frictional heat at the pick point or as the machine performs its task. Second, the removal of energy from the compressed air will result in a reduction of the temperature of that air at the exhaust ports of the machine.

If we assume that the change in potential energy of the air is negligible and that there is no significant heat transfer across the machine casing then the steady flow energy for the compressed air motor is

$$\frac{u_1^2 - u_2^2}{2} + W = C_p(\theta_2 - \theta_1) \quad \text{J/kg} \tag{15.49}$$

where u_1 and u_2 are the velocities of the air in the supply pipe and the exhaust ports, respectively, W is the mechanical work done on the air (this is numerically negative as mechanical energy is leaving the system; J/kg), C_p is the specific heat of air (1005 J/(kg °C) for dry air), θ_1 is the temperature of the supply air (°C) and θ_2 is the temperature of air at the outlet ports (°C).

In the majority of cases the temperature of the compressed air supplied to the motor is equal to the ambient dry bulb temperature, θ_{amb}. Furthermore, the velocity of the compressed air in the supply pipe, u_1, is small compared with that at the exhaust ports, u_2. Hence,

$$\frac{-u_2^2}{2} + W = C_p(\theta_2 - \theta_{amb}) \quad \text{J/kg} \tag{15.50}$$

Now consider the subsequent process during which the cold exhaust air mixes with the ambient air, increasing in temperature, extracting heat from the general ventilating airstream and, hence, cooling that airstream. Again, the velocity of the ambient airstream will be small compared with that at the outlet ports, u_2. The steady-flow energy equation for this process becomes

$$\frac{u_2^2}{2} = C_p(\theta_{amb} - \theta_2) - q \quad \text{J/kg} \tag{15.51}$$

where q is the heat added to the machine exhaust from the ambient air.

The overall effect may be quantified by adding equations (15.50) and (15.51), giving

$$W = -q$$

Hence, the mechanical work output of the machine (which is subsequently transferred by friction into heat) is balanced precisely by the cooling effect on the ventilating airstream. If the moisture content of the compressed air is similar to that of the ambient airstream, then there is no net heating or cooling arising from the expansion of the air through the compressed air motor.

Despite this analysis, a cooling effect is certainly noticeable when standing immediately downstream from a compressed air motor. There are two reasons for this. First, the cooling of the ambient airstream, q, by a motor exhausting at subzero temperatures is immediate and direct, while part of the balancing heat produced by friction may be stored temporarily within the body of the machine, in the solid or broken rock, and distributed over the working length of the machine. Secondly, the compressed air supply will normally be drier and have a much lower sigma heat than the ventilating air. The increase in sigma heat of the exhaust air will then exceed the numerical equivalent of the work output and, hence, result in a true net cooling of the ambient air. This will be of consequence only if the mass flow of compressed air is significant compared with the mass flow of the ventilating airstream. The dryness of compressed air, coupled with a very local kinetic energy effect ($u^2/2 = C_p(\theta_2 - \theta_1)$ also explains why air issuing from a leaky compressed air pipe appears cool even though no energy has been extracted from it.

In calculating the heat load for a complete district or mine, it errs on the side of safety to assume that there is no net cooling effect from compressed air motors. On the other hand, if a detailed analysis in the immediate vicinity of compressed air devices is required then the local cooling effects should be considered.

Air compressors are large sources of heat. Indeed, the local cooling effect at the outlet ports of the compressed air motors is reflected by an equivalent amount of heat produced at the compressors, even if the latter are 100% efficient. This need be of little import if the compressors are sited on surface and the compressed air is cooled before entering the shaft pipes. If, however, compressors are sited in intake airways close to hot workings, then it would be beneficial to investigate the circulation of hot water from the compressor coolers to heat exchangers in return airways.

15.3.3 Fissure water and channel flow

Groundwater migrating through strata towards a subsurface opening can add very considerably to the transfer of geothermal heat through the rock (section 15.2.10). Such water may continue to add heat to the ventilating airstream after it has entered the mined opening.

Fissure water is often emitted at a temperature close to the virgin rock temperature (VRT). In circumstances of local geothermal activity or radioactive decay, it may even be higher. The total heat load on the mine environment can be calculated from the flowrate and the drop in temperature of the water between the points of emission and effective exit from the mine ventilation system.

Example A mine produces 5 million litres of water per day, emitted at an average temperature of 42 °C. In the shaft sump where the water is collected for pumping, the water temperature is 32 °C. Determine the heat load from the water on the mine ventilation system.

Solution

heat load = mass flow × specific heat × drop in temperature

$$= \frac{5\,000\,000}{24 \times 60 \times 60} \times \frac{4187}{1000} \times (42 - 32) \quad \frac{\text{kg}}{\text{s}}\,\frac{\text{kJ}}{\text{kg}\,^\circ\text{C}}\,^\circ\text{C}$$

$$= 2423 \text{ kW}$$

The rate at which heat is emitted depends on the difference between the temperature of the air and the water, and whether the water is piped or in open channels. In the latter case, cooling by evaporation will be the major mode of heat transfer and will continue while the temperature of the water exceeds the wet bulb temperature of the air. Hot fissure water should be allowed no more than minimum direct contact with intake air. The most effective means of dealing with this problem are (a) to transport the water in closed pipes and (b) to restrict water flow routes to return airways. A less effective but expedient measure is to restrict air–water contact by covering drainage channels by boarding or other materials.

Where chilled service water is employed, the run-off may be at a temperature below that of the local wet bulb temperature. While this continues, the water will continue to absorb both sensible and latent heat, thus providing a cooling effect on the airflow.

In, or close to, the working areas of many mines, it is often inevitable that open drainage channels will be used. Similarly, the footwalls or floors may have areas that are covered with standing, or slowly moving, water.

The transfer of sensible and latent heat between the surface of the water and the air may be determined from equations (15.34) and (15.35), i.e.

$$q_{sen} = h_c(\theta_{ws} - \theta_d) \quad \text{W/m}^2$$

and

$$q_L = 0.0007\, h_c L_{ws} \frac{e_{ws} - e}{P} \quad \mathrm{W/m^2}$$

where subscript ws means water surface and d means dry bulb.

The heat transfer coefficient for the water surface, h_c, is the least certain of the variables, depending on the geometry of the airway, the condition of the airflow (in particular, the local air velocity) and the degree of turbulence existing in the flowing water. An approximation may be obtained from equations (15.22) and (15.23) using the Reynolds' number for the airway and a value of coefficient of friction, f, that represents the degree of water turbulence on the liquid surface. This may vary from 0.002 for a calm flat surface to 0.02 for a highly turbulent surface.

Example Water flows at a rate of 60 l/s along an open drainage channel that is 0.5 m wide and 0.25 m deep. The channel is located in an airway of cross-sectional area 12 m^2, perimeter 14 m and which passes an airflow of 50 m^3/s. Over a 10 m length of channel, the following mean values are given:

water temperature, θ_{ws}	32 °C
air dry bulb temperature, θ_d	30 °C
air wet bulb temperature, θ_w	25 °C
barometric pressure, P	112 kPa

Determine the sensible and latent heat emitted from the water to the air and the corresponding drop in water temperature over the 10 m length. Assume that no heat is added to the water from the strata.

Solution

1. *Determine the heat transfer coefficient for the water surface.*

$$\text{volume flow of water} = \frac{60}{1000} = 0.06\ \mathrm{m^3/s}$$

$$\text{velocity of water} = \frac{0.06}{0.5 \times 0.25} = 0.48\ \mathrm{m/s}$$

At this velocity, it is estimated that ripples on the water surface will have an equivalent coefficient of friction, $f = 0.008$. The Reynolds' number for the airway is

$$\mathrm{Re} = 268\,000 \frac{Q}{\mathrm{per}}$$

$$= 268\,000 \times \frac{50}{14} = 957\,140$$

Equation (15.22) gives the Nusselt number to be

$$N_u = 2127$$

(this can be estimated directly from Fig. 15.5). Hydraulic mean diameter of the airway,

$$d = \frac{4A}{\text{per}} = \frac{4 \times 12}{14} = 3.429 \text{ m}$$

The convective heat transfer coefficient is then given by equation (15.23). Using $k_a = 0.026$ W/(m °C),

$$h_c = \frac{2127 \times 0.026}{3.429} = 16.1 \text{ W/(m}^2 \text{ °C)}$$

2. *Determine the psychrometric conditions.* Using the psychrometric relationships given in section 14.6, the following values are determined: for the water surface at 32 °C,

$$L_{ws} = 2426 \times 10^3 \text{ J/kg}$$
$$e_{ws} = 4753 \text{ Pa}$$

and for the air at $\theta_w = 25$ °C, $\theta_d = 30$ °C and $P = 112$ kPa,

$$e = 2805 \text{ Pa}$$

3. *Determine the heat transfers from the water to the air.* Sensible heat,

$$q_{sen} = h_c(\theta_{ws} - \theta_d)$$
$$= 16.1\,(32 - 30)$$
$$= 32.2 \text{ W/m}^2$$

Latent heat,

$$q_L = 0.0007\, h_c L_{ws} \frac{e_{ws} - e}{P}$$

$$= 0.0007 \times 16.1 \times 2426 \times 10^3 \times \frac{4753 - 2805}{112\,000}$$

$$= 475.5 \text{ W/m}^2$$

As the width of the drainage channel in the 10 m length is 0.5 m, the heat flows may be stated as

$$\text{sensible heat} = \frac{32.2 \times 10 \times 0.5}{1000} = 0.161 \text{ kW}$$

$$\text{latent heat} = \frac{475.5 \times 10 \times 0.5}{1000} = 2.378 \text{ kW}$$

These values illustrate the dominant effect of latent heat transfer and why it is important to prevent direct contact between the air and the water.

The total heat gained by the air from the water is the sum of the sensible and latent heat transfers,

$$q = 2.378 + 0.161 = 2.539\,\text{kW}$$

4. *Determine the drop in temperature of the water.* As 1 l of water has a mass of 1 kg, the mass flow of water, $m = 60$ kg/s. The total heat lost by the water, then becomes

$$q = m\Delta\theta C_w$$

where $\Delta\theta$ = change in water temperature (°C) and C_w = specific heat of water (4187 J/(kg °C), giving

$$\Delta\theta = \frac{q}{mC_w}$$

$$= \frac{2.539 \times 10^3}{60 \times 4187} = 0.0101\,°\text{C} \quad \text{in the 10 m length}$$

15.3.4 Oxidation

Coal and sulphide ore mines are particularly liable to suffer from a heat load arising from oxidation of fractured rock. An estimate of the heat produced can be determined from the rate of oxygen depletion.

Assuming complete combustion of carbon

$$C + O_2 \rightarrow CO_2$$

$$12\,\text{kg} + 32\,\text{kg} \rightarrow 44\,\text{kg} \qquad (15.52)$$

Each 1 kg of oxygen consumed oxidizes 12/32 kg of carbon. Taking the calorific value of carbon as 33 800 kJ/kg gives a corresponding heat production of (12/32) × 33 800 = 12 675 kJ of heat per kg of oxygen used.

Similarly, for complete oxidation of sulphur,

$$S + O_2 \rightarrow SO_2$$

$$32\,\text{kg} + 32\,\text{kg} \rightarrow 64\,\text{kg} \qquad (15.53)$$

Each 1 kg of oxygen consumed will oxidize 1 kg of sulphur (calorific value 9304 kJ/kg) to produce 9304 kJ of heat.

Example 10 m^3/s of air enters a working district in a coal mine at an oxygen content of 21% by volume, and leaves at 20.8%. Calculate the heat generated by oxidation assuming complete combustion.

Solution The equivalent volume flow of oxygen at entry,

$$0.21 \times 10 = 2.1\,\text{m}^3/\text{s}$$

The equivalent volume flow of oxygen at exit,

$$0.208 \times 10 = 2.08\,m^3/s$$

Taking the density of oxygen to be 1.3 kg/m^3 gives the oxygen depletion rate to be

$$(2.1 - 2.08) \times 1.3 = 0.026\,kg/s$$

Heat produced,

$$0.026 \times 12\,675 = 330\,kW$$

The type of calculation illustrated by this example is useful in determining the heat produced by oxidation in an existing mine or section of a mine where air samples can be taken for analysis in intake and return airways. The extent to which oxidation takes place depends on the mineralogical content of the rock, the psychrometric condition of the air and the surface area exposed. These factors make it very difficult to predict heat loads from oxidation by other than empirical means.

In the case of open surfaces in operating airways and working places, the heat of oxidation will be an immediate and direct load on the mine ventilation system. On the other hand the heat that is produced by oxidation in waste areas, old workings or caved stopes will initially be removed only partially by leakage air. The remainder will be retained causing a rise in the temperature of the rock. This, in turn, will enhance the rate of heat transfer to the leakage air and, in most cases, an equilibrium is reached when the heat removed by the air balances the heat of oxidation. In minerals liable to spontaneous combustion, however, the increased temperature of the rock will accelerate the oxidation process to an extent that the balance between heat produced and heat removed may not be attained. In such cases the temperature will continue to rise until the rock becomes incandescent resulting in a spontaneous fire. (section 21.4).

15.3.5 Explosives

The heat that is produced during blasting varies with the type of explosives used and charge density. The amount of heat released for most explosives employed in mining falls within the range of 3700 kJ/kg for ANFO to 5800 kJ/kg for nitroglycerine (Whillier, 1981). This heat is dispersed in two ways. First, a fraction of it will appear in the blasting fumes and cause a peak heat load on the ventilation system. In mines where a re-entry period is enforced following blasting, this peak load will have cleared prior to personnel being readmitted to the area.

Secondly, the remainder of the heat will be stored in the broken rock. The magnitude of this will depend on the mining method. If the rock is blasted into a free space through which the ventilating airstream passes, such as in an open stope or on a longwall face, and the fragmentation is high, then as much as 40–50% of the heat produced by the explosive may be removed rapidly as a peak load with the blasting fumes. On the other hand, if the blast occurs within a region through which there is little or no ventilation such as in sublevel or forced caving techniques, or if the fragmentation is poor, then a much larger proportion of the heat will be retained in the rock.

Example In a 2000 t blast, the charge density of ANFO is 0.8 kg/t. It is estimated that 20% of the blast heat will be removed within 1 h with the blasting fumes. (1) Calculate the mean value of the rate of heat removed by the airflow during this hour. (2) If the specific heat of the rock is 950 J/(kg°C) determine the average increase in temperature of the rock due to blasting.

Solution

1. Mass of explosive used:

$$2000 \times 0.8 = 1600 \text{ kg ANFO}$$

Heat produced by ANFO,

$$3700 \times 1600 \frac{\text{kJ}}{\text{kg}} \text{kg}$$
$$= 5\,920\,000 \text{ kJ}$$

of which 20% is removed in the blasting fumes over 1 h. Rate of heat removal with blasting fumes,

$$\frac{5\,920\,000}{3600} \times 0.2 \text{ kJ/s}$$
$$= 329 \text{ kW}$$

2. Heat retained in rock,

$$5\,920\,000 \times 0.8 = 4\,736\,000 \text{ kJ}$$

Rise in rock temperature,

$$\frac{\text{heat retained}}{\text{mass} \times \text{specific heat}} = \frac{4\,736\,000}{2000 \times 1000 \times 0.950} \frac{\text{kJ}}{\text{kg}} \frac{\text{kg}}{\text{kJ}} {}^\circ\text{C}$$
$$= 2.49\,^\circ\text{C}$$

15.3.6 Falling rock

When strata or fractured rock moves downward under gravitational effects then the reduction in potential energy will ultimately produce an increase in temperature through fragmentation, impact, braking or other frictional effects. The fraction of the resulting heat that produces a load on the ventilation system again depends on the exposure of the broken rock to a ventilating airstream. Hence, for example, a large amount of heat is produced by the mass of superincumbent strata subsiding through to the surface over a period of time. Fortunately, the vast majority of this is retained within the rock mass, raising its temperature slightly, and little enters the mine ventilation system.

On the other hand, mineral descending through an ore pass or vertical bunker may immediately, or subsequently, be exposed to a ventilating airstream. The loss of potential energy of the rock will then appear as a heat load on the ventilation system. This can be significant.

The rise in temperature of falling rock, $\Delta\theta_r$, is given by the expression

$$\Delta\theta_r = \frac{\Delta z g}{C} \quad {}^\circ\mathrm{C} \tag{15.54}$$

where Δz is the distance fallen (m), g is the gravitational acceleration (9.81 m/s^2) and C is the specific heat of rock (J/(kg °C)), or

$$\Delta\theta_r = 981/C \quad {}^\circ\mathrm{C}\ \text{per 100 m fall} \tag{15.55}$$

15.3.7 Fragmented rock

When fragmented rock is exposed to a ventilating airstream and there is a temperature difference between the rock and the air, then heat transfer will take place. This will occur on working faces, drawpoints, conveyors and along other elements of a mineral transportation system.

The heat load from broken rock is given by:

$$mC(\theta_1 - \theta_2) \quad \mathrm{kW} \tag{15.56}$$

where m is the mass flow of rock (kg/s), C is the specific heat of rock (J/(kg °C)), θ_1 is the temperature of the broken rock immediately after fragmentation (°C) and θ_2 is the temperature of the rock at exit from the ventilation system (°C).

In most mining methods the temperature of the solid rock immediately prior to fragmentation will be less than VRT because of cooling of the rock surface. On the other hand, the process of fragmentation, whether by blasting or mechanized techniques, will raise the temperature of the broken rock. As shown in earlier sections, estimates may be made of these effects. However, in practice, the procedure may be simplified by assuming that the temperature of the newly broken rock, θ_1, is equal to VRT.

The mean temperature of the broken rock at exit from the system, θ_2, will depend on the degree of fragmentation, the exposure of rock surfaces to the airstream, and the velocity and psychrometric condition of the air. Mineral transported along a conveyor system will cool to a much greater extent than in a locomotive system. Furthermore, wetted material will yield up its heat more readily owing to evaporative cooling. Although the rate of heat transfer from a given particle size of rock to a known quantity and quality of airflow can be calculated, the many variables in any actual transportation system enforce the use of empirical measurements of temperature along the mineral transportation system. Such measurements are facilitated by the use of infrared thermometry.

Example Fragmented ore of specific heat 900 J/(kg °C) enters the top of an ore pass at a temperature of 35 °C and at a rate of 500 t/h. At an elevation 200 m below, the ore is discharged onto a conveyor and reaches the shaft bottom at a temperature of 32 °C. Calculate the heat transferred from the broken ore to the ventilation system between the top of the ore pass and the shaft.

Solution From equation (15.54), temperature rise of the rock in the ore pass

$$\frac{200 \times 9.81}{900} = 2.18\,^{\circ}\mathrm{C}$$

Temperature of rock at bottom of ore pass

$$35 + 2.18 = 37.18\,^{\circ}\mathrm{C}$$

Mass flow of ore,

$$m = \frac{500 \times 1000}{3600} = 138.9 \text{ kg/s}$$

Heat load from ore on the conveyor (from equation (15.56))

$$mC(\theta_1 - \theta_2) = 138.9 \times 0.900 \times (37.18 - 32)$$
$$= 647 \text{ kW}$$

15.3.8 Metabolic heat

The rate at which the human body produces metabolic heat depends on a number of factors including rate of manual work, physical fitness and level of mental stress. The question of heat transfer between the human body and the surrounding environment is of great importance in ascertaining the risk of heat stress in mine workers (Chapter 17). However, metabolic heat makes only a small contribution to mine heat load. It may, nevertheless, become significant where labour-intensive activities take place in a location of limited throughflow ventilation such as a poorly ventilated heading or a barricaded refuge chamber.

Like other heat engines, the human body emits heat by three mechanisms. The most significant of these in physiological processes is heat loss from the body surface. Secondly, the large, wet surface area of the lungs provides an effective heat exchanger and 'exhaust' heat is emitted through respiration. Third, any mechanical work done by the individual on the external surroundings will produce frictional heat, unless that work is done against gravity. As the human body is an inefficient heat engine, this latter mechanism is the least significant and is often omitted in calculations of physiological heat transfer.

The heat produced by a fit worker who is acclimatized to the environment will produce metabolic heat at rates varying from about 100 W for sedentary work, through 400 W for a medium level of activity such as walking, to over 600 W during intermittent periods of strenuous manual work. (see Table 17.1).

REFERENCES

Carrier, W. H. (1940) Air cooling in the gold mines on the Rand *Trans. AIME* **141**, 176–287.

Carslaw, H. and Jaeger, J. C. (1956) *Conduction of Heat in Solids*, Oxford University Press.

Danko, G., *et al.* (1987) Development of an improved method to measure *in situ* thermal rock properties in a single drill hole. *Proc. 3rd US Mine Ventilation Symp., Penn State*, 33–52.

Danko, G., *et al.* (1988) Heat, mass and impulse transport for underground airways. *Proc. 4th Int. Mine Ventilation Congr., Brisbane*, 237–47.

Deen, J. B. (1988) Laboratory verification of heat transfer analogies. *MS Report*, University of California, Berkeley.

Gibson, K. L. (1976) The computer simulation of climatic conditions in underground mines. *PhD Thesis*, University of Nottingham.

Goch, D. C. and Patterson, H. S. (1940) Heat flow into tunnels. *J. Chem., Metall. Min. Soc. S. Afr.* **41**, 117–28.

Hemp, R. (1965) Air temperature increases in airways. *J. Mine Vent. Soc. S. Afr.* **38**(2).

Kibble, J. D. (1978) Some notes on mining diesels. *Min. Technol.* (October), 393–400.

McPherson, M. J. (1986) The analysis and simulation of heat flow into underground airways. *Int. J. Min. Geol. Eng.*, **4**, 165–196.

Mousset-Jones, P. and McPherson, M. J. (1984) Measurement of *in situ* thermal parameters in an underground mine. *Proc. 2nd Annu. Workshop, Generic Mineral Technology Center (Mine System Design and Ground Control), USA*, 113–31.

Mousset-Jones, P., *et al.* (1987) Heat transfer in mine airways with natural roughness. 3rd *US Mine Ventilation Symp., Penn State.*

Nunner, W. (1956) *Z. Ver. Dtsch. Forsch.* **455**.

Stroh, R. (1979) A note on the downcast shaft as a thermal flywheel. *J. Mine Vent. Soc. S. Afr.* **32**, 77–80.

Whillier, A. (1981) Predicting cooling requirements for caving and sublevel stoping in hot rock. *Int. Conf. on Caving and Sublevel Stoping, AIME, Denver.*

Whillier, A. (1982) *Environmental Engineering in South African Mines*, Chapter 19.

FURTHER READING

Bluhm, S. J. (1987) Chamber of Mines of South Africa, personal communication.

Hartman, H. L. (1982) *Mine Ventilation and Air Conditioning*, Chapter 20, Wiley.

Hemp, R. (1982) *Environmental Engineering in South African Mines*, Chapter 22, Mine Ventilation Society of South Africa.

McPherson, M. J. (1976) Refrigeration in South African gold mines. *Min. Eng.* (February), 245–258.

Robinson, G., *et al.* (1981) Underground environmental planning at Boulby Mine, Cleveland Potash Ltd. *Trans. Inst. Min. Met.* (July).

Stewart, J. (1982) Fundamentals of heat stress. *Environmental Engineering in South African Mines* Chapter 20, 495–533.

Verma, Y. K. (1981) Studies in virgin strata temperatures, with special reference to the NCB's Western Area Mines. *Min. Eng.* **140**(234), 655–63

A15 APPENDICES: MATHEMATICAL BACKGROUND

15A.1 Solution of the three-dimensional transient heat conduction equation (15.13) as obtained by Carslaw and Jaeger

The solution to the radial heat conduction equation 16 given by Carslaw and Jaeger (1956) is

$$G\frac{4}{\pi^2}\int_0^\infty \frac{\exp(-V^2T)\mathrm{d}V}{[I_0(V)+V/DI_1(V)]^2+[Y_0(V)+V/DY_1(V)]^2V}$$

where I_0, I_1, Y_0 and Y_1 are Bessel functions and V is the variable of integration.

15A.2 Gibson's algorithm for computation of dimensionless temperature gradient, *G*

$F=\alpha t/r_s^2$ (Fourier number)

$B=hr_s/k$ (Biot number)

$x=\log_{10}(F)$

$y=\log_{10}(B)$

$c=x(0.000\,104x+0.000\,997)-0.001\,419$

$c=-\{x[x(xc-0.046\,223)+0.315\,553]+0.006\,003\}$

$d=y-[x(4x-34)-5]/120$

$d=0.949+0.1\exp(-2.69035d^2)$

$$m=\left\{(y-c)^2+\frac{216+5x}{70}\left[0.0725+0.01\tan^{-1}\left(\frac{x}{0.7048}\right)\right]\right\}^{1/2}$$

$n=(y+c-m)/2$

$G=10^n/d$

15A.3 Background to equations for the heat transfer coefficient, *h*

Consider the transfer of heat, $q(\mathrm{W/m^2})$, across the boundary layers shown on Fig. 15A.1. In the laminar sublayer, there are no cross-velocities and Fourier's law of heat conduction applies

$$q=-k_a\frac{\mathrm{d}\theta}{\mathrm{d}y}\quad \mathrm{W/m^2} \tag{15A.1}$$

where k_a = thermal conductivity of air (W/m °C)), θ = fluid temperature (°C) and y = distance from the wall. In the turbulent boundary layer (and within the main-

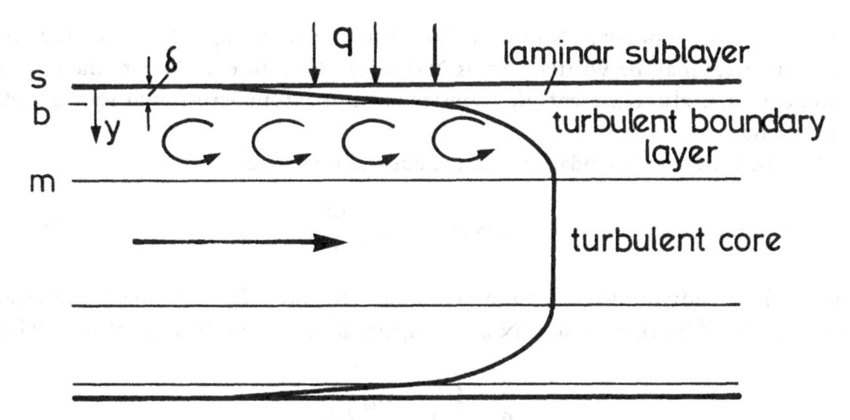

Figure 15A.1 Heat is transported by conduction and molecular diffusion across the laminar sublayer, and by eddy diffusion across the turbulent boundary layers.

stream), heat transfer is assisted by eddy action and the equation becomes

$$q = -(k_a + \rho_a C_p E_h)\frac{d\theta}{dy} \quad W/m^2 \tag{15A.2}$$

where ρ_a = air density (kg/m^3), C_p = specific heat (1005 J/(kg °C) for dry air) and E_h = eddy diffusivity of heat (m^2/s). The term $\rho_a C_p$ (J/(m^3 °C)) is the amount of heat transported in each m^3 for each Centigrade degree of temperature difference while E_h (m^2/s) represents the rate at which this heat is transported by eddy action. The product $\rho_a C_p E_h$ is much larger than k_a.

Equation (15A.2) can be rewritten as

$$\frac{q}{\rho_a C_p} = -\left(\frac{k}{\rho_a C_p} + E_h\right)\frac{d\theta}{dy} \quad \frac{m\,°C}{s}$$

$$= -\left(\alpha + E_h\right)\frac{d\theta}{dy}$$

where $\alpha = k_a/\rho_a C_p$ is the thermal diffusivity of air (m^2/s).

Figure 15A.1 illustrates two boundary layers only. In fact, a third transitional or **buffer** layer exists between the laminar and turbulent layers, within which both conduction and eddy diffusion are significant. This is similar to the transitional region between laminar flow and fully developed turbulence shown on Fig. 2.6 for pipes.

Let us now consider the transfer of momentum across the boundary layers. For the laminar sublayer where the flow is viscous, Newton's equation (2.22) applies

$$\tau = \mu\frac{du}{dy} \quad N/m^2 \tag{15A.3}$$

where τ = shear stress transmitted across each lamina of fluid (N/m^2 or Pa), μ = coefficient of dynamic viscosity (N s/m^2) and u = fluid velocity (m/s). Note that the units of shear stress may be written also as N s/(m^2 s). Therefore, as N s are the units of momentum, τ also represents the transfer of momentum across each square metre per second.

For the turbulent boundary layer, the equation becomes

$$\tau = (\mu + \rho_a E_m)\frac{du}{dy} \tag{15A.4}$$

where E_m = eddy diffusivity of momentum (m^2/s) and $\rho_a E_m$ = momentum transfer across each m^2 by eddy action (N s/m^2). Again, $\rho_a E_m$ is much larger than μ. Then

$$\frac{\tau}{\rho_a} = \left(\frac{\mu}{\rho_a} + E_m\right)\frac{du}{dy}$$

$$= (\nu + E_m)\frac{du}{dy}\ \frac{\text{J}}{\text{kg}} \tag{15A.5}$$

where $\nu = \mu/\rho_a$ = kinematic viscosity or momentum diffusivity (m^2/s).

In order to combine heat transfer and momentum transfer, Reynolds divided equations (15A.2) and (15A.5), giving

$$\frac{q}{\rho_a C_p}\frac{\rho_a}{\tau} = -\frac{\alpha + E_h}{\nu + E_m}\frac{d\theta}{dy}\frac{dy}{du}$$

or

$$\frac{q}{C_p \tau} = -\frac{\alpha + E_h}{\nu + E_m}\frac{d\theta}{du}$$

Reynolds argued that as the eddy components predominate and the eddy diffusivity for heat, E_h, must be closely allied to the eddy diffusivity for momentum, E_m, we may equate those terms, leaving

$$\frac{q}{C_p \tau} = -\frac{d\theta}{du}$$

Reynolds integrated this equation directly, making the erroneous assumption that u remained linear with respect to y across the composite boundary layers. However, both Taylor and Prandtl, working independently, later realized that the integration had to be carried out separately for the laminar sublayer and the turbulent boundary layer.

Laminar sublayer

From equation (15A.1)

$$q\int_0^\delta dy = -k_a\int_{\theta_s}^{\theta_b} d\theta$$

$$q\delta = k_a(\theta_s - \theta_b) \tag{15A.7}$$

where δ = thickness of laminar sublayer (m), θ_s = temperature at the surface (°C) and θ_b = temperature at the edge of the laminar sublayer. Similarly, integrating equation (15A.3) from zero velocity at the wall to u_b at the edge of the laminar sublayer.

$$\tau = \mu \frac{u_b}{\delta} \tag{15A.8}$$

Dividing equation (15A.7) by (15A.8) gives

$$\frac{q}{\tau} = \frac{k_a}{\mu} \frac{\theta_s - \theta_b}{u_b} \tag{15A.9}$$

Turbulent boundary layer

Here we employ the same equation (15A.6)

$$\frac{q}{C_p \tau} du = -d\theta$$

and integrate from the edge of the laminar sublayer (subscript b) to the mainstream (subscript m) (see Fig. 15A.1)

$$\frac{q}{C_p \tau} \int_{u_b}^{u_m} du = -\int_{\theta_b}^{\theta_m} d\theta$$

giving

$$\frac{q}{C_p \tau} = \frac{\theta_b - \theta_m}{u_m - u_b} \tag{15A.10}$$

A difficulty here is that neither θ_b nor u_b is amenable to accurate location or measurement. We can eliminate θ_b by rewriting equation (15A.9) as

$$\theta_b = \theta_s - \frac{q}{\tau} \frac{\mu}{k_a} u_b \tag{15A.11}$$

Substituting for θ_b in equation (15A.10) and engaging in some algebraic manipulation leads to

$$\frac{q}{C_p \tau} = \frac{\theta_s - \theta_m}{u_m} \frac{1}{1 + \frac{u_b}{u_m}\left(\frac{\mu C_p}{k_a} - 1\right)}$$

Now the combination $\mu C_p / k_a$ is another dimensionless number known as the Prandtl number P_r. Hence

$$\frac{q}{C_p \tau} = \frac{\theta_s - \theta_m}{u_m} \frac{1}{1 + \frac{u_b}{u_m}(P_r - 1)} \tag{15A.12}$$

Furthermore, the shear stress, τ, is related to the coefficient of friction, f, and the mainstream velocity, u_m, by equation (2.41):

$$\tau = f\rho_a \frac{u_m^2}{2}$$

Also, $q = h(\theta_s - \theta_m)$ from equation (15.16), where $\theta_m = \theta_d$ = dry bulb temperature of the mainstream. Substituting for τ and q in equation (15A.12),

$$\frac{h(\theta_s - \theta_m)}{C_p (f/2)\rho u_m^2} = \frac{\theta_s - \theta_m}{u_m} \frac{1}{1 + (P_r - 1)u_b/u_m}$$

$$\frac{h}{C_p \rho_a u_m} = \frac{f}{2} \frac{1}{[1 + (P_r - 1)u_b/u_m]}$$

The left-hand side of this equation can be separated into three dimensionless groups

$$\frac{hd}{k_a} \frac{\mu}{\rho_a u_m d} \frac{k_a}{C_p \mu} = \frac{N_u}{\text{Re}P_r}$$

Nusselt number (N_u) 1/(Reynolds' number (Re)) 1/(Prandtl number (P_r))

Hence

$$N_u = \frac{f}{2} \text{Re}P_r \frac{1}{1 + (P_r - 1)\, u_b/u_m} \tag{15A.13}$$

This is sometimes known as the Taylor or Taylor–Prandtl equation and is the basic relationship that has been used by numerous other workers. These varied in the manner in which they treated the u_b/u_m ratio as u_b remains elusive to measure. Taylor himself used an empirical value of $u_b/u_m = 0.56$. Several other relationships are listed in Table 15A.1. A more comprehensive listing is given by Danko (1988).

A theoretical and experimental study of flow over rough surfaces by Nunner (1956) produced a more sophisticated equation:

$$N_u = \frac{f}{2} \text{Re}\, P_r \frac{1}{1 + \dfrac{1.5}{\text{Re}^{1/8} P_r^{1/6}} \left(P_r \dfrac{f}{f_0} - 1 \right)} \tag{15A.14}$$

where f_0 = friction coefficient for a smooth tube at the same value of Reynolds' number. This is known as Nunner's equation.

Table 15A.1 Values of the u_b/u_m ratio

Authority	*Value of u_b/u_m*
Taylor	0.56
Von Kármán	$9.77\sqrt{f/2}$
Rogers and Mayhew	$1.99\ \text{Re}^{-0.125}$

An analysis of hydraulically smooth pipes by Colburn gave

$$N_u = 0.023\,\mathrm{Re}^{0.8} P_r^{\,0.4} \tag{15A.15}$$

For air in the atmospheric range, the Prandtl number is near constant. Using the values at 20 °C,

$$\mu = (17 + 0.045 \times 20) \times 10^{-6} = 17.9 \times 10^{-6}\,\mathrm{N\,s/m^2}$$

$$k_a = 2.2348 \times 10^{-4}(273.15 + 20)^{0.8353} = 0.0257\,\mathrm{W/(m\,°C)}$$

$$C_p = 1005\,\mathrm{J/(kg\,°C)}$$

gives

$$P_r = \frac{C_p \mu}{k_a} \tag{15A.16}$$

$$= \frac{1005 \times 17.9 \times 10^{-6}}{0.0257} = 0.700$$

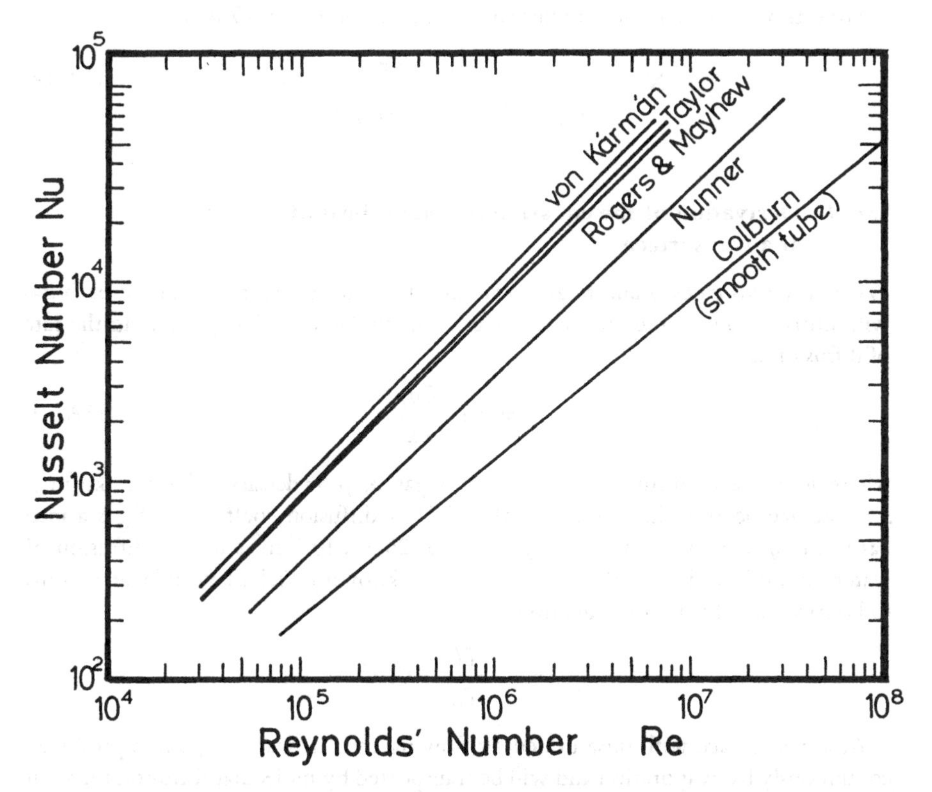

Figure 15A.2 Relationships between Nusselt number and Reynolds' number proposed by various authorities for Pr = 0.7 and $f = 0.02$.

In order to illustrate the differences between the relationships quoted, Fig. 15A.2 has been produced for values of $P_r = 0.7$ and $f = 0.02$. There is reasonable agreement within the Von Kármán, Taylor and Rogers equations. The Nunner relationship produces Nusselt numbers a little less than one-half those given by the other authorities. As might be expected, the Colburn smooth tube equation gives much lower Nusselt numbers and, hence, heat transfer coefficients. Experimental evidence in mine airways (Mousset-Jones *et al.*, 1987; Danko *et al.*, 1988) and also in scale models under controlled conditions (Deen, 1988) has indicated a better correlation with the Nunner equation than the others. Hence, this is the relationship suggested for use in underground openings at the present time. Ongoing research will no doubt produce further equations or algorithms for N_u.

One further problem remains with equation (15A.14), the evaluation of f_0, the coefficient of friction for a hydraulically smooth tube at the same Reynolds' number. However, the approximation given in section 2.3.6 gives acceptable results for the range of Reynolds' numbers common in mine airways:

$$f_0 = 0.046\,\mathrm{Re}^{-0.2} \qquad (15A.17)$$

Substituting for f_0 in the Nunner equation and using $P_r = 0.7$ gives

$$N_u = \frac{0.35 f\,\mathrm{Re}}{1 + \dfrac{1.592}{\mathrm{Re}^{0.125}}(15.217 f\,\mathrm{Re}^{0.2} - 1)} \qquad (15A.18)$$

15A.4 Derivation of the equation for latent heat of evaporation at a wet surface

When any two gases, a and b, are brought into contact then molecules of each gas will diffuse into the other at the interface. The fundamental law governing the rate of diffusion is

$$m = D_v \frac{\partial \rho_a}{\partial x} \qquad (15A.19)$$

where m = rate of diffusion of gas a into gas b, ρ_a = density of gas a (kg/m^3), x = distance beyond the interface (m) and D_v = diffusion coefficient of gas a into gas b (m^2/s) (experimental results give this as 25.5×10^{-6} m^2/s for the diffusion of water vapour into air (ASHRAE, 1985)). This is known as Fick's law and is analogous to Fourier's law for heat conduction:

$$q = -k \frac{\partial \theta}{\partial x} \quad \mathrm{W/m^2}$$

At a wet surface with unsaturated air flowing over it, water vapour is produced continuously by evaporation and will be transported by molecular diffusion through the laminar sublayer and then, additionally, by eddy action through the transitional (buffer) and turbulent layers. The transfers of heat, momentum and mass are all affected in the same way by eddies at any given surface. Hence, the governing equations have the same form for all three. Just as we had (see equation (15.6))

$$q = h(\theta_s - \theta_d) \quad \mathrm{W/m^2}$$

for heat flow across the boundary layers, so we can write an analogous equation for the migration of water molecules:

$$m = h_m (\rho_{vws} - \rho_v) \quad \mathrm{kg/(s\,m^2)} \tag{15A.20}$$

where m = mass of water vapour migrating from each square metre, h_m = mass transfer coefficient (m/s), ρ_{vws} = density of water vapour at the wet surface and ρ_v = density of water vapour in the mainstream. The density difference $\rho_{vws} - \rho_v$ provides the potential or driving mechanism for the diffusion of mass.

Now from the general gas law (see equation (3.11))

$$e = \rho_v R_v T \quad \mathrm{Pa}$$

where e = partial pressure of water vapour (Pa), R_v = gas constant for water vapour (461.5 J/(kg K) and T = absolute temperature (K). Hence, equation (15A.20) becomes

$$m = h_m \frac{1}{R_v T}(e_{ws} - e) \quad \frac{\mathrm{kg}}{\mathrm{s\,m^2}} \tag{15A.21}$$

e is the vapour pressure within the main airstream and e_{ws} the vapour pressure at the wet surface. It is reasonable to assume that, immediately adjacent to the liquid surface, the space will be saturated. e_{ws} is then also the saturated vapour pressure at the wet surface temperature, θ_{ws}.

Now, the general gas law for a mixture of air and water vapour gives

$$\rho_m = \frac{P}{R_m T} \quad \mathrm{Pa}$$

where ρ_m = density of moist air (kg/m^3), P = barometric pressure (Pa) and R_m = gas constant for moist air (J/(kg K)). However, for the atmospheric range in mines, R_m differs negligibly from the gas constant for dry air, R_a = 287.04J/(kg K) (section 14.2.4). Hence, we can express equation (15A.21) as

$$\begin{aligned} m &= h_m \frac{\rho_m}{R_v T} \frac{1}{\rho_m}(e_{ws} - e) \\ &= h_m \frac{\rho_m}{R_v T} \frac{R_a T}{P}(e_{ws} - e) \\ &= h_m \rho_m 0.622 \frac{e_{ws} - e}{P} \quad \frac{\mathrm{kg}}{\mathrm{s\,m^2}} \end{aligned} \tag{15A.22}$$

as

$$\frac{R_a}{R_v} = \frac{287.04}{461.5} = 0.622$$

Now let us turn our attention to the mass transfer coefficient, h_m. In 1934, Chilton and Colburn, investigating the analogy between heat transfer and mass diffusion, produced a relationship between the convective heat transfer coefficient, h_c, and the mass transfer coefficient, h_m. In its most compact form (ASHRAE, 1985), this can be expressed as

$$h_m = \frac{h_c}{C_p \rho_m} \left(\frac{P_r}{S_c} \right)^{2/3} \quad \text{m/s} \tag{15A.23}$$

where C_p = specific heat of air (J/(kg K)), P_r = is the Prandtl number for air (section 15A.3),

$$P_r = \frac{\mu_a C_p}{k_a}$$

where μ_a = dynamic viscosity (Ns/m^2) and, k_a = thermal conductivity (W/(m °C)) and S_c is yet another dimensionless number (Schmidt number):

$$S_c = \frac{\mu_a}{\rho_a D_v}$$

Both of these dimensionless numbers remain reasonably constant over the range of air pressures and temperatures found in subsurface facilities.

Using the values for air at 20 °C,

$$\mu_a = 17.9 \times 10^{-6} \, \text{N s/m}^2$$
$$k_a = 0.0257 \, \text{W/(m °C)}$$
$$C_p = 1010 \, \text{J/(kg K)}$$

mid-range value for moist air,

$$\rho_a = 1.2 \, \text{kg/m}^3$$

and

$$D_v = 25.5 \times 10^{-6} \, \text{m}^2\text{/s}$$

for the diffusion of water vapour into air, gives

$$P_r = 0.703$$

and

$$S_c = 0.585$$

Combining equations (15A.22) and (15A.23) gives

$$m = \frac{h_c}{C_p \rho_m} \left(\frac{P_r}{S_c} \right)^{2/3} \rho_m \, 0.622 \frac{e_{ws} - e}{P}$$

and, inserting the numerical values for P_r and S_c,

$$m = h_c \, 1.130 \times \frac{0.622}{C_p} \frac{e_{ws} - e}{P} \quad \frac{\text{kg}}{\text{s m}^2} \tag{15A.24}$$

Now the latent heat emitted when m kg of surface water is evaporated is

$$q_L = m\,L_{ws} \qquad \frac{\text{kg}}{\text{s}\,\text{m}^2}\frac{\text{J}}{\text{kg}} = \frac{\text{W}}{\text{m}^2} \tag{15A.25}$$

where

$$L_{ws} = (2502.5 - 2.386\,\Theta_{ws})\,10^3 \quad \text{J/kg}$$

is the latent heat of evaporation at the temperature of the wet surface (see equation (14.6)). Substituting for m from equation (15A.24) gives

$$q_L = h_c\;1.130 \times \frac{0.622\,L_{ws}}{C_p}\frac{e_{ws} - e}{P} \quad \text{W/m}^2 \tag{15A.26}$$

or, using the value $C_p = 1010\,\text{J/(kg K)}$ for moist air,

$$q_L = 0.0007\,h_c L_{ws} \frac{e_{ws} - e}{P} \quad \frac{\text{W}}{\text{m}^2} \tag{15A.27}$$

Equation (15A.27) may also be written as

$$q_L = h_e\,(e_{ws} - e) \quad \text{W/m}^2 \tag{15A.28}$$

where the evaporative heat transfer coefficient, h_e, is given as

$$h_e = 0.0007\,h_c \frac{L_{ws}}{P} \quad \frac{\text{W}}{\text{m}^2\,\text{Pa}} \tag{15A.29}$$

The ratio h_e/h_c is known as the Lewis ratio (LR). Using the latent heat of evaporation of water at 20 °C (2455×10^3 J/kg) and a barometric pressure of 100×10^3 Pa gives

$$\text{LR} = \frac{h_e}{h_c} = 0.017 \quad \frac{^\circ\text{C}}{\text{Pa}} \tag{15A.30}$$

The Lewis ratio changes slowly through the atmospheric range of pressures and temperatures.

Another correlation with psychrometric relationships occurs in equation (15A.26). The group $0.622\,L_{ws}/C_p$ is the reciprocal of the psychrometric 'constant' introduced in section 14.4.2.

References for appendices

ASHRAE (1985) *Fundamentals Handbook in SI Units,* Chapters 3 and 5, American Society of Heating, Ventilating and Air Conditioning Engineers.

Carslaw, H. and Jaeger, J. C. (1956) *Conduction of Heat in Solids*, Oxford University Press.

Danko, G. *et al.* (1988) Heat, mass and impulse transport for underground airways. *Proc. 4th Int. Mine Ventilation Congr., Brisbane*, 237–47.

Deen, J. B. (1988) Laboratory verification of heat transfer analogies. *MS Report*, University of California, Berkeley.

Mousset-Jones, P., *et al.* (1987) Heat transfer in mine airways with natural roughness. *3rd US Mine Ventilation Symp. Penn State.*
Nunner, W. (1956) *Z. Ver. Dtsch. Forsch.* **455**.

Further reading for appendices

Chilton, T. H. and Colburn, A. P. (1934) Mass transfer (absorption) coefficients—prediction from data on heat transfer and fluid friction. *Ind. Eng. Chem.* (November), 1183.

16

Simulation of climatic conditions in the subsurface

16.1 BACKGROUND

The complexities of the relationships that govern heat flow from the strata into ventilated underground openings are illustrated by the analyses given in section 15.2. Indeed, the routine use of those relationships become practical only through the availability of computer assistance. The first computer programs to simulate heat flow into mine workings were developed in South Africa (Starfield, 1966a,b). The early programs estimated strata heat flow from the Goch and Patterson tables (section 15.2.6), either by interpolation or from regression-fitted equations that approximated those tables. Since that time, simulation programs of increasing sophistication have been developed in a number of countries. Current programs recognize the influence of boundary layers close to the rock–air interface, allow for heat sources other than the strata (section 15.3) and predict the psychrometric effects of heat and moisture additions on the mine climate.

The common feature of mine climate simulation models is that they are based on solutions of the fundamental equation for heat conduction (equation (15.13)), and on utilization of the dimensionless Fourier and Biot numbers. However, the programs may vary in the manner in which they determine rock surfaces temperatures and heat transfer coefficients, and in the characterization and treatment of wet surfaces (Mousset-Jones, 1988).

Individual programs may be constrained to particular geometries of stopes or working faces, while others have been written for airways or headings. Some involve empirical relationships that are applicable only to specified layouts or methods of working. Again, some program packages allow combinations of airways, headings and working faces within a network structure while others are essentially 'single airway' simulators that must either be run separately for each branch or used in conjunction with a ventilation network analysis package (section 16.3.5).

In this chapter, we shall examine the essential features of mine climate simulation programs and how such programs may be utilized in the design of subsurface ventilation and air conditioning systems.

16.2 ELEMENTS OF MINE CLIMATE SIMULATION PROGRAMS

16.2.1 Organization of the programs

All mine climate simulations commence with the psychrometric condition of the air at the inlet end of the airway (or face) being defined by the user. This is normally accomplished by specifying the inlet wet bulb temperature, dry bulb temperature and barometric pressure. The program then divides the airway into incremental lengths, each of which is sufficiently short that wet and dry bulb temperatures may be assumed to be constant within the increment for the calculation of strata heat flows.

Each increment is traversed in the direction of airflow and the following parameters are calculated:

1. sensible and latent heat flows from the strata and other sources
2. change in moisture content of the air
3. change in dry bulb temperature
4. conversions between potential and thermal energies for shafts or inclined openings (autocompression)
5. change in barometric pressure
6. change in wet bulb temperature
7. other psychrometric parameters and indices of heat stress at the exit end of the increment

The conditions for the start of the next successive increment are then defined. Each incremental length is treated in this way until the complete airway has been traversed.

The following subsections outline the computational procedures involved in each of the steps listed.

16.2.2 Incrementation of airway length

The length of airway increment, Y_i, over which changes in temperature have no significant impact on strata heat flux will vary according to the magnitude of heat additions, the airflow and inclination of the airway. The value of Y_i may be (i) fixed at some small value (say 10 or 20 m) within the program, (ii) a fixed fraction of the total length or (iii) chosen by the user. Some programs may accept all three but use the smallest of those values of Y_i during the computational procedures.

16.2.3 Heat additions

In any incremental length of airway there will, in general, be transfers of both sensible and latent heat from one or more sources. In order to determine the corresponding psychrometric changes in the airflow, the sensible and latent heat components are each accumulated separately.

Strata heat

Let us consider the rock surface to be divided into wet and dry areas. We can then define a **wetness fraction**, w, as that fraction of the total surface area that is covered or coated with liquid water. The concept of wetness fraction is discussed further in section 16.3.1.

The area of wet surface within the incremental length Y_i now becomes

$$A_w = 2\pi r_a Y_i w \quad \text{m}^2 \tag{16.1}$$

where r_a = the effective radius of the airway (perimeter/2π). The remaining dry surface area is

$$A_d = 2\pi r_a Y_i(1 - w) \quad \text{m}^2 \tag{16.2}$$

Sensible heat transfer will take place on both the dry and wet surfaces while latent heat transfer occurs at the wet surface only. The normalized strata heat flows (per square metre) may be calculated using the methods described in sections 15.2.8 and 15.2.9.

Let us denote these heat flows as follows:

sensible heat from dry surface $q_{sen,d}$ (W/m^2)

sensible heat from wet surface $q_{sen,w}$ (W/m^2)

latent heat from wet surface $q_{L,w}$ (W/m^2)

The strata heat flowing into the increment, Y_i, is then the combination of

dry surface $$q_{sen,d}A_d = 2\pi r_a q_{sen,d} Y_i(1 - w) \quad \text{W} \tag{16.3}$$

wet surface $$q_{sen,w}A_w = 2\pi r_a q_{sen,w} Y_i w \quad \text{W} \tag{16.4}$$

wet surface $$q_{L,w}A_w = 2\pi r_a q_{L,w} Y_i w \quad \text{W} \tag{16.5}$$

where A = surface area (m) and subscripts w and d refer to wet and dry areas respectively.

Machines and other sources of heat

Here again, heat additions from operating equipment must be separated into components of sensible and latent heat. In the case of electrical plant, all of the power supplied may be considered as a sensible heat addition to the airflow, excepting any work that is done against gravity (section 15.3.2). The user must supply two values representing

1. the full power rating (kW) of the motor or device, FPR, reduced, if necessary, by the rate of work done against gravity, and

2. the machine utilization factor, MUF, defined as that fraction of time over which, if the machine was running at full load, would consume the same amount of energy as the actual intermittent operation of the device. Hence, the machine utilization factor for a motor running continuously at full load would be 1.0.

The value of sensible heat produced by the electrical device is then given as

$$\text{FPR} \times \text{MUF} \quad \text{W} \tag{16.6}$$

Diesel engines produce both sensible and latent heat (section 15.3.2). Here again, the user must supply a full power rating, FPR, and a machine utilization factor, MUF, for each piece of diesel equipment. The average machine load given as the product FPR × MUF (kW) is then converted to fuel consumption (l/h) using a mean empirical value (typically 0.3 l of fuel per kW engine rating per hour). The total heat produced is then simply the fuel consumption multiplied by the calorific value of the fuel (typical value 34 000 kJ/l).

Using the values quoted, the fuel consumed, FC, becomes

$$\text{FC} = \text{FPR} \times \text{MUF} \times \frac{0.3}{3600} \quad \text{l/s} \tag{16.7}$$

and

$$\text{the total heat produced} = \text{FC} \times 34\,000 \times 10^3 \quad \text{W} \tag{16.8}$$

The user may, of course, specify the rate of fuel consumption directly rather than through the factors on the right-hand side of equation (16.7). However, practical experience has indicated that most mine ventilation engineers find it more convenient to assess machine rating and utilization data than to acquire fuel consumption for individual diesel units.

In order to separate out the amount of water vapour produced by the diesel engine, the user must supply a third item of data, namely, the water/fuel ratio, WFR, defined as the number of litres of water (liquid equivalent) produced for each litre of fuel consumed. The combustion of 1 l of diesel fuel will produce between 1.1 and 1.5 l of water. However, this may be multiplied several times by engine cooling systems and exhaust scrubbers. Values as high as 9 l of water per litre of fuel have been reported (Mousset-Jones, 1988).

Water vapour is then added to the airstream at a rate of

$$\text{FC} \times \text{WFR} \quad \frac{\text{l fuel}}{\text{s}} \frac{\text{l water}}{\text{l fuel}} \tag{16.9}$$

i.e. kg of water vapour per second. The equivalent value of latent heat is given as

$$\text{FC} \times \text{WFR} \times L \quad \text{W} \tag{16.10}$$

where L = latent heat of evaporation of water in J/kg, while the sensible heat produced by the diesel becomes

$$\text{total diesel heat} - \text{latent diesel heat}$$

from equations (16.8) and (16.10) respectively.

Compressed air equipment adds no overall heat to a ventilating airstream and, indeed, can produce a small net cooling effect (section 15.3.2). Such equipment is usually ignored in mine climate simulations.

Other sources of heat (sections 15.3.3 to 15.3.8) may be specified by the user. In the general-purpose climate simulators, it is left to the user to select the sensible and latent heat components. However, special-purpose programs have been developed to handle cases such as ducts, pipes, cables, water channels or spray chambers located within the airway. Air coolers may be specified simply as negative heat sources.

A climate simulation program will compute the components of strata heat for each incremental length of airway. In the case of machines or other sources of heat, the user must identify the locations of those machines or other sources. This can be accomplished by specifying either (a) a spot source or (b) a distributed source.

A spot source will produce its heat at a fixed location and, hence, will appear in a single element of airway length. This is the type of heat source chosen for stationary pieces of equipment such as transformers, conveyor gearheads or mobile devices that move over short distances only. A distributed source, as the term implies, will spread its heat load over a distance specified by the user (e.g. length of a conveyor, hot or cold pipes, drainage channels). In a simple case, the heat will be distributed linearly over the selected distance. A more sophisticated program will allow the heat to be distributed according to a function specified by the user.

For each increment of airway length, the simulation program will determine and sum both the sensible and latent heat components from all sources relevant to that increment. We can now refer to the corresponding summations as $\sum q_{sen}$ and $\sum q_L$, respectively, for each increment.

16.2.4 Change in mositure content

The apparent density, ρ_{app}, of the air at inlet to the airway is calculated from the psychrometric condition of the air specified by the user for that starting location:

$$\rho_{app} = \frac{P - e}{287.04(\theta_d + 273.15)} \quad \frac{\text{kg of dry air}}{\text{m}^3 \text{ of air}}$$

(see equation (14.51)), where P = barometric pressure (Pa), e = actual vapour pressure (Pa) and θ_d = dry bulb temperature (°C). The mass flow, M, of the dry air component of the air–vapour mixture is then gives as

$$M = Q\rho_{app} \quad \text{kg dry air/s} \tag{16.11}$$

where Q is the known rate of airflow at inlet (m^3/s). The value of M remains constant along any given airway.

The rate at which water vapour is added to an increment of airway length is simply

$$\frac{\sum q_L}{L} \quad \frac{\text{J}}{\text{s}}\frac{\text{kg}}{\text{J}} = \frac{\text{kg}}{\text{s}}$$

where L = latent heat of evaporation (J/kg). The increase in moisture content of the air becomes

$$\Delta X = \frac{\sum q_L}{LM} \quad \text{kg/(kg dry air)} \tag{16.12}$$

Hence, if we use subscripts 1 and 2 for entry and exit from the incremental length, then

$$X_2 = X_1 + \Delta X \quad \text{kg/(kg dry air)} \tag{16.13}$$

16.2.5 Change in dry bulb temperature and autocompression

The steady-flow energy equation (3.25) gives

$$H_2 - H_1 - q_{12} = \frac{u_1^2 - u_2^2}{2} + (Z_1 - Z_2)g \quad \frac{\text{J}}{\text{kg}} \tag{16.14}$$

where H = enthalpy (J/kg), q_{12} = heat added (J/kg), u = air velocity (m/s) and Z = height above datum (m). However, equation (14.40) shows that the enthalpy term comprises a sensible heat component, $C_{pa}\,\Theta_d$, and a latent heat component, just as q_{12} involves both sensible and latent heat. If we subtract the latent heat component from both the H and the q terms, then equation (16.14) can be rewritten as

$$C_{pa}(\theta_{d,2} - \theta_{d,1}) - q_{12,sen} = \frac{u_1^2 - u_2^2}{2} + (Z_1 - Z_2)g \quad \frac{\text{J}}{\text{kg}}$$

where C_{pa} = specific heat for dry air (1005 J/(kg °C)). We must remember that $q_{12,sen}$ is the sensible heat addition in J/kg while our earlier sensible heat summation for the increment, $\sum q_{sen}$, was in watts. However,

$$q_{12,sen} = \frac{\sum q_{sen}}{M} \quad \frac{\text{J}}{\text{kg dry air}} \tag{16.15}$$

The increase in dry bulb temperature then becomes

$$\Delta\theta_d = \theta_{d,2} - \theta_{d,1} = \left[\frac{u_1^2 - u_2^2}{2} + (Z_1 - Z_2)g + \frac{\sum q_{sen}}{M}\right]\frac{1}{C_{pa}} \tag{16.16}$$

All terms on the right-hand side of equation (16.16) are known except the change in kinetic energy. This is caused by the change in density only, in an airway of fixed cross-sectional area and, hence, is very small. However, it can be estimated as follows:

$$\frac{u_1^2 - u_2^2}{2} = \frac{1}{2}\frac{Q_1^2 - Q_2^2}{A^2}$$

where Q = volume flowrate (m^3/s) and A = cross-sectional area (m^2). From equation (16.11), $Q = M/\rho_{app}$, giving

$$\frac{u_1^2 - u_2^2}{2} = \frac{1}{2}\frac{M^2}{A^2}\left(\frac{1}{\rho_{1,app}^2} - \frac{1}{\rho_{2,app}^2}\right) \quad \frac{\text{J}}{\text{kg}} \tag{16.17}$$

A difficulty arises here in that $\rho_{2,\mathrm{app}}$ requires a knowledge of the psychrometric conditions at the exit end of the increment. These are still unknown. However, as the change in kinetic energy is a very weak parameter, satisfactory results are obtained by using the change in air density established for the previous increment in equation (16.17). For the first increment in the airway, the kinetic energy correction may be ignored without significant loss of accuracy.

Equation (16.16) then enables the change in dry bulb temperature, $\Delta\theta_d$, to be determined, having taken into account both the sensible heat added and the change in elevation. The dry bulb temperature of the air leaving the increment is simply

$$\theta_{d,2} = \theta_{d,1} + \Delta\theta_d \quad ^\circ\mathrm{C} \tag{16.18}$$

16.2.6 Change in barometric pressure

The variation in absolute pressure of the air, P, is caused by (a) the conversion of mechanical energy into heat causing a frictional pressure drop, p, and (b) changes in elevation $Z_1 - Z_2$. The effect of changes in moisture content through any single element remains small provided that the incremental length is not excessive.

The frictional pressure drop is given by Atkinson's equation (5.2):

$$p = kY_i \frac{\mathrm{per}}{A} u^2 \quad \mathrm{Pa} \tag{16.19}$$

where k = Atkinson's friction factor ($\rho f/2\,\mathrm{kg/m^3}$) and per = airway perimeter (m). This can be converted into the work done against friction

$$F_{12} = p/\rho_m \quad \mathrm{J/kg} \tag{16.20}$$

where ρ_m = mean air density in the increment ($\mathrm{kg/m^3}$). Here again, ρ_1 may be used for the first increment and then, subsequently, ρ_m extrapolated from the previous increment.

Assuming polytropic flow through the incremental length of airway, the steady flow energy equation written in the form of equation (8.1) gives

$$R_m(T_2 - T_1)\frac{\ln(P_2/P_1)}{\ln(T_2/T_1)} = \frac{u_1^2 - u_2^2}{2} + (Z_1 - Z_2)g - F_{12} \quad \mathrm{J/kg} \tag{16.21}$$

The gas constant is given as (see equation (14.14))

$$R_m = \frac{287.04 + 461.5X}{1 + X} \quad \mathrm{J/(kg\,K)}$$

and T = absolute dry bulb temperature (θ_d + 273.15 K) where X and θ_d are the mean values for the increment.

The outlet pressure, P_2, remains the only unknown in equation (16.21) and, hence, may be calculated.

16.2.7 Change in wet bulb temperature

The psychrometric condition of air is completely specified if any three of its psychrometric properties are defined. At the exit from the incremental length of airway, the moisture content, X_2, the dry bulb temperature, $\theta_{d,2}$, and the barometric pressure, P_2, have now all been determined. All other psychrometric parameters can then be calculated from the equations given in section 14.6.

In particular, the wet bulb temperature, $\theta_{w,2}$, may be found from the following equations:

$$X_2 = \frac{S_2 - 1005\theta_{d,2}}{L_{w,2} + 1884(\theta_{d,2} - \theta_{w,2})} \quad \frac{\text{kg}}{\text{kg dry air}}$$

where

$$S_2 = L_{w,2}X_{s,2} + 1005\theta_{w,2} \quad \text{J/(kg dry air)}$$

$$L_{w,2} = (2502.5 - 2.386\theta_{w,2})1000 \quad \text{J/kg}$$

$$X_{s,2} = 0.622\frac{e_{sw,2}}{P - e_{sw,2}} \quad \frac{\text{kg}}{\text{kg dry air}}$$

and

$$e_{sw,2} = 610.6 \exp\left(\frac{17.27\theta_{w,2}}{237.3 + \theta_{w,2}}\right) \quad \text{Pa}$$

In this sequence of equations, $\theta_{w,2}$ is the only unknown independent parameter. Hence, by assuming an initial value, the equations may be cycled iteratively until a value of $\theta_{w,2}$ is found that satisfies all the equations simultaneously.

16.2.8 Relative humidity and saturation conditions

One of the psychrometric variables that may be calculated for each incremental length of airway is the relative humidity, defined as (see equation (14.54))

$$\text{rh} = \frac{e}{e_{sd}}$$

where e = actual vapour pressure (Pa) and e_{sd} = saturated vapour pressure at dry bulb temperature (Pa). If the result of this calculation is a value exceeding unity (supersaturation) then this indicates that condensation will take place. Such condensation will occur on all surfaces with a temperature less than that of the air wet bulb temperature and also as a fog within the airstream. Indeed, the former will occur even if the relative humidity is less than 1.0. If the computed psychrometric variables indicate supersaturation, then the heat of condensation released causes the dry bulb temperature to rise until saturation is attained at the same level of enthalpy as the original supersaturation condition. This is approximated closely by setting the revised dry bulb temperature equal to the current value of wet bulb temperature.

16.3 USING A MINE CLIMATE SIMULATOR

The first stage of commissioning a mine climate simulation system is to obtain a program package that provides the required features. The pointers given in section 7.4.5 for network simulation packages apply equally well here. Secondly, for any given airway or network of airways, the data must be assembled. There is a much greater variety of information required for the prediction of psychrometric conditions along each airway than for the ventilation network analyses described in Chapter 7. Furthermore, some of those data may, initially, be of uncertain precision. Hence, the correlation exercises that must precede employment of a climate simulator for long-term planning will normally involve modifications of the initial data.

This section deals with data preparation, correlation trials, the procedures involved in running a mine climate simulator, and design exercises that can be carried out with the assistance of such program packages.

16.3.1 Data preparation

The parameters that influence heat flow into a subsurface airway were introduced in section 15.2.1. The following subsections deal more specifically with the data that must be quantified for each airway in preparation for a mine climate simulation.

Physical description of airway

Geometry

The parameters required are airway length, cross-sectional area, perimeter, and levels of both the inlet and exit ends of the airway. It is normally assumed that the cross-sectional area, perimeter and gradient each remain constant throughout the length of the airway. Should there be significant deviations in any of these parameters then the airway should be divided into two or more sublengths for simulation.

Friction factor

The roughness of the airway lining is normally quantified as the Atkinson friction factor (section 5.2).

Age

The program package may allow the age of each end of the airway to be specified separately. The age of each simulated increment of airway length is then computed assuming a uniform rate of drivage. This feature takes into account the time taken to develop the airway and is particularly useful for advancing headings.

Wetness fraction

This parameter was introduced in section 16.2.3 as the fraction of airway surface that is covered or coated with liquid water. However, we have not yet discussed the actual distribution of water over the surface. Furthermore, although the strata heat flow analysis of section 15.2 considered both dry and wet surfaces, the solution of the basic equation for heat conduction assumed radial heat flow. In a simple case of a square airway with a completely wetted floor but dry sides and roof, the wetness fraction would be 0.25. However, the heat flow through the rock would no longer be radially symmetric.

The concept of wetness fraction, as originally introduced by Starfield in 1966, envisaged the wetness to be spread uniformly over the surface. With a surface which is damp but not completely covered in water, asperities on the rough surface may be dry on their peaks but with liquid water in the troughs. This water may be emitted through capillaries in the rock matrix or from discrete fractures before spreading out by a combination of surface capillarity and gravitational forces. The temperature of this type of surface will then take a value lying between those of the completely dry and completely wet surfaces analysed in section 15.2 and, hence, be dependent on the degree of wetting.

In practice, the wetting of airway surfaces is usually far from being uniform. Within a single airway there may be surface areas varying from completely wetted to visually dry. Starfield and Bleloch (1983) and Mack and Starfield (1985) proposed an 'equivalent wetness factor' for such circumstances, this being defined as the wetness fraction over a partially uniformly wetted surface that would give the same rates of heat and moisture transfer as the actual surface of non-uniform wetness.

Values of wetness fraction can be established for existing airways through correlation exercises (section 16.3.2). Furthermore, ventilation engineers who are experienced in running a climate simulation package become adept at estimating wetness fractions by visual inspection of airways—in much the same way that they can estimate friction factors from the appearance of surface roughnesses. Wetness fractions seldom fall below 0.04, even in airways that appear quite dry.

For planned but yet unconstructed airways, the wetness fraction may be estimated from previous experience of mining within that same geologic formation and at similar depths. If no such data are available then hydrologic studies and the projected use of service water can provide an indication of the potential average wetness of future workings. For particularly important projects, test drivages within the relevant horizon(s) will provide invaluable data regarding the migration of moisture from the strata.

Condition of airflow at inlet

Most mine climate simulators assume that the airflow and psychrometric condition of the air at inlet have remained constant since the airway was driven. In the majority of cases this gives acceptable results because of the thermal flywheel effect (section 15.2.2). Variations in the surface climate are usually well damped by the time the air reaches the working areas. If the transients that occur along intake airways are of concern then methods that employ the principles of superposition are available for such analyses (Hemp, 1982).

The inlet data required for conventional non-transient simulators are simply airflow, barometric pressures, and wet and dry bulb temperatures. Satisfactory results for hot mines can be obtained by assuming mean summer (not extreme) atmospheric conditions on surface.

Thermal parameters and other heat sources

The data in this category comprise the following.

Rock thermal conductivity
It is the effective thermal conductivity of the strata that should be specified. It is preferable that this parameter should have been established through *in situ* tests as described in section 15.2.10. If values are available from laboratory samples only, then corrections may be necessary to obtain the effective thermal conductivity that actually pertains underground (Mousset-Jones, 1988). Here again, correlation trials with a climate simulator are valuable in establishing values of effective thermal conductivity.

Rock thermal diffusivity
This can be measured directly *in situ* as well as by laboratory methods. However, it is normally satisfactory to determine thermal diffusivity indirectly as (see equation (15.13))

$$\alpha = \frac{k_r}{\rho_r C_r} \quad \mathrm{m^2/s}$$

where k_r = effective thermal conductivity (W/(m °C)), ρ_r = rock density (kg/m^3) and C_r = specific heat of the rock (J/(kg °C)).

Virgin rock temperature and geothermic step (section 15.2.4)
The combination of VRT at the inlet of the airway and the rate at which rock temperature increases with depth allows the VRT to be determined for each increment of length along a non-horizontal airway. These parameters may be determined from temperature logs taken from boreholes. In the case of a downcast shaft, the inlet VRT should actually be the rock temperature sufficiently far below the surface to be unaffected by variations in the surface climate. This is usually between 10 and 20 m.

Heat sources other than the strata
The location of every machine or other source of heating or cooling and the distance over which it extends must be specified. The magnitudes of both the sensible and latent heat components should be quantified either directly or indirectly by identifying a machine type (section 16.2.3).

16.3.2 Correlation tests

All simulation packages should be subject to correlation tests in order to verify that they do, indeed, simulate the real system to the required accuracy prior to their being employed for planning purposes. The accuracy of data must also be verified by site-specific correlation trials. This is true for ventilation network analysis (section 9.2.3) and is also the case for mine climate simulation studies.

Because of the uncertainty that may be associated with some of the data, correlation trials of mine climate simulators should include sensitivity runs involving those input parameters of uncertain accuracy. Experience has shown that such correlations conducted for a few airways in any mine can not only highlight previously unidentified sources of heat and humidity but also provide a range of typical data values that may be tested against other airways until sufficient confidence has been established to embark upon planning studies.

There is a definite procedure to follow in conducting sensitivity studies and correlations of climatic simulation output with actual conditions in a mine. The airways chosen for initial correlation should each be well established and continuous with no intermediate additions or losses of airflow. Any gradient from horizontal to vertical is acceptable provided that it remains uniform along the length of the airway. The primary trials should seek to provide correlation between computed results and the effects of strata heat and, perhaps, autocompression. Additional sources of heat that can be quantified easily, such as metered electrical equipment, may also be included. However, the initial trials should avoid airways that contain diesel equipment or open drainage channels. Such sources of heat should be subject to secondary correlation runs.

Careful measurements of airflow, barometric pressure, and wet and dry bulb temperatures should be made at the intake end of the correlation airway. Additional wet and dry bulb temperatures should also be taken at about 100 m intervals along the airway. All of the other parameters required as input (section 16.3.1) should be ascertained or ascribed initial estimated values.

In comparing the computed output with observed temperatures, attention should be focused first on the wet bulb temperatures. If there is a consistent divergent trend between the computed and measured values then it is probable that a continuous heat source has been over or underestimated, or perhaps even omitted entirely. A check should be carried out on the depths of the airway ends and their corresponding ages (if less than two years). Sensitivity runs should also be made to test the effect of thermal conductivity.

If the oberved wet bulb temperatures do not show a smooth trend then the reasons for discontinuities should be investigated. There may be occurrences such as leakage of air or other fluids from old workings, machine heat, or increased inflow of fissure water. Correlation exercises provide a valuable educational experience in tracing sources of heat in the mine.

When reasonable correlation ($\pm 1\,°C$) has been obtained on the wet bulb temperature, attention should be turned to the dry bulb temperature. Any remaining deviations of this parameter will almost certainly be due to the evaporation or condensation of water. If the deviation shows a consistent trend then it is likely that the wetness factor has been wrongly assessed. Sensitivity runs on wetness factor will test for such a condition. More localized deviations may be caused by inaccurate assessment of the water vapour produced by diesel equipment, or the effects of dust suppression sprays.

Having followed this correlation procedure over a series of airways, the ventilation engineer will have built up a store of information on the values and dispersion of heat sources in his mine. He will also have determined a range of *in situ* values for thermal conductivities, wetness factors and contributions of heat and humidity from mechanized equipment. At that stage, forward planning studies may be initiated with confidence levels that have been established through the correlation procedures. Climatic simulations then provide an invaluable tool of unprecedented detail in planning environmental control for hot underground facilities.

16.3.3 Case study

Figure 16.1 illustrates a 500 m inclined airway descending from 450 to 530 m below a surface datum. The airway contains a diesel unit of output rating 175 kW working at a mean rate of 60% full load, an air cooler of 200 kW cooling capacity and a conveyor expending 230 kW running from the midpoint to the exit. The complete data for the airway have been assembled under the categories given in section 16.3.1 and are shown in Table 16.1. The order and format of the input data depend on the particular software package being used.

The tabulated results of the simulation are shown in Table 16.2. Here again, the form of the output will depend upon the program employed. The package used for this case study allowed any of the computed variables to be produced in graphical form. Figures 16.2 and 16.3 are computer plots of the variations in temperature and moisture content of the air respectively. The effects of the diesel unit, cooler and conveyor are shown clearly. Despite the influence of autocompression and strata heat, evaporation in this rather wet airway results in a decrease of dry bulb temperature except where affected by diesel or conveyor heat.

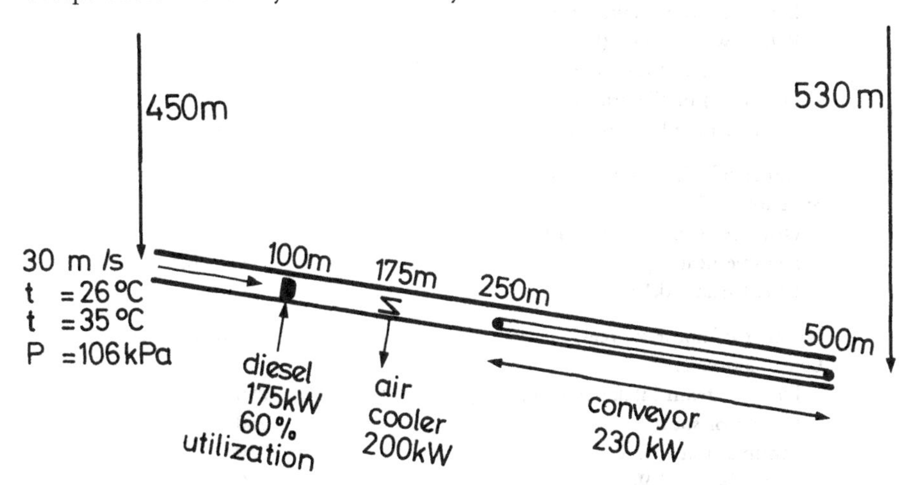

Figure 16.1 Ramp used in case study of mine climate simulation.

Table 16.1 Input data for the case study requested by a mine climate simulator

Computer prompt	*User response*
Name of airway	Ramp 530–540
Physical description of Ramp 530–450	
Length (m)	500
Cross sectional area (m^2)	8.75
Perimeter (m)	12
Depth at inlet (m)	450
Depth at outlet (m)	530
Friction factor (kg/m^3)	0.014
Age at inlet (days)	350
Age at outlet (days)	70
Wetness fraction	0.2
Ventilation at intake	
Airflow (m^3/s)	30
Pressure (kPa)	106
Wet bulb temperature (°C)	26
Dry bulb temperature (°C)	35
Rock thermal parameters	
Thermal conductivity (W/m°C)	4.5
Thermal diffusivity (m^2/s)	2.153×10^{-6}
VRT at inelt (°C)	42
Geothermal step (m/°C)	40
Heat sources	
Number of diesel and electrical spot sources	1 (diesel)
Spot source 1:	
Distance from intake (m)	100
Full power rating (kW)	175
Percentage utilization at equivalent full load	60
Diesel (D) or Electric (E)	D
Water emitted (litres/litre fuel)	7.0
Number of other spot sources	1
Spot source 2:	
Distance from intake (m)	175
Sensible heat (kW)	− 200 (cooler)
Latent heat (kW)	0
Number of linear sources	1 (conveyor)
Linear source 1:	
Distance from intake end to start (m)	250
Length of source (m)	250
Sensible heat (kW)	230
Latent heat (kW)	0
Airway length increments between output lines (m)	100

Table 16.2 Tabulated output produced by a mine climate simulator for the case study

Predicted Environment: ramp 530–450

dist (m)	*dry blb (C)*	*wet blb (C)*	*moist cont (g/kg)*	*rel hum (%)*	*press (kPa)*	*den (kg/m³)*	*sigma heat (kJ/kg)*	*vrt (C)*	*ACP (W/m²)*	*eff tmp (C)*
0	35.00	26.00	16.43	48.53	106.000	1.186	75.52	42.0	948	25.80
50	34.49	26.06	16.74	50.88	106.088	1.188	75.75	42.2	943	25.51
100	34.04	26.13	17.03	53.11	106.175	1.191	76.00	42.4	939	25.25
After spot source no. 1 sensible heat = 147.78 kW, latent heat = 149.37 kW										
150	37.55	28.21	19.10	48.98	106.262	1.177	84.56	42.6	742	28.57
After spot source no. 2 sensible heat = − 200.00 kW, latent heat = 0.00 kW										
200	31.65	26.92	19.38	68.95	106.350	1.201	79.20	42.8	881	23.91
250	31.55	27.00	19.55	70.04	106.438	1.202	79.51	43.0	874	23.87
Start 250 m linear source no. 1 sensible heat = 230 kW, latent heat = 0 kW										
300	32.69	27.40	19.73	66.28	106.527	1.200	81.11	43.2	833	25.03
350	33.77	27.80	19.94	63.07	106.616	1.197	82.70	43.4	793	26.06
400	34.79	28.18	20.16	60.30	106.704	1.193	84.29	43.6	754	26.98
450	35.76	28.56	20.41	57.90	106.792	1.190	85.88	43.8	714	27.83
500	36.68	28.94	20.68	55.81	106.879	1.188	87.46	44.0	675	28.60

Sensible heat flow from rock surface to air = − 144.2 kW
Latent heat flow from rock surface to air = 217.7 kW
Total heat flow from rock surface to air = 73.5 kW

16.3.4 Organization of simulation exercises

The flow chart given on Fig. 16.4 illustrates the procedure for investigating variations in the air temperatures, humidities and indices of heat stress throughout a ventilation network. From the results of a ventilation network analysis, one or more routes should be selected through the system. These may commence at an air intake point on surface, proceed through shafts, intake airways, work areas and return passages back to surface. Routes should be chosen that may be expected to produce the severest psychrometric conditions. Climatic simulations may then be conducted along successive branches throughout the selected routes.

The input shown on Fig. 16.4 may be supplied by manual interaction with the climate simulation package as illustrated in the case study of section 16.3.3. Alternatively, the input may be drawn from data files that have been assembled over a period of time, again, either manually or as a result of previous simulations. Such files may also be used or amended by other program packages. For example, the airflow for each individual branch may have been determined from a ventilation network analysis package (sections 7.4 and 9.1). Similarly, branch inlet temperatures and pressures will have been produced by climate simultions on upstream airways.

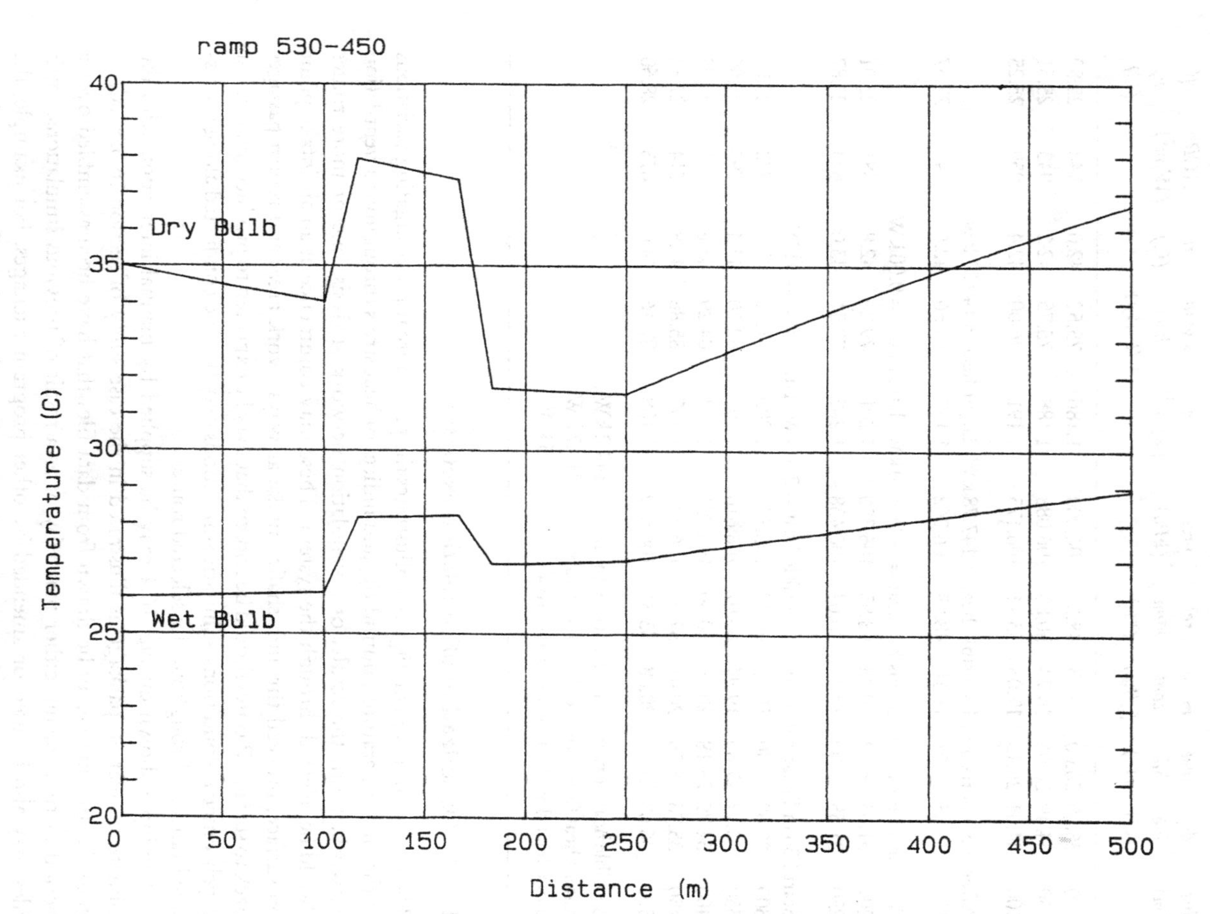

Figure 16.2 Computer plot of temperatures along the airway for the case study.

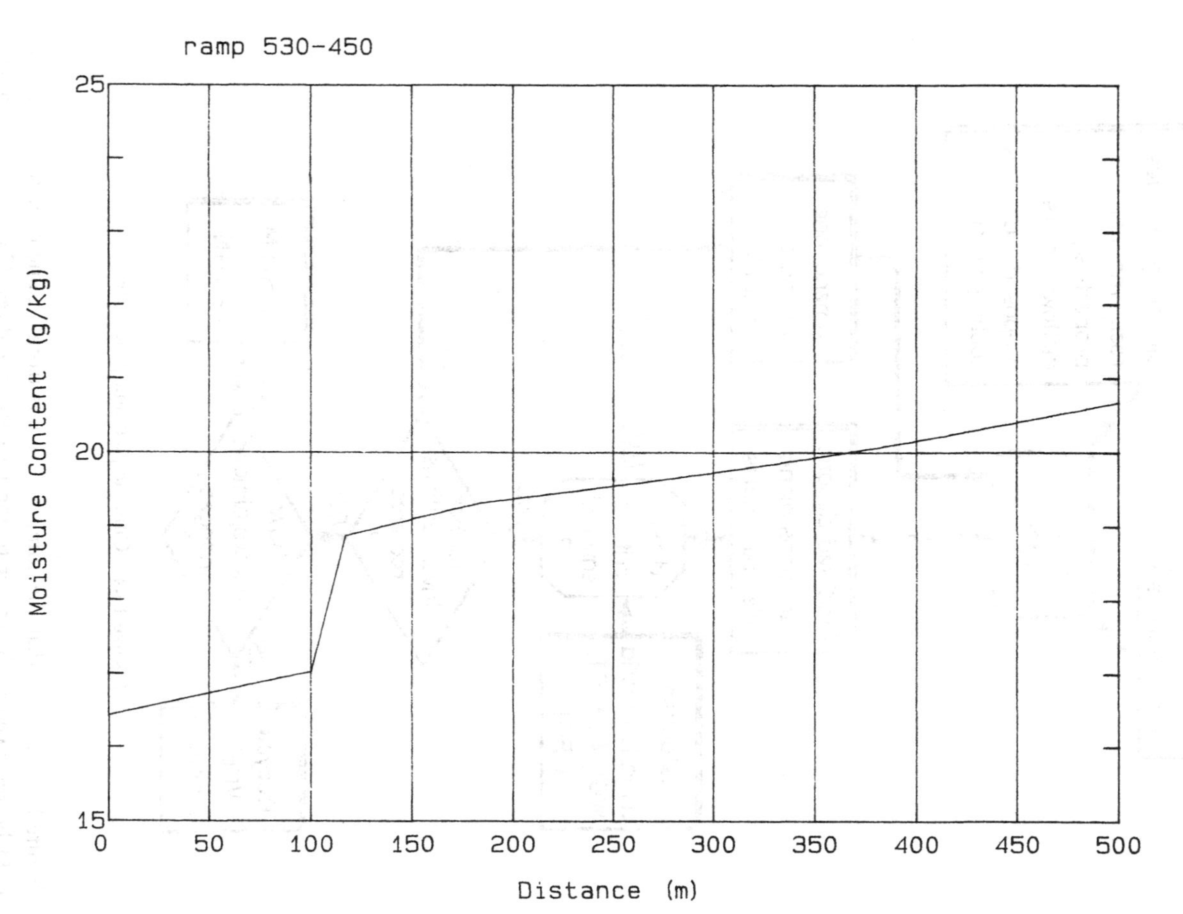

Figure 16.3 Computer plot of air moisture content along the airway for the case study.

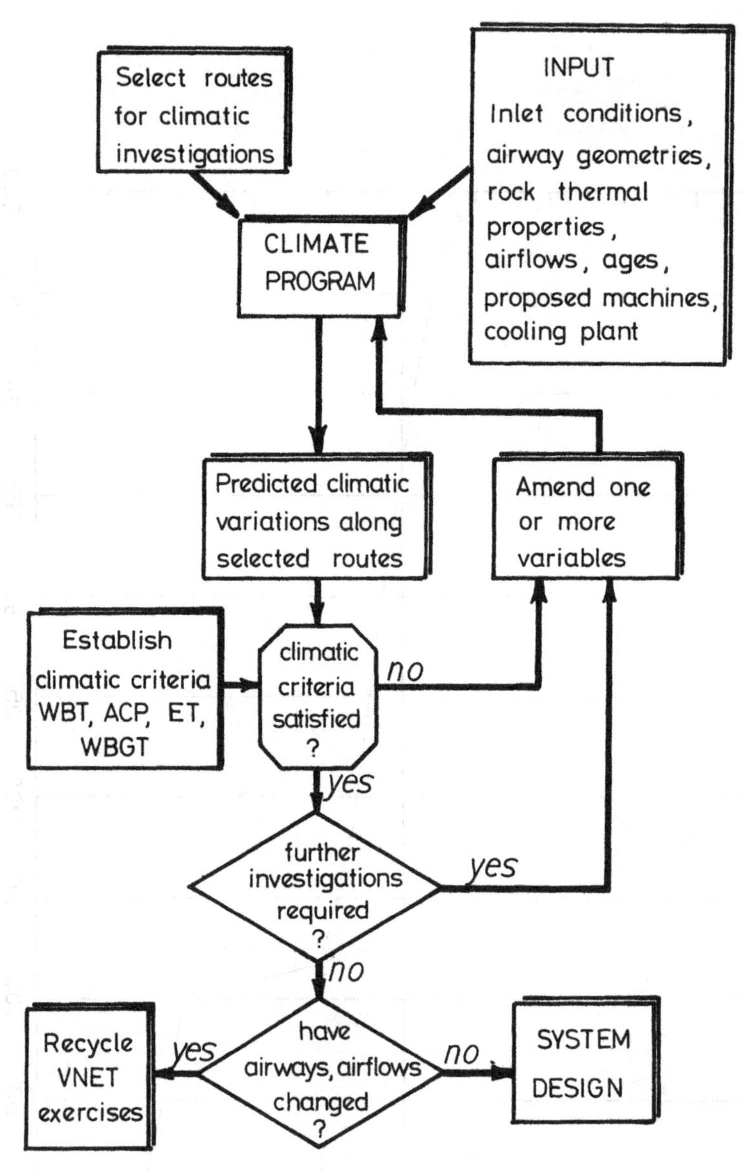

Figure 16.4 Climatic planning procedure.

Figure 16.4 indicates that the computed results from a mine climate simulation should be reviewed with respect to the heat strain indices (Chapter 17) or other climatic criteria relevant to the particular study. Should these criteria not be met then the program should be re-run successively in order to investigate means of reaching the approved standards. Parameters that may typically be amended are airflow, airway size or lining, reduction in machine power (for example, replacing diesels by electrical equipment) and the installation, siting and cooling capacities of heat exchangers. Sensitivity studies should also be run to test the influence of any input parameters that may be of questionable accuracy.

The convenient editing features of modern climate simulators allow individual items of input data to be updated with very few keystrokes and the program re-run within a matter of seconds.

There will usually be several ways of meeting the climatic criteria. Again, the procedure may be cycled successively to investigate alternative arrangements. This sometimes necessitates going back to re-evaluate an upstream airway.

16.3.5 Interaction with ventilation network analysis packages

Figure 9.1, introduced earlier to illustrate the systems analysis of subsurface ventilation planning, shows the relationships of climate simulations to ventilation network exercises. While the initial airflows used in climate simulations may have been produced by VNET exercises, it might have been necessary to amend those airflows during the climate investigations in order to meet set criteria. The sizes and linings of some airways may also have been altered, affecting their resistances. In those cases, it becomes necessary to re-run the network analysis package in order to revise fan duties and other consequential effects throughout the ventilation system.

Where the installation of cooling plant has proved to be necessary during the simulation exercises, the airflow requirements should be reviewed most carefully in order to approach an optimized combination of ventilating and cooling costs while, at the same time, providing airflows capable of diluting other airborne contaminants (section 9.3) (Anderson and Longson, 1986).

Programs have been developed that combine the functions of ventilation network analysis and climate simulation. However, the volume and variety of both input data and output can become daunting to the user. Practical experience of ventilation engineers leads to a preference for separate program packages for network analysis and climate simulation while, at the same time, maintaining the efficiencies of rapid transfer of information between data files. This also allows manual modification of those files whenever required and establishes a firmer control over the design process by the planning engineers.

REFERENCES

Anderson, J. and Longson, I. (1986) The optimisation of ventilation and refrigeration in British coal mines. *Min. Eng.* (146), 115–20.

Hemp, R. (1982) Sources of heat in mines. *Environmental Engineering in South African Mines*, Chapter 22, Mine Ventilation Society of South Africa.

Mousset-Jones, P. (1988) Determination of *in situ* rock thermal properties and the improved simulation of the underground climate. *PhD Thesis*, Royal School of Mines, University of London.

Starfield, A. M. (1966) The flow of heat into the advancing stope. *J. Mine Vent. Soc. S. Afr.* **19**, 13–28.

Starfield, A. M. (1966) The computation of air temperature increases in advancing stopes. *J. Mine Vent. Soc. S. Afr.* **19**, 189–99.

Starfield, A. M. and Bleloch, A. L. (1983) A new method for the computation of heat and moisture transfer in a partly wet airway. *J. S. Afr. Inst. Min. Met.* **84**, 263–9.

FURTHER READING

Amano, K., *et al.* (1980) Variation of temperature and humidity of air current passing through a partially wetted airway. *J. Min. Metall. Inst. Jpn.* **96**, 13–7.

Banerjee, S. P. and Chernyak, V. P. (1985) A simplified method for prediction of temperature and moisture content changes of air during its passage in a mine airway. *Int. Bur. Min. Thermophys.* **1** (May).

Boldizsar, T. (1960) Geothermal investigations concerning the heating of air in deep and hot mines. *Mine Quarry Eng.* (June), 253–8.

Cifka, I. and Eszto, M. (1978) Investigation of the method of determining *in situ* characteristics of underground climatic conditions and the thermal properties of rocks. *Report No. 16-5/78.K*, Mining Research Institute, Budapest.

Gibson, K. L. (1977) The computer simulation of climatic conditions in mines. *Proc. 15th APCOM Symp., Brisbane*, 349–54.

Hall, A. E. (1975) Computer techniques for calculating temperature increases in stopes and mine airways. *J. Mine Vent. Soc. S. Afr.* **28**(4), 55–9.

Hall, A. E. (1976) The design and analysis of ventilation systems in deep metal and non-metal mines. *MSHA Contract No. SO231095.*

Hemp, R. (1985) Air temperature increases in airways. *J. Mine Vent. Soc. S. Afr.*

Lambrechts, J. de V. (1967) Prediction of wet bulb temperature gradients in mine airways. *J. Mine Vent. Soc. S. Afr.* **67**, 595–610.

Longson, I. and Tuck, M. A. (1985) The computer simulation of mine climate on a longwall coal face. *Proc. 2nd US Mine Ventilation Symp., Reno, NV*, 439–48.

Mack, M. G. and Starfield, A. M. (1985) The computation of heat loads in mine airways using the concept of equivalent wetness. *Proc. 2nd US Mine Ventilation Symp., Reno, NV*, 421–8.

McPherson, M. J. (1975) The simulation of airflow and temperature in the stopes of S. African gold mines. *Proc. 1st Int. Mine Ventilation Congr., Johannesburg*, 35–51.

McPherson, M. J. (1986) The analysis and simulation of heat flow into underground airways. *Int. J. Min. Geol. Eng.* **4**, 165–96.

Mine Ventilation Services (1986) CLIMSIM 2.0. *Climatic Simulation Program User's Manual*, 2nd edn.

Voss, J. (1975) Control of the mine climate in deep coal mines. *Proc. 1st Int. Mine Ventilation Congr., Johannesburg*, p. 331.

Vost, K. R. (1982) The prediction of air temperatures in intake haulages in mines. *J. Mine Vent. Soc. S. Afr.* (November), 316–28.

Whillier, A. and Thorpe, S. A. (1971) Air temperature increases in development ends. *Research Report 18/71*, Chamber of Mines of South Africa.

17

Physiological reactions to climatic conditions

17.1 INTRODUCTION

In conventional mining operations, the need to control air temperatures and humidities arises primarily because of the relatively narrow range of climatic conditions within which the unprotected human body can operate efficiently. In some areas of underground repositories or where remotely controlled equipment is in use, a different spectrum of atmospheric criteria may be allowed or even required. However, in this chapter, we shall concentrate on the effects of climatic variations on the human body, and alternative means of quantifying the ability of a given environment to maintain an acceptable balance between metabolic heat generation and body cooling.

17.2 THERMOREGULATION OF THE HUMAN BODY

Within the human body, chemical and biological reactions act upon consumed nutrients to produce metabolic energy. A little of this is in the form of mechanical energy expended against external forces. However, the majority of metabolic energy results in the generation of heat. If the body is to remain in thermal equilibrium then this metabolic heat must be transferred to the ambient surroundings at a rate equal to that at which it is produced. Section 17.3 analyses and attempts to quantify the various elements of physiological heat exchange, namely by respiration, convection, radiation and evaporation, whose net sum must equal the rate of metabolic heat generation for thermal equilibrium to exist. First, however, let us discuss, in general terms, the mechanisms employed by the human body in adjusting to variations in climatic conditions.

A conceptual model that is commonly envisioned is one in which the body is made up of a central core of average temperature, t_c, surrounded by an outer layer of tissue that has an average temperature t_{sk}. The latter is referred to simply as the skin temperature. Contrary to popular belief, the core temperature is not constant but varies normally between 36 and 38 °C with respect to both location within the body and muscular activity for a healthy individual experiencing no heat strain. The skin temperature is even more variable, depending on position on the body surface, clothing, temperature and velocity of the ambient air, rate of perspiration and, to a lesser extent, metabolic activity. The average temperature and wettedness of the skin are parameters of major importance in physiological heat exchange. For a resting person located within an environment that gives no sensation of either heat or cold, average skin temperature approximates to 34 °C. This may be termed the **neutral skin temperature**.

When the surrounding environment removes heat from the human body at a rate that is not equal to the metabolic heat generation then one or both of two types of response will occur. A *behavioural* response consists of conscious steps taken to alleviate the situation. These might consist of shedding or donning clothing, and of decreasing or increasing physical activity. The second type of response comprises the involuntary actions taken by the body to reestablish a stable heat balance with the surroundings. This is known as body **thermoregulation**.

Temperature-sensitive receptors exist throughout both the core and skin elements of the body. These react to departures of t_c and t_{sk} away from their neutral values. The predominant mode of heat conductance from the core to skin tissues is by the flow of blood. A 'high' signal from the core receptors will produce **vasodilation**, the relevant blood vessels expand to allow a greater flow of blood and, hence, heat from the core to the skin. Further increases in the 'high' signals from core receptors will result in perspiration and evaporative cooling, the most powerful mode of surface heat transfer from the body.

As the skin temperature approaches its neutral value of some 34 °C the skin receptors react to excite the sweat glands into additional activity. However, some areas are more efficient at producing sweat than others to the extent that unevaporated run-off may occur in some zones before other areas are fully wetted. There is also considerable variation in sweat rates between individuals.

'Low' signals from the skin receptors (cold conditions) cause **vasoconstriction**. The blood vessels that supply skin tissues contract. The vital organs in the core of the body continue to be supplied by warm blood and are thus protected preferentially while the skin is allowed to cool. This can cause extremities such as fingers, toes or ears to suffer tissue damage by frostbite in very cold conditions. Reduced skin temperature may also produce the involuntary muscular movement of shivering in an attempt to generate additional metabolic heat.

In order to follow the thermoregulatory reactions of the human body to hot conditions, let us conduct an imaginary experiment in which a volunteer undertakes physical work at a steady but unhurried rate within an environment that is initially fairly cool, say at 10 °C. In this situation, heat loss from the body will be composed, mainly, of convection with lesser amounts of cooling by respiration and radiation. The sweat glands are effectively inactive and evaporation of water by direct diffusion through the skin contributes only a little evaporative cooling. The procedure of the experiment is to increase the air temperature in increments, maintaining all other parameters constant, including the rate of work and, hence, metabolic heat generation. The incremental changes in air temperature are made sufficiently slowly to allow the body to maintain itself in or near thermal equilibrium. The purpose of the experiment is to observe the changes that take place in the modes of heat transfer.

As the air temperature rises from its initial low value, the skin temperature, t_{sk}, also increases in an attempt to retain the differential between t_{sk} and the temperature of the surrounding environment and, hence, to maintain convective and radiant cooling. However, the influence of respiratory cooling decreases because the heat content of the expired air rises at a lesser rate than that of the inspired air.

At some point, as the air temperature continues to climb and depending on the rate of work, signals from the core receptors cause sweating to commence, resulting in partial wetting of the body surface. This initiates some significant changes in the heat transfer processes. First, throughout the period of increased sweating and until full surface wetness occurs, evaporative cooling moderates the rate at which skin temperature continues to rise with respect to ambient temperature. Also, throughout this stage, the combined effects of convective and radiative cooling are progressively replaced by evaporative heat transfer. When the skin temperature approaches and exceeds some 34 °C, the skin receptors react to produce further enhanced sweating. The increased area that is coated in perspiration and the corresponding effectiveness of evaporative cooling both grow rapidly. However, when the skin becomes fully wetted, further cooling can be gained only by additional rises in skin temperature. Hence, the skin temperature resumes a higher rate of increase.

If the temperature of the surroundings exceeds skin temperature then the convective and radiant heat transfers will reverse. Evaporative cooling must then balance the heat gain from convection and radiation in addition to metabolic heat.

When skin temperature exceeds about 36 °C the subject is likely to begin exhibiting the effects of heat strain (section 17.5) unless he has been well acclimatized to hot working environments. At skin temperatures above 37 °C the person is in danger of excessive and, perhaps, even fatal core temperatures. This might be a good time to terminate the experiment.

17.3 PHYSIOLOGICAL HEAT TRANSFER

A great deal of research has been devoted to quantification of the various modes of heat exchange associated with the human body and the corresponding physiological reactions. Such work has involved laboratory and field experiments, analytical studies and numerical simulation. The development of a numerical thermoregulation model involves a combination of relationships that approximate the modes of heat transfer, in order to predict one or more of the parameters that are deemed to represent physiological reaction such as skin temperature, core temperature or sweat rate. One such model is described in section 17.3.6. First, however, we must develop relationships that quantify the individual modes of physiological heat transfer.

17.3.1 The metabolic heat balance

In common with many other living organisms, the human body is a biological heat engine of low mechanical efficiency. Fuel is consumed in the form of nutrients and combines with oxygen to produce

1. metabolic heat,
2. mechanical work, and
3. changes in mass (body growth).

The rate of body growth is negligible compared with other physiological changes and can be ignored for the purposes of this analysis. The mechanical work output, shown as W on Fig. 17.1, is seldom more than 20% of the total metabolic energy even for strenuous efforts by trained athletes and is usually very much less than this. It is considered only when the work involves significant lifting or lowering of a mass, including the human body itself.

metabolic energy (proportional to oxygen consumption)
= metabolic heat + work done against gravity (W)

For example, a person of mass 70 kg walking up an incline of 4 in 100 at a speed of 1 m/s would do work against gravity at a rate of

$$70 \times 9.81 \times \frac{4}{100} = 27.5 \text{ W}$$

The metabolic energy produced by the person should be reduced by this amount to give the corresponding metabolic heat generation. However, in walking uphill, the rate of oxygen consumption and, hence, metabolic energy are increased. In practice, it is usual to ignore mechanical work in analyses of physiological response to climatic conditions.

At equilibrium, the generation of metabolic heat, M, is balanced by heat transfer from the body to the surroundings. However, a fraction of the metabolic heat, Ac, may be accumulated within the body resulting in a transient rise in core temperature, Δt_c, and (usually) also skin temperature. By assuming that 80% of the body mass is at core temperature, t_c, and 20% at skin temperature, t_{sk}, the rate of heat accumulation may be approximated as

$$\text{Ac} = \frac{m_b C_b (0.8\,\Delta t_c + 0.2\,\Delta t_{sk})}{\text{time (seconds)}} \quad \text{W} \tag{17.1}$$

where m_b = mass of body (kg) and C_b = average specific heat of the body (3470 J/(kg K)) (Stewart, 1982).

For sedentary persons at thermal equilibrium with the surroundings, average oral temperature is approximately 36.9 °C. Skin temperature varies over the body surface and with the ambient air temperature but has an average value of 34 °C for non-stressed personnel experiencing a sensation of thermal comfort. The effect of heat storage, Ac, is to increase core temperature, t_c, and, hence, skin temperature. This

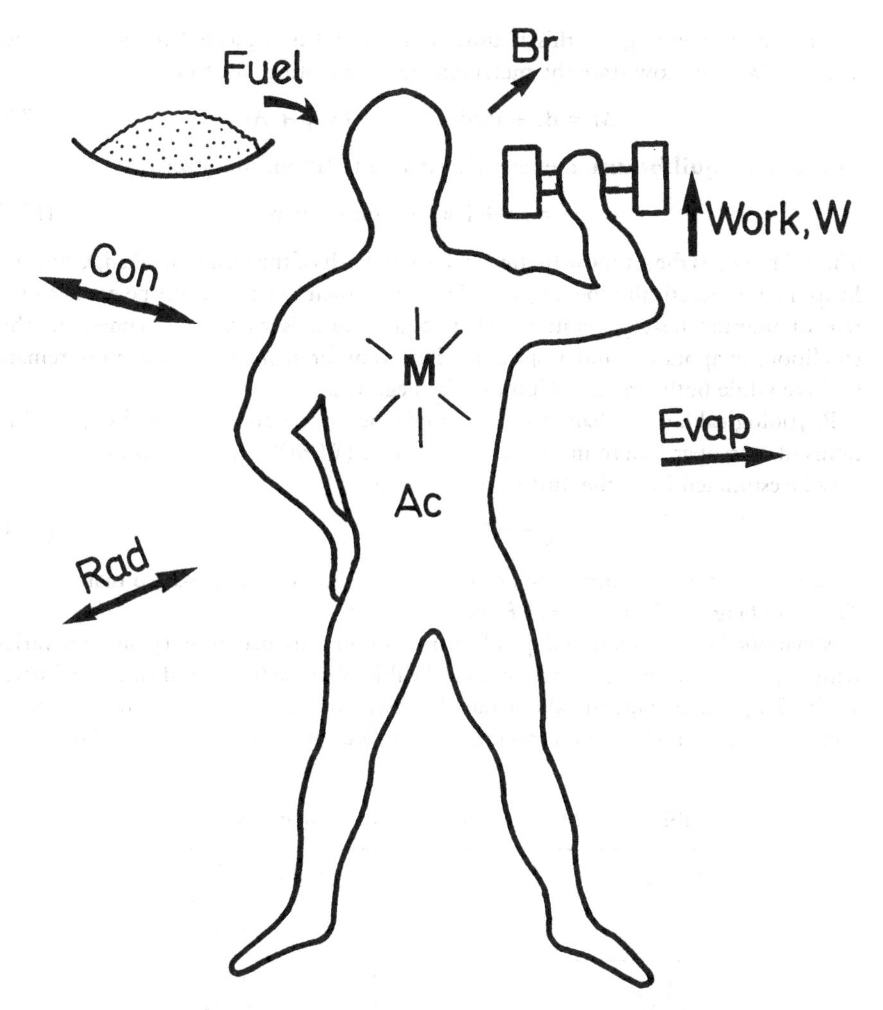

Figure 17.1 The human body is a heat engine producing both work and heat, *M*. M = Br + Con + Rad + Evap + Ac.

normally results in enhanced body cooling in order to attain a new equilibrium at which the rate of heat accumulation declines to zero.

Figure 17.1 shows that heat loss from the body occurs through a combination of heat transfer processes:

1. respiratory heat exchange (breathing), Br,
2. convection, Con,
3. radiation, Rad, and
4. evaporation, Evap.

Conductive heat transfer may also take place at any areas of contact with solid surroundings. Here again, this is usually small and is neglected in the following analyses. We can now state the metabolic heat generation, M, to be

$$M = \mathrm{Br} + \mathrm{Rad} + \mathrm{Con} + \mathrm{Evap} + \mathrm{Ac} \qquad (17.2)$$

At **thermal equilibrium**, the rate of heat accumulation, Ac, is zero. Then

$$M = \mathrm{Br} + \mathrm{Rad} + \mathrm{Con} + \mathrm{Evap} \qquad (17.3)$$

This is known as the **metabolic heat balance**. Each of the terms Br, Rad, Con, and Evap may, theoretically, be negative. However, their net sum must be equal to the rate of internal heat generation, M, if equilibrium is to be maintained. In this condition, evaporative and respiratory heat transfer from the body must remain positive while both Rad and Con may be negative.

Physiological heat exchange and metabolic heat generation are usually quoted in terms of watts per square metre of body surface ($\mathrm{W/m^2}$) where the **skin area**, A_{sk}, may be estimated from the duBois relationship

$$A_{sk} = 0.202 m_b^{0.425} h_b^{0.725} \quad \mathrm{m^2} \qquad (17.4)$$

where h_b = height (m) and m_b = body mass (kg). For an average-sized man of mass 70 kg and height 1.73 m, $A_{sk} = 1.83\ \mathrm{m^2}$.

Metabolic heat generation depends primarily on muscular activity but also varies with respect to the condition of individual health, physical fitness and emotional state. Table 17.1 gives average metabolic rates for a series of activities both in terms of total energy production (W) and normalized ($\mathrm{W/m^2}$) on the basis of $A_{sk} = 1.83\ \mathrm{m^2}$.

Table 17.1 Metabolic rates for various activities

Activity	*Metabolic heat production*	
	(W)	*M(W/m²)*
Sleeping	73	40
Seated	107	58.5*
Standing but relaxed	128	70
Walking on the level at		
1 m/s	238	130
1.4 m/s	320	175
1.8 m/s	403	220
Manual work		
Very light	174	95
Light	265	145
Moderate	448	245
Heavy	622	340

* 1 met = 58.5 $\mathrm{W/m^2}$ and is sometimes used as a unit of metabolic rate.

17.3.2 Respiratory heat transfer

The large internal area and wetness of a human lung give good efficiency of heat exchange. The actual degree of respiratory cooling depends on the rate of breathing and the wet bulb temperature of the ambient air:

$$Br = m_{resp}(S_{out} - S_{in})/A_{sk} \quad W/m^2 \tag{17.5}$$

where m_{resp} = mass rate of breathing (kg air/s), S_{out} = sigma heat of exhaled air (J/kg) and S_{in} = sigma heat of inhaled (ambient) air (J/kg).

Experimental evidence (Fanger, 1970) has indicated that the rate of respiration, m_{resp}, is proportional to the total metabolic energy production:

$$\begin{aligned} m_{resp}, &= 1.7 \times 10^{-6} \times \text{metabolic energy} \\ &= 1.7 \times 10^{-6} \times M \times A_{sk} \quad kg/s \end{aligned} \tag{17.6}$$

Hence, equation (17.5) becomes

$$Br = 1.7 \times 10^{-6} M(S_{out} - S_{in}) \quad W/m^2 \tag{17.7}$$

where the metabolic rate, $M(W/m^2)$, may be estimated from Table 17.1.

The sigma heat terms S_{out} and S_{in} are calculated from the equations given in section 14.6 at the corresponding wet bulb temperatures for exhaled air, t_{ex}, and ambient inhaled air, t_w, respectively. The latter can be measured directly. The wet bulb temperature of the near-saturated exhaled air depends on body temperature and the psychrometric condition of the ambient air. Again, empirical data (Fanger, 1970) have produced the relationship

$$t_{ex} = 32.6 + 0.066 t_d + 0.0002 e \quad °C$$

where t_d = ambient dry bulb temperature (°C) and e = actual vapour pressure of the ambient air (Pa).

Example Estimate the rate of respiratory cooling for a man of average size ($A_{sk} = 1.83\ m^2$) walking at 1.8 m/s in an atmosphere of wet bulb temperature, $t_w = 18\ °C$, dry bulb temperature, $t_d = 25\ °C$ and barometric pressure, $P =$ 110 kPa.

Solution From Table 17.1 the metabolic rate, M, is selected as 220 W/m². Applying the given climatic conditions to the psychrometric equations of section 14.6 leads to sigma heat,

$$S_{in} = 47\,334\ J/kg$$

and actual vapour pressure,

$$e = 1566\ Pa$$

Then, temperature of exhaled air (equation (17.8))

$$\begin{aligned} t_{ex} &= 32.6 + (0.066 \times 25) + 0.0002 \times 1566 \\ &= 34.56\ °C \end{aligned}$$

Applying this temperature of saturated air to the same equations of section 14.6 gives the sigma heat of the expired air to be

$$S_{out} = 113\,756 \text{ J/kg}$$

Equation (17.5) then gives the heat removed by respiration as

$$\text{Br} = 1.7 \times 10^{-6} \times 220\,(113\,756 - 47\,334)$$
$$= 24.8 \text{ W/m}^2$$

This represents some 11% of the metabolic rate of 220 W/m^2. We shall continue with this example as we introduce the other modes of heat transfer.

The influence of respiratory cooling decreases as the ambient wet bulb temperature rises towards that of the expired air. At ambient wet bulb temperatures in excess of t_{ex}, condensation will occur within the lungs and respiratory tracts, accompanied by a rapid onset of heat strain.

17.3.3 Convective heat transfer, Con

The convective heat transfer equation

The convective, radiative and evaporative modes of heat transfer from the human body are all affected by the degree of covering and the thermal properties of clothing. In the case of an unclothed body, the convective heat transfer is given by equation (15.16):

$$\text{Con} = h_c(t_{sk} - t_d) \quad \text{W/m}^2 \tag{17.8}$$

where h_c = convective heat transfer coefficient (W/(m^2 °C)) and t_{sk} = skin temperature (°C).

To take the effect of clothing into account, let us consider the heat transfer in two parts:

1. from the skin at temperature t_{sk} to the outside of the clothing at temperature t_{cl};
2. from the clothing to the surrounding air of dry bulb temperature t_d.

For each square metre of skin surface, we can write

$$\text{Con} = h_{cl}(t_{sk} - t_{cl}) \quad \text{W/m}^2 \tag{17.9}$$

where h_{cl} = effective heat transfer coefficient through the clothing (W/(m^2 °C)) and

$$\text{Con} = h_c f_{cl}(t_{cl} - t_d) \quad \text{W/m}^2 \tag{17.10}$$

h_c remains the surface convective heat transfer coefficient from the (now clothed) body to the surroundings (W/(m^2 °C)) while

$$f_{cl} = \frac{\text{surface area of clothed body}}{\text{surface area of unclothed body}}$$

The latter factor accounts for the increase in overall surface area caused by clothing.

At equilibrium, equations (17.9) and (17.10) represent the same value of Con.

As t_{cl} is particularly difficult to establish independently, we can eliminate it by rewriting equation (17.9) as

$$t_{cl} = t_{sk} - \frac{\text{Con}}{h_{cl}} \tag{17.11}$$

For practical application, $1/h_{cl}$ may be expressed as

$$R_{cl} = 1/h_{cl} \quad \text{m}^2\,{}^\circ\text{C/m} \tag{17.12}$$

and is known as the thermal resistance of the clothing. Substituting for t_{cl} in equation (17.10) and re-arranging gives

$$\text{Con} = \frac{t_{sk} - t_d}{R_{cl} + 1/(f_{cl}h_c)} \quad \frac{\text{W}}{\text{m}^2} \tag{17.13}$$

Note that, for an unclothed body, $R_{cl} = 0$ and $f_{cl} = 1$. The relationship then simplifies to equation (17.8).

The only variable in equation (17.13) that is amenable to direct measurement is the ambient dry bulb temperature. The remaining parameters may be estimated from the following empirical data and relationships.

Clothing factors

The effective thermal resistances, R_{cl}, of clothing ensembles commonly worn in underground workings are given in Table 17.2 together with typical corresponding area factors, f_{cl}. This table has been derived from data reported in the ASHRAE *Fundamentals Handbook* (1985).

Convective heat transfer coefficient, h_c

A number of authorities have reported experimental data and proposed empirical relationships for the convective heat transfer coefficient applicable to the human body (Stewart, 1982; ASHRAE, 1989). This parameter depends on the psychrometric condition of the air, the localized (microclimate) relative velocity, and the size and shape of the body. A simplified relationship that gives acceptable agreement with experimental data for the velocity range 0.2 to 5 m/s is

$$h_c = 0.008\,78\; P^{0.6} u^{0.5} \quad \text{W/(m}^2\,{}^\circ\text{C)} \tag{17.14}$$

where P = barometric pressure (Pa) and u = local (effective) relative velocity (m/s). The local (effective) relative velocity is not the same as the difference between the velocities of the body and the general airstream. Migration of heat and water vapour creates an envelope around the body within which the velocity and psychrometric properties of the air differ from those in the general airstream. This envelope is known

Table 17.2 Clothing factors for ensembles commonly worn in mining

*Ensemble**	*Thermal resistance* R_{cl}*(°C m²/W)*	*Area factor* f_{cl}	$h_{cl} = 1/R_{cl}$ *(W/(m² °C))*
Unclothed	0	1.00	Infinite
Walking shorts	0.051	1.05	19.6
Shorts and short-sleeved shirt	0.076	1.11	13.2
Thin trousers and short-sleeved shirt	0.085	1.14	11.8
Thick trousers and long sleeved shirt	0.110	1.28	9.1
Long sleeved coveralls and long sleeved shirt	0.143	1.28	7.0

*All ensembles include boots and stockings.

as the body microclimate. The velocity within the microclimate is affected by the corresponding temperature gradient and body movements as well as the mainstream air velocity. However, a practical approximation may be obtained for the range 0 to 3 m/s as

$$u = 0.8u_r + 0.6 \quad \text{m/s} \tag{17.15}$$

where u_r = relative velocity between the main airstream and the general motion of the body. For values of u_r in excess of 3 m/s, the correction becomes negligible and we may then accept that $u = u_r$.

Base skin temperature, t_{sk}

The only parameter in equation (17.13) that remains to be investigated is the average skin temperature. This is a factor of prime importance, controlling convective, radiant and evaporative modes of heat transfer. While neutral skin temperature is of the order of 34 °C (section 17.2) it may rise to 26 °C in hot conditions without the subject necessarily suffering from heat strain. Well-acclimatized individuals may be able to tolerate skin temperatures of up to 37 °C without harm. However, the probability of progressive symptoms of heat strain increases rapidly after skin temperatures have reached 36 °C. Skin temperatures of more than 37 °C are likely to be indicative of dangerously high core temperatures. A close correlation has been observed between average skin temperature and heat strain. Hence, average skin temperature is one of the physiological response parameters that may be used as an indicator of heat strain.

Skin temperature may vary considerably over the surface of the body depending on the temperature and velocity of the ambient airstream, the degree of covering and thermal resistance of clothing, and the wetness of the skin.

Figure 17.2 shows an experimental curve obtained by Gagge *et al.* (1969) for unclothed subjects within an airflow of 40% relative humidity and low velocity. Fanger (1970) and other authorities have reported some indication that skin tem-

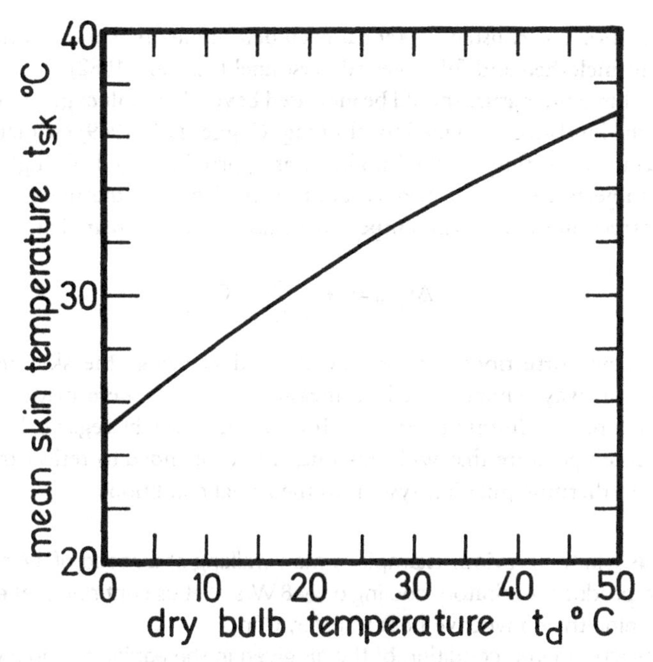

Figure 17.2 Average skin temperature for unclothed subjects as a function of dry bulb temperature in air of low velocity and relative humidity 40% (after Gagge *et al.*, 1969). $t_{sk} = 24.85 + 0.322\,t_d - 0.00165\,t_d^2$

perature reduces as metabolic rate increases. However, those experiments were related to the measurement of thermal 'comfort' in surface buildings rather than for industrial conditions.

The relationship obtained by regression analysis on the curve shown on Fig. 17.2 is

$$t_{sk} = 24.85 + 0.322\,t_d - 0.001\,65 t_d^2 \quad °C \tag{17.16}$$

where t_d = dry bulb temperature of the ambient air (°C). The reducing slope of this curve as t_d and t_{sk} increase indicates the moderating effect of evaporative cooling.

The average skin temperature obtained from equation (17.16) should be corrected for air velocity and the effects of clothing. As the velocity increases the base skin temperature will, in general, fall provided that the air temperature is less than that of the skin. The effect is mitigated by the presence of clothing. It is assumed here that heavy clothing having a thermal resistance, R_{cl}, of 0.15 will provide full protection against the effect of air velocity on skin temperature. The correction to be applied to the 'low velocity' skin temperature, t_{sk}, of equation (17.16) is

$$\Delta t_{sk,vel} = -0.009 h_c (t_{sk} - t_d) \frac{0.15 - R_{cl}}{0.15} \quad °C \tag{17.17}$$

The heat transfer coefficient, h_c, is calculated as a function of air velocity from equation (17.14). The constant 0.009 has been determined by curve fitting to results reported for unclothed and fully wetted personnel (Stewart, 1982).

The base skin temperature should be increased beyond the values given on Fig. 17.2 because of the insulation provided by clothing. Gagge *et al.* (1969) estimated that the skin temperature for fully clothed bodies was approximately 1 °C higher than for unclothed subjects. In order to take account of the degree of thermal insulation, the clothing correction to base skin temperature may be approximated as

$$\Delta t_{sk,cl} = +\frac{R_{cl}}{0.15} \quad ^{\circ}\mathrm{C} \tag{17.18}$$

Despite these corrections for air velocity and clothing, the skin temperature obtained in this way remains based on measurements that were made in a low air velocity and a relative humidity of 40%. It may, therefore, be regarded as an **initial** or **base** skin temperature that will subsequently be amended to reflect the response of the human thermoregulation system to the actual conditions.

Example In the previous example, a man walking at a speed of 1.8 m/s was found to produce respiratory cooling of 24.8 W/s². Let us continue that example to determine the convective heat transfer, Con.

The psychrometric condition of the air given in the earlier example was

$$t_w = 18\,^{\circ}\mathrm{C},\ t_d = 25\,^{\circ}\mathrm{C},\ P = 110\,\mathrm{kPa}$$

The additional data now supplied are as follows.

1. The man is clothed in thick trousers, long-sleeved shirt and boots. From Table 17.2, the corresponding values are obtained for

$$R_{cl} = 0.110^{\circ}\mathrm{C\,m^2/W} \text{ and } f_{cl} = 1.28$$

2. The ambient air velocity is 0.5 m/s and the man is walking at 1.8 m/s in the same direction as the airflow. The relative velocity is, therefore, $u_r = 1.3$ m/s.

Solution The microclimate velocity is given by equation (17.15)

$$u = (0.8 \times 1.3) + 0.6 = 1.64\,\mathrm{m/s}$$

Equation (17.14) then gives the convective heat transfer coefficient to be

$$h_c = 0.008\,78(110\,000)^{0.6}(1.64)^{0.5}$$
$$= 11.91\,\mathrm{W/(m^2\,^{\circ}C)}$$

In the absence of any skin wetness data (for the moment), we use equation (17.16) to determine the base skin temperature, i.e. for an unclothed subject in a low velocity air stream

$$t_{sk} = 24.85 + (0.322 \times 25) - (0.001\,65 \times 25^2)$$
$$= 31.869\,^{\circ}\mathrm{C}$$

Equation (17.17) gives the correction for air velocity to be

$$\Delta t_{sk,vel} = -0.009 \times 11.91(31.87 - 25)\frac{0.15 - 0.11}{0.15} - 0.196\,°C$$

The clothing correction is given by equation (17.18) as

$$\Delta t_{sk,cl} = \frac{0.11}{0.15} = 0.733\,°C$$

The corrected base skin temperature now becomes

$$t_{sk} = 31.869 - 0.196 + 0.733 = 32.406\,°C$$

The convective cooling is then given by equation (17.13)

$$\text{Con} = \frac{32.406 - 25}{0.11 + 1/1.28 \times 11.91}$$
$$= 42.2\,\text{W/m}^2$$

17.3.4 Radiant heat transfer

The exchange of radiant heat takes place between the outer surface of the body or clothing and any surrounding surfaces or ambient water vapour. The temperature of the outer surface of clothing is given by equations (17.11) and (17.12) as

$$t_{cl} = t_{sk} - \text{Con} \times R_{cl} \quad °C \tag{17.19}$$

This reduces to t_{sk} if the body is unclothed, i.e. $R_{cl} = 0$. If the radiant temperature of the surroundings is t_r, then the absolute average radiant temperature becomes

$$T_{av} = \frac{t_{cl} + t_r}{2} + 273.15 \quad \text{K} \tag{17.20}$$

The Stefan–Boltzmann relationship (equation (15.27)) then gives the radiant heat transfer coefficient to be

$$h_r = 4 \times 5.67 \times 10^{-8} T_{av}^3 \quad \text{W/(m}^2\,°\text{C)} \tag{17.21}$$

For most mining circumstances, h_r lies within the range 5 to 7 W/m^2. The radiant heat transfer equation (15.26) gives

$$\text{Rad} = h_r f_r (t_{cl} - t_r) \quad \text{W/m}^2 \tag{17.22}$$

The emissivity of the body may be taken as unity unless specially reflective clothing is worn.

The view (or posture) factor, f_r, has been reported in the range 0.7 to 0.75 for seated, standing or crouching personnel (ASHRAE, 1989; Stewart, 1982).

Example To continue with the previous example, the man is walking along an airway where the temperature of the rock walls is 26 °C. Calculate the radiant heat exchange.

Solution From equation (17.19) the temperature of the outer surface of the clothing is

$$t_{cl} = 32.406 - 42.2 \times 0.11$$
$$= 27.76\,°\text{C}$$

As $t_r = 26\,°\text{C}$, the absolute average radiant temperature (equation (17.20)) is

$$T_{av} = \frac{(27.76 + 26)}{2} + 273.15$$
$$= 300.03\ \text{K}$$

Equation (17.21) then gives the radiant heat transfer coefficient to be

$$h_r = 4 \times 5.67 \times 10^{-8}(300.03)^3 = 6.13\ \text{W/(m}^2\,°\text{C)}$$

Taking the posture factor, f_r, to be 0.73 for a walking person, equation (17.22) gives the radiant cooling as

$$\text{Rad} = 6.13 \times 0.73 \times (27.76 - 26)$$
$$= 7.9\ \text{W/m}^2$$

17.3.5 Evaporative heat transfer

Evaporation from unclothed and clothed bodies

An analysis of the heat and mass transfers at a wet surface was give in section 15A.4 of the appendix following Chapter 15. Equation (15A.28) gave the evaporative or latent heat transfer to be

$$\text{Evap} = q_L = h_e(e_{ws} - e) \quad \text{W/m}^2 \tag{17.23}$$

where

$$h_e = 0.0007\, h_c \frac{L_{sk}}{P} \quad \frac{W}{\text{m}^2\,\text{Pa}} \tag{17.24}$$

is the evaporative heat transfer coefficient at the outer surface, e_{ws} = saturated vapour pressure at the wet surface temperature (Pa), e = actual vapour pressure in the main airstream (Pa) and L_{sk} = latent heat of evaporation of water at wet surface temperature (J/kg). (The ratio h_e/h_c (known as the Lewis ratio) is then

$$\frac{h_e}{h_c} = 0.0007 \frac{L_{sk}}{P} \quad °\text{C/Pa} \tag{17.25}$$

Employing mid-range values of $L_{sk} = 2455 \times 10^3$ J/kg and $P = 100\,000$ Pa gives a Lewis ratio of 0.017 °C/Pa.)

In the case of physiological heat transfer, e_{ws} becomes e_{sk}, the saturated vapour pressure at the wet skin temperature. Furthermore, if the skin is only partially wetted then equation (17.23) becomes

$$\text{Evap} = wh_e(e_{sk} - e) \quad \text{W/m}^2 \tag{17.26}$$

where w = fraction of the surface that is wet (see, also, the wetness fraction described in section 16.3.1).

In order to take the effect of clothing into account, we employ the analogy between heat and mass transfers by expanding equation (17.26) to the same form as equation (17.13), i.e.

$$\text{Evap} = \frac{w(e_{sk} - e)}{R_{e,cl} + 1/(f_{cl}h_e)} \quad \text{W/m}^2 \tag{17.27}$$

where $R_{e,cl}$ = the resistance to latent heat through the clothing (m^2 Pa/W) (and is analogous to the thermal resistance, R_{cl}) and f_{cl} = the clothing area ratio introduced in section 17.3.3.

In order to evaluate $R_{e,cl}$, let us recall from equation (17.12) that $R_{cl} = 1/h_{cl}$. We might expect, therefore, that the heat and mass transfer analogies would give $R_{e,cl} = 1/h_{e,cl}$ where $h_{e,cl}$ is the evaporative heat transfer coefficient through the clothing. Practical observations on sweating mannequins have determined that $R_{e,cl}$ is proportional but not equal to $1/h_{e,cl}$ for any given clothing ensemble. We can write

$$R_{e,cl} = \frac{1}{i_{cl}h_{e,cl}} \quad \text{m}^2\,\text{Pa/W} \tag{17.28}$$

where i_{cl} may be called the vapour permeation efficiency of the clothing. This varies from 0 for a completely vapour-proof garment to 1.0 for an unclothed person. There appears to be a scarcity of i_{cl} data for a variety of clothing types. However, Oohori *et al.*(1984) found that most indoor clothing systems have i_{cl} values of approximately 0.45. Direct experimentation with sweating mannequins or other means may be employed to determine i_{cl} values for protective or other specially designed garments.

The evaporate heat transfer coefficient for the outer surface of the body or clothing was given by equation (17.24). By replacing the coefficients h_e and h_c by their counterparts for evaporative heat transfer through the clothing, we obtain

$$h_{e,cl} = 0.0007\, h_{cl}\frac{L_{sk}}{P} \quad \text{W/(m}^2\,\text{Pa)} \tag{17.29}$$

Hence, as $h_{cl} = 1/R_{cl}$ and $h_{e,cl} = 1/R_{e,cl}\, i_{cl}$,

$$R_{e,cl} = \frac{R_{cl}P}{0.0007\, L_{sk}\, i_{cl}}$$

i.e.

$$R_{e,cl} = \frac{h_c}{h_e}\frac{R_{cl}}{i_{cl}} \quad \frac{m^2\,Pa}{W} \tag{17.30}$$

Substituting into equation (17.27) (from equation (17.25)) in order to eliminate $R_{e,cl}$ gives

$$\text{Evap} = \frac{wh_e(e_{sk} - e)}{h_c R_{cl}/i_{cl} + 1/f_{cl}} \quad W/m^2 \tag{17.31}$$

Skin wetness fraction

The skin wetness fraction, w, or 'sweat fraction', is a measure of the proportion of skin surface area from which evaporation of sweat takes place. Diffusion of water vapour through the skin occurs at a rate normally equivalent to about $w = 0.06$ and this may be taken as a base value. (Individuals who are dehydrated or who are acclimatized to living in arid regions may reduce this base value to about 0.03).

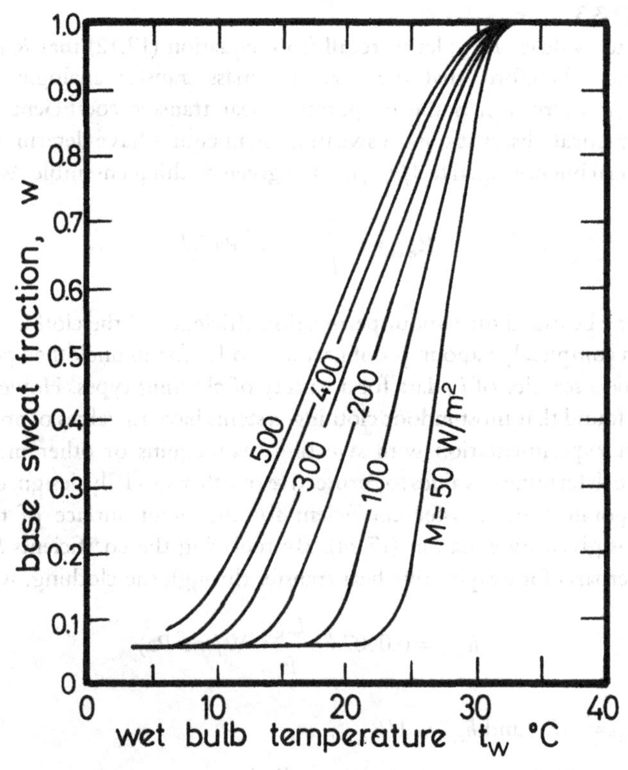

Figure 17.3 Base wetness fraction as a function of ambient wet bulb temperature and metabolic rate.

Signals from temperature receptors within the body activate sweat glands in order to increase the size of the wetted area. The sweat rate is closely related to the core temperature and skin temperature. Each of these three may be employed as an indicator of potential heat strain. Furthermore, as core temperature is affected by metabolic energy, M, it follows that sweat rate must also be influenced by the metabolic rate.

An analysis of the effective limits of evaporative cooling given in the ASHRAE *Fundamentals Handbook* (1985) leads to the estimated curves shown on Fig. 17.3. These form a **base wetness fraction** for any given metabolic heat generation and ambient wet bulb temperature. The curves converge to give a fully wetted body surface ($w = 1$) at a wet bulb temperature of 33 °C, irrespective of the metabolic heat production. The curves are described numerically by the algoithm contained within the sweat fraction subroutine of the computer program listed in the appendix at the end of this chapter.

There is, however, an additional effect when the temperature receptors indicate a high skin temperature. Figure 17.4 illustrates the value to be added to the base wetness fraction when skin temperature exceeds 32.5 °C. This curve has been obtained from a consideration of sweat production as a function of the thermoregulatory signals generated by body temperature receptors. The curve is represented adequately by the equation

$$\Delta w_{sk} = \frac{\sin[(31.85\, t_{sk} - 762) \times 0.017\,45]}{2} + 0.5 \qquad (17.32)$$

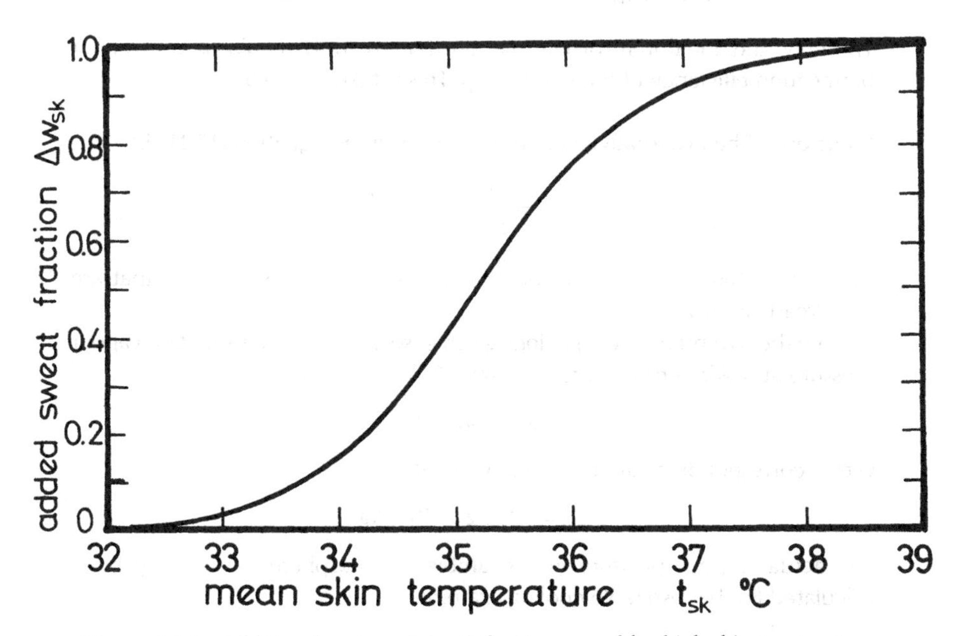

Figure 17.4 Additional wetness (sweat) fraction caused by high skin temperature.

(where the sine function is computed in radians) for skin temperatures between 32.5 and 39.0 °C. The actual value of skin wetness fraction is then calculated as

$$w = \text{base wetness fraction} + \Delta w_{sk} \tag{17.33}$$

with the maximum value set at $w = 1$.

There is considerable variation in sweat rate between individuals and according to the degree of acclimatization. Furthermore, the production of perspiration is by no means constant over the body surface. A run-off of sweat may occur in some zones before others are fully wetted. Light absorbent clothing may assist in holding and distributing perspiration that might otherwise be lost as run-off, so improving the evaporative cooling process. This can result in a distinct chilling effect on wetted skin during rest periods or when entering a cooler or drier atmosphere.

Example Let us continue once again with the earlier example to find the evaporative heat transfer. As a reminder,

metabolic heat generation, M	220 W/m^2
base skin temperature, t_{sk}	32.406 °C
ambient wet bulb temperature, t_w	18 °C
ambient dry bulb temperature, t_d	25 °C
barometric pressure, P	110 000 Pa
convective heat transfer coefficient, h_c	11.91 W/(m^2 °C)
thermal resistance of the clothing, R_{cl}	0.11 °C m^2/W
area factor, f_{cl}	1.28

The only additional item of information that we now require is the vapour permeation efficiency of the clothing, i_{cl}. This is taken as 0.45.

Solution The evaporative heat transfer is given by equation (17.31), i.e.

$$\text{Evap} = \frac{wh_e(e_{sk} - e)}{h_c R_{cl}/i_{cl} + 1/f_{cl}}$$

We must determine the values of those parameters in this equation that are not already known.

From the psychrometric equations given in section 14.6 the saturation vapour pressure at a skin temperature of 32.406 °C is calculated as

$$e_{sk} = 4863 \text{ Pa}$$

with a corresponding latent heat of evaporation

$$L_{sk} = 2425.2 \times 10^3 \text{ J/kg}$$

The actual vapour pressure in the ambient atmosphere has already been calculated (in the respirable cooling example) as

$$e = 1566 \text{ Pa}$$

Equation (17.24) gives the evaporative heat transfer coefficient to be

$$h_e = 0.0007 \times 11.91 \times \frac{2425.2 \times 10^3}{110\,000}$$

$$= 0.1838 \text{ W/(m}^2\text{ Pa)}$$

The base wetness fraction, w, at a metabolic rate of $M = 220$ W/m^2 and a wet bulb temperature $t_w = 18$ °C may be read from Fig. 17.3 or calculated from the sweat factor subroutine given within the program listed in the appendix as $w = 0.268$. Had the skin temperature been greater than 32.5 °C then an addition to the wetness fraction would have been required from Fig. 17.4.

Equation (17.31) then gives

$$\text{Evap} = \frac{0.268 + 0.1838(4863 - 1566)}{11.91 \times 0.11/0.45 + 1/1.28}$$

$$= 44.0 \text{ W/m}^2$$

17.3.6 A thermoregulation model

The analyses in the previous subsections have produced a set of equations that describe the heat transfers from a human body. A numerical model of the human thermoregulation system involves organizing those equations into a logical sequence; then, after comparing the net heat transfer with a known metabolic heat generation, adjust one or more of the physiological parameters in a manner that simulates the thermoregulatory response of an average, healthy person. The process is repeated iteratively until either thermal equilibrium is reached (when net heat transfer equals metabolic heat) or the model indicates untenable conditions of heat strain.

It must be appreciated that considerable variation occurs between individuals in their reaction to climatic variations. Empirical data gained from physiological tests represent average response. Combined with the approximations that are inherent in the analyses of surface heat and mass transfer theory, this indicates that the results of any thermoregulation model provide guidelines to average reaction rather than precise predictions for individual workers.

The thermoregulation model developed here employs average skin temperature as the physiological indicator of heat strain (section 17.3.3). The model calculates an initial or base skin temperature on the basis of ambient dry bulb temperature from Gagge *et al.*'s equation (17.16) or from Fig. 17.2. The heat transfer from respiration, Br, convection, Con, radiation, Rad, and evaporation, Evap, are calculated from equations (17.7), (17.13), (17.22) and (17.31) respectively. The difference between net heat transfer and metabolic heat then indicates the heat accumulated, Ac (see equation (17.2)):

$$\text{Ac} = M - (\text{Br} + \text{Con} + \text{Rad} + \text{Evap}) \quad \text{W/m}^2 \qquad (17.34)$$

If Ac is positive then heat is temporarily accumulated within the body. The reaction of an actual person may be to rest, to reduce the rate of work, or to discard clothing (the behavioural response). However, the simulation model assumes that the metabolic rate, M, is maintained. The skin temperature and, consequently, wetness fraction are increased as a function of Ac. In the model given as a program listing in the appendix, the correction to t_{sk} may be taken as

$$0.02\,\mathrm{Ac}$$

The heat transfer equations are re-evaluated iteratively until Ac is within 1 W/m^2 of zero. If the final equilibrium skin temperature is over 36 °C then the worker may exhibit the progressive symptoms of heat strain, while skin temperatures in excess of 37 °C should be considered to be dangerous.

Example The example that was built up sequentially through sections 17.3.2 to 17.3.5 is continued here to illustrate the reaction of the model and iterative convergence towards thermal equilibrium. The initial or base skin temperature was calculated to be 32.406 °C. The heat transfers associated with this skin temperature were calculated to be

respiration, Br,	24.8 W/m^2
convection, Con,	42.2 W/m^2
radiation, Rad,	7.9 W/m^2
evaporation, Evap,	44.0 W/m^2
total (net cooling)	118.9 W/m^2

The metabolic rate was, however, 220 W/m^2. Hence, the initial rate of heat accumulation is

$$\mathrm{Ac} = 220 - 118.9 = 101.1\ \mathrm{W/m^2}$$

This causes the simulated skin temperature to increase to $32.406 + (0.02 \times 101.1) = 34.428$ °C. The heat transfer equations are then re-evaluated iteratively until Ac approaches zero.

Table 17.3 illustrates the convergence of the model (listed in the appendix) towards equilibrium when net cooling equals metabolic rate. These results show that the person will be perspiring at a medium rate ($w = 0.65$) and will attain an equilibrium skin temperature of 34.8 °C if he maintains his work rate at 220 W/m^2. This skin temperature indicates acceptable conditions with negligible risk of heat strain.

By running the thermoregulation model listed in the appendix, the equilibrium skin temperature for any given combination of psychrometric conditions, air velocity, type of clothing and metabolic rate can be computed. This may then be used as a measure of potential heat strain. As indicated earlier, the probability of unacceptable heat strain is small for skin temperatures up to 36 °C in acclimatized workers. However, for even greater stringency, Fig. 17.5 has been developed from the work of Stewart (1982) and Wyndham (1974). This gives

Table 17.3 Convergence of the thermoregulation model towards thermal equilibrium

	Skin temperature (°C)	*Respiration Br (W/m²)*	*Convection Con (W/m²)*	*Radiant Rad (W/m²)*	*Evaporation Evap (W/m²)*	*Wetness Fraction w*	*Net cooling (W/m²)*
1	32.41	25	42	8	44	0.27	119
2	34.43	25	54	11	106	0.55	196
3	34.90	25	56	12	135	0.68	229
4	34.73	25	55	12	124	0.63	216
5	34.80	25	56	12	129	0.65	221

maximum allowable values of equilibrium skin temperature for fully wetted and acclimatized persons after four hours of continuous work, if there is to be a probability of no more than 10^{-6} of dangerous core temperatures.

For the purposes of illustrating the behaviour of equilibrium skin temperature, cyclic runs of the thermoregulation model produced the results that are depicted on Fig. 17.6 for differing degrees of clothing. In these runs the barometric pressure was maintained at 100 kPa, metabolic rate at 100 W/m², and the dry bulb temperature and radiant temperature both set at 5 °C above the wet bulb temperature. For those conditions, the equilibrium skin temperature can be

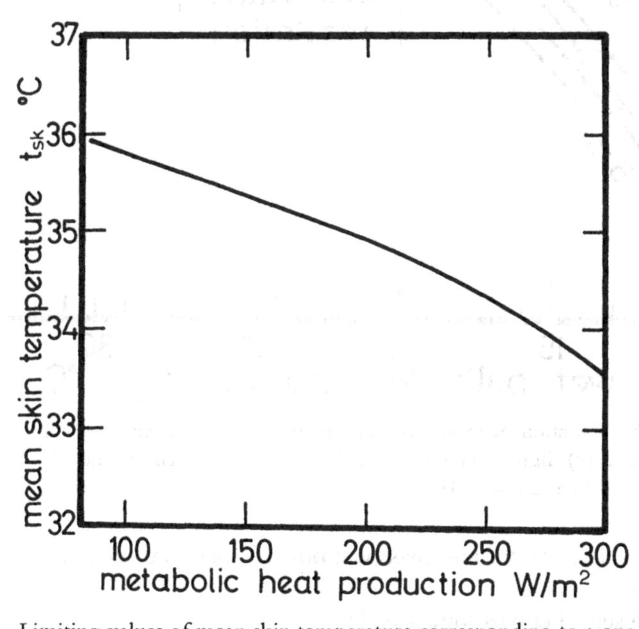

Figure 17.5 Limiting values of mean skin temperature corresponding to a one in a million risk of core temperature rising above 40° C in acclimatized men.

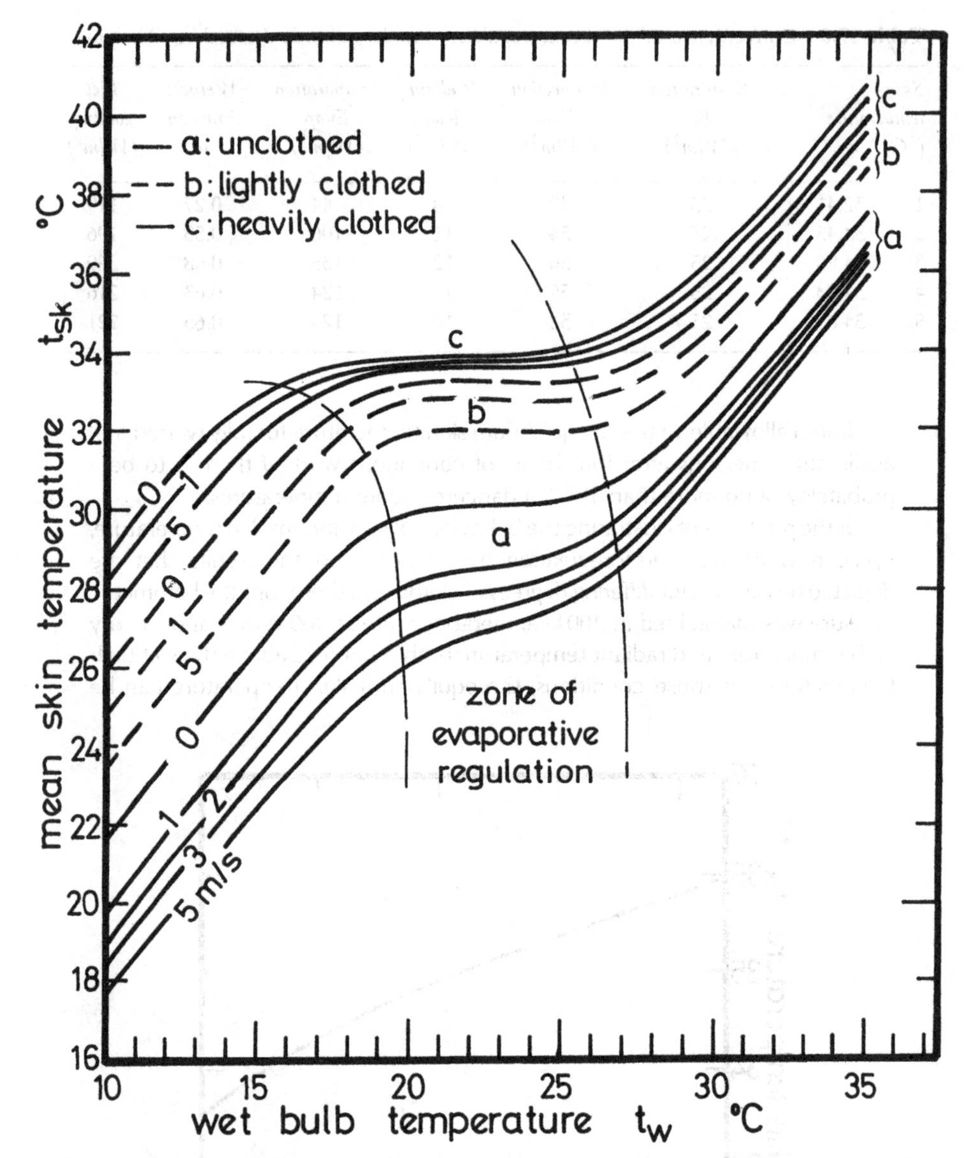

Figure 17.6 Variation of mean skin temperature with ambient wet bulb temperature for (a) unclothed, (b) lightly clothed and (c) heavily clothed personnel. $M = 100\,\mathrm{W/m^2}$, $t_d = t_r = t_w + 5°$ C and $P = 100\,\mathrm{kPa}$.

read from Fig. 17.6 for any given wet bulb temperature, air velocity, and type of clothing.

These sets of curves show clearly

1. the increases in skin temperature both below and above the zone of evaporative regulation,

2. the levelling of skin temperatures during evaporative regulation (when wetness factor is increasing),
3. the increased skin temperatures caused by clothing, and
4. the reduced influence of air velocity as clothing is added.

17.4 INDICES OF HEAT STRESS

17.4.1 Purpose and types of heat stress indices

It is clear from the preceding sections that the reaction of the human body to climatic variations involves a complex mix of psychrometric parameters and physiological responses. For detailed investigation, a thermoregulation model such as that described in section 17.3.6 may be employed. This requires a specification of all relevant data pertaining to work rate, clothing and psychrometric condition of the air, and the availability of the relevant program and computer hardware.

For less rigorous applications such as routine checking of the cooling power of an existing environment or for initial conceptual designs, it is convenient to employ some measure of air cooling power or physiological reaction that can be quoted as a single number or **index of heat stress**. For manual use, such an index should be capable of direct measurement, or calculated easily from only a few types of observation. Over 90 indicates of heat stress have been developed during the twentieth century (Hanoff, 1970). This reflects the large number of variables involved, the complexity of the human thermoregulation system and the range of climatic conditions to be encompassed. Furthermore, the lack of computing power until the 1960s rendered the use of sophisticated numerical models impractical. The availability of microcomputers in engineering offices has diminished the need for simplified indices of heat stress.

We may classify heat stress indices into three types:

1. single measurements,
2. empirical methods, and
3. rational indices.

The indices within each of these groups that have been recommended for subsurface ventilation systems are introduced in the following subsections.

17.4.2 Single measurements

No single psychrometric parameter, by itself, gives a reliable indication of physiological reaction. In hot and humid environments where the predominant mode of heat transfer is evaporation, the wet bulb temperature of the ambient air is the most powerful variable affecting body cooling. Indeed, many mines retain the wet bulb temperature as a sole indicator of climatic acceptability. A psychrometric (aspirated) wet bulb temperature of 27 or 28 °C may be employed as a criterion above which work rates or shift hours are reduced, while 32 °C may be regarded as an upper limit of acceptability.

The second most important variable in hot conditions is the air velocity. Although air velocity, by itself, gives little indication of climatic acceptability, it can easily be combined with wet bulb temperature. This may be achieved by measuring the **natural** wet bulb temperature, that is, the temperature indicated by a non-aspirated wet bulb thermometer held stationary within the prevailing airstream.

Dry bulb temperature alone has limited effect on climatic acceptability in hot mine environments. However, dry bulb temperatures in excess of 45 °C can give a burning sensation on exposed skin facing the airstream. Hot dry environments may also result in skin ailments (section 17.5.4). Heating and ventilating engineers concerned with comfort conditions in surface buildings often employ a combination of dry bulb temperature and relative humidity.

For cold conditions where convection and radiation are the primary modes of heat exchange, dry bulb temperature and air velocity become the dominant factors. These may be combined into a wind chill index (section 17.6).

17.4.3 Empirical methods

These techniques produce indices of heat stress that have either evolved from a statistical treatment of observations made on volunteers working under conditions of a controlled climate or are based on simplified relationships that utilize measurable parameters but have not been derived through a rational or theoretical analysis.

Effective temperature, ET

Effective temperature is one of the older and, probably, most widely used indices of heat stress. In the mid-1920s, F. C. Houghton and C. P. Yaglou of the then American Society of Heating and Ventilating Engineers (ASHVE) conducted a series of experiments in which three volunteers were asked to pass between two adjoining rooms. One room was maintained at saturated conditions and negligible air velocity during any given experiment while the wet bulb temperature, dry bulb temperature and air velocity were varied in the second room. The instantaneous and subjective thermal sensation of each volunteer on passing between rooms was recorded simply as 'hotter', 'cooler' or 'the same'. The experiments were conducted first on men stripped to the waist then repeated with the same subjects wearing lightweight suits.

The **effective temperature** was defined as the temperature of still saturated air that would give the same instantaneous thermal sensation as the actual environment under consideration. Charts were produced that allowed the effective temperature to be read for given wet bulb temperature, dry bulb temperature and air velocity. An example is shown on Fig. 17.7.

The concept of effective temperature has been employed widely both as a comfort index for office workers and also as a heat stress index for industrial and military occupations. The effect of radiant heat exchange may be taken into account in the corrected effective temperature in which the dry bulb temperature is replaced by the

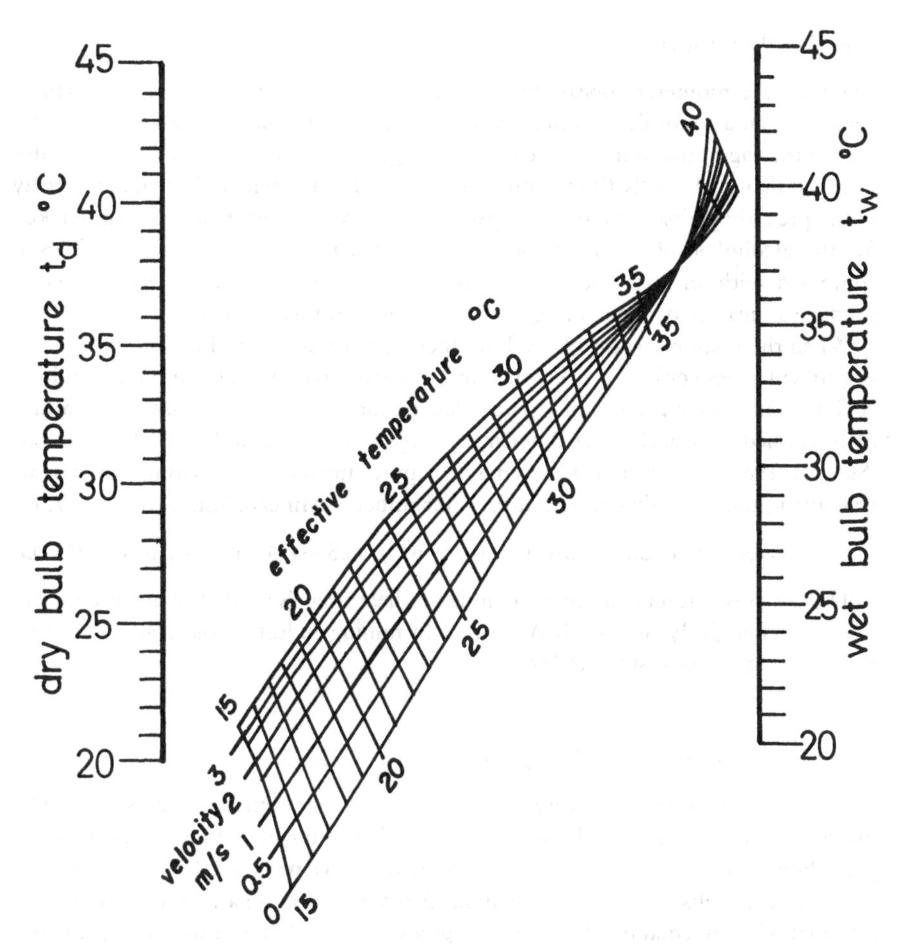

Figure 17.7 Effective temperature chart (normal scale) for lightly clothed personnel.

globe temperature measured at the centre of a blackened 150 mm diameter hollow copper sphere.

The use of effective temperature for the prediction of physiological strain has been shown to have several shortcomings (Wyndham, 1974) and most mining industries have abandoned it. Effective temperature remains in use in the United Kingdom and Germany. Some mining companies in the United States also still utilize effective temperature although it is no longer ratified by any US organization. Where effective temperature remains in use, it is recommended that periods of continuous work should be reduced at effective temperatures in excess of 28 °C and work should be terminated if it rises above 32 °C.

Wet Kata thermometer

The Kata thermometer appears to have been introduced by Dr. L. Hill in 1916. It consists of an alcohol thermometer with an enlarged bulb at the base and a smaller bulb at the top of the stem. To use it, the enlarged bulb is submerged in warm water until the alcohol partially fills the upper reservoir. The instrument is then hung freely in the prevailing airstream and a stopwatch employed to determine the time taken for the alcohol level to fall between two marks inscribed on the stem. This is combined with an instrument factor given with each individual thermometer to produce a measure of the cooling power of the air in mcal/(cm^2/s).

When the main reservoir of the Kata thermometer is covered in wet muslin, the instrument gives cooling powers that are representative of a 20 mm wetted bulb at 36.5 °C. These have been found to correlate reasonably well with core temperatures of unclothed and acclimatized workers subjected to hot and humid conditions (Stewart, 1982). Wet Kata readings may be approximated from measurements of air velocity, u, and wet bulb temperature, t_w (Chamber of Mines of South Africa, 1972):

$$\text{wet Kata cooling power} = (0.7 + u^{0.5})(36.5 - t_w) \quad \text{mcal/(cm}^2\text{ s)} \qquad (17.35)$$

The use of wet Kata readings as an index of heat stress is limited to hot and humid environments. Only the South African gold mining industry has retained the wet Kata reading as a heat stress index.

Wet bulb globe temperature and wet globe temperature

The wet bulb globe temperature (WBGT) relies on two measurements only. The first is the reading indicated by a wet bulb thermometer held stationary in the prevailing airstream. This is sometimes termed the **natural** wet bulb temperature, t_{nw}, and at air velocities of less than about 3 m/s will indicate a temperature greater than that of the normal psychrometric (aspirated) wet bulb thermometer, t_w (section 14.3.5). Secondly, the temperature at the centre of a matte black hollow sphere or globe temperature, t_g, is measured. The two are combined to give

$$\text{WBGT} = 0.7t_{nw} + 0.3t_g \qquad (17.36)$$

When a significant source of radiant heat is visible then

$$\text{WBGT} = 0.7t_{nw} + 0.2t_g + 0.1t_d \qquad (17.37)$$

where t_d = dry bulb temperature. This latter form of the equation is normally employed only in sunlight.

The wet bulb globe temperature is a function of the major climatic parameters that affect physiological reaction, i.e. wet and dry bulb temperatures, air velocity and radiant temperature. However, it has the advantage that it does not require a separate measurement of air velocity. The National Institute of Occupational Safety and Health of the United States (NIOSH, 1986) has employed WBGT as a heat stress

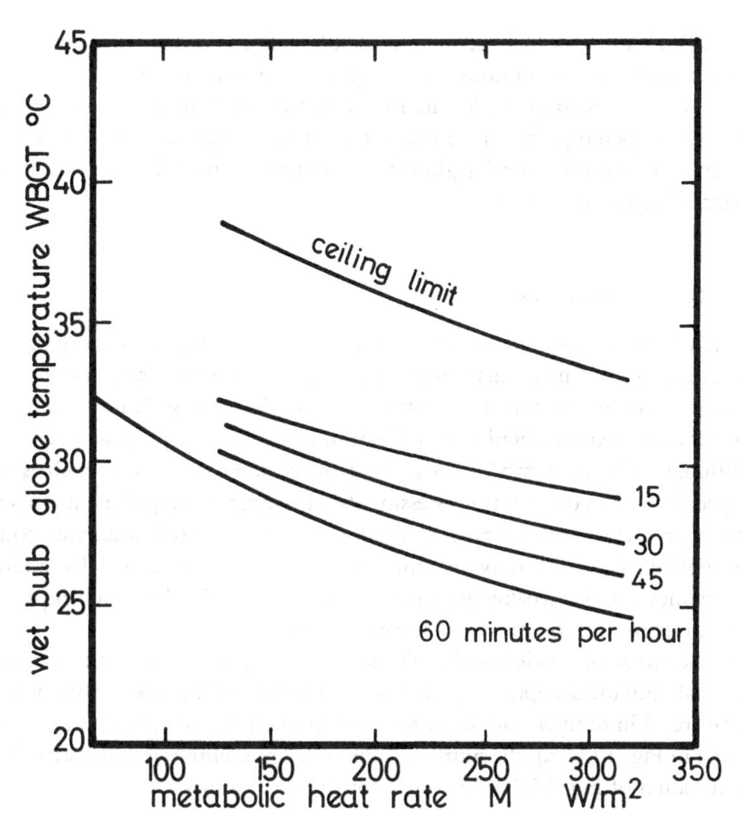

Figure 17.8 Wet bulb globe temperature exposure limits for acclimatized workers (recommended by the US National Institute of Occupational Safety and Health).

standard and recommended the exposure limits shown on Fig. 17.8. Heat stress levels based on the WBGT have also been subject to an international standard (ISO, 1982).

A simplified version of the wet bulb globe temperature was developed by J. H. Botsford in 1971 and is known as the **wet globe temperature** (WGT). The instrument, now called a Botsball, consists of a 6 cm diameter blackened copper sphere covered by a wetted black fabric. A dial thermometer indicates the temperature at the middle of the hollow sphere.

Regression equations may be employed to relate WGT to WBGT for known psychrometric conditions. For the particular range of dry bulb temperatures 20 to 35 °C, relative humidities greater than 30% and air velocities less than 7 m/s, Onkaram *et al.* (1980) have suggested the experimentally derived equation

$$\mathrm{WBGT} = 1.044\,\mathrm{WGT} - 0.187 \quad {}^{\circ}\mathrm{C} \tag{17.38}$$

where WGT is measured in °C.

The WBGT and its simplified companion, WGT, are both well suited to the rapid assessment of climatic conditions in existing locations where instrumentation may be located. They are, at present, less useful for the prediction of air cooling power in underground openings that have not yet been constructed, as neither natural wet bulb temperature nor wet or dry globe temperatures are included as output in current subsurface climate simulators.

17.4.4 Rational indices

A rational index of heat stress is one that has been established on the basis of the physiological heat balance equation (17.3) and that conforms to heat transfer relationships such as those developed in section 17.3. A thermoregulation model (section 17.3.6) may be used for detailed investigation of existing or proposed facilities. The establishment of such a model on a mine office microcomputer enables it to be employed for day to day routine assessments. However, for rapid manual assessment or where predictions of average cooling power for a work area are required, a thermoregulation model may be simplified into charts or tables by establishing specific values for the weaker parameters, or defining fixed relationships between those parameters and the more dominant variables.

A choice must be made on the physiological response that is to be used as an indicator of climatic acceptability. This is usually one of the linked parameters, core temperature, skin temperature or sweat rate. In the thermoregulation model output illustrated in Fig. 17.6, equilibrium skin temperature is employed in association with the limit values of Fig. 17.5 as the indicator of acceptability.

Air cooling power (M scale), ACPM

Figure 17.6(a) gives a series of curves relating equilibrium skin temperature to wet bulb temperature and air velocity for unclothed personnel producing 100 W/m^2 of metabolic heat, and for specified values of

$$P = 100\,\text{kPa and } t_d = t_r = t_w + 5\,°\text{C} \tag{17.39}$$

Figure 17.5 indicates that at $M = 100\,\text{W/m}^2$ an equilibrium skin temperature of 35.8 °C is associated with a probability of no more than one in a million that core temperature will exceed 40 °C, i.e. negligible risk of unacceptable heat strain. Scanning across the $t_{sk} = 35.8\,°\text{C}$ line on Fig. 17.6(a) allows us to read the corresponding limiting wet bulb temperatures for air velocities 0 to 5 m/s.

If the procedure were repeated with a sufficiently large and detailed number of charts of the type illustrated on Fig. 17.6 then tables relating metabolic heat, wet bulb temperature and air velocity could be compiled for each clothing ensemble. This exercise has been carried out, not graphically, but by cyclic operation of the thermoregulation model described in section 17.3.6. The results are shown on Fig. 17.9.

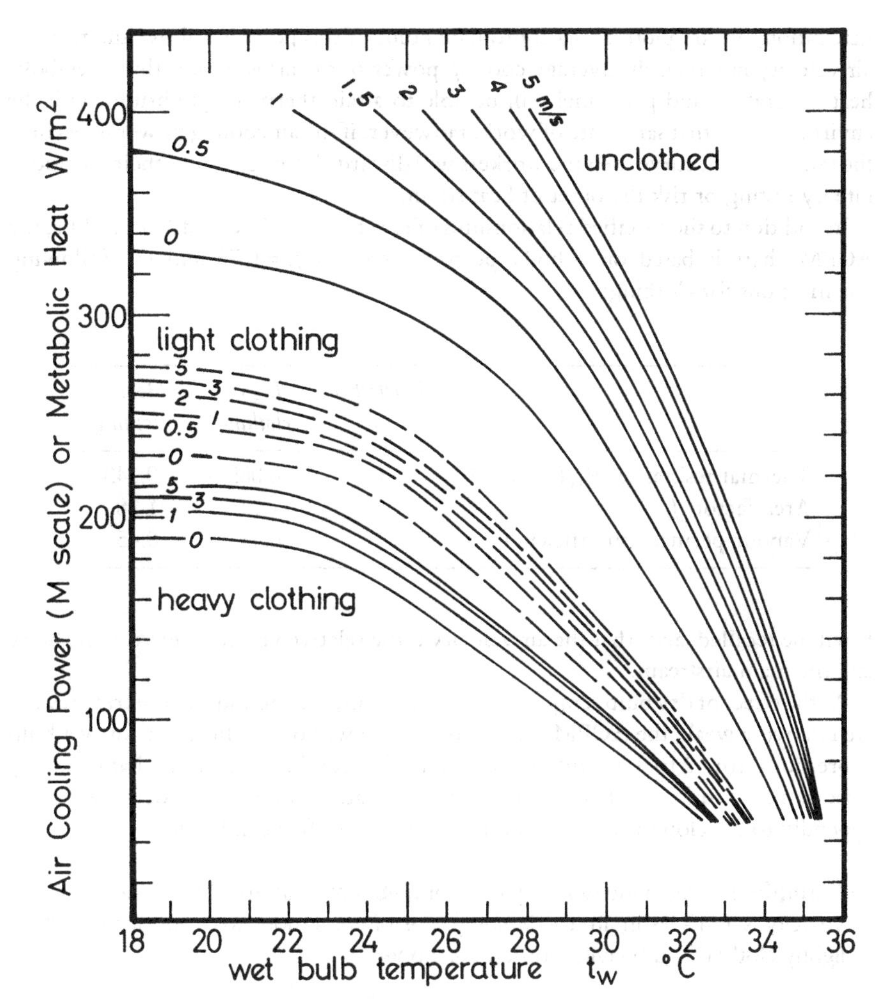

Figure 17.9 Air cooling power (M scale) or ACPM chart (based on $t_d = t_r = t_w + 5°$ C and $P = 100$ kPa).

The lines on this chart have been established from equilibrium conditions when the ambient environment removes metabolic heat at the same rate as it is generated. Hence, they may be regarded as limit lines of climatic acceptability. For this reason, the vertical axis has been renamed air cooling power (M scale) as a reminder that the lines represent equilibrium with metabolic heat, M, and to distinguish it from earlier versions of air cooling power.

In order to use the chart as a rapid means of assessing climatic acceptability of any given environment, the psychrometric wet bulb temperature, t_w, and metabolic heat generation, M, are plotted as a coordinate point. If this point lies above the relevant air velocity line then the average cooling power of the air is greater than metabolic heat generation and personnel will be able to attain thermal equilibrium with the environment at that same rate of work. However, if the air cooling power is less than the relevant limit curve then the workers will discard clothing, reduce their metabolic rate by resting or risk the onset of heat strain.

In addition to the specified relationships of $t_d = t_r = t_w + 5\,°C$ and $P = 100$ kPa, the ACPM chart is based on a body posture factor of $f_r = 0.75$ and the following specifications for clothing:

	Unclothed	*Light clothing*	*Heavy clothing*
Thermal resistance, R_{cl} (°C m^2/W)	0	0.085	0.143
Area factor, f_{cl}	1	1.14	1.28
Vapour permeation efficiency	1	0.45	0.45

It will be recalled, also, that the air velocity is the relative velocity between the body and the main airstream.

As the effect of dry bulb temperature, radiant temperature and barometric pressure are relatively weak, the ACPM chart may be used without undue error for wet bulb depressions from 2 to 8 °C and for the range of atmospheric pressures that normally exist in mine workings. It is a straightforward matter to use the model listed in the appendix to develop similar charts for any other specified conditions.

Example 1 In a mine working, the wet bulb temperature is 28 °C and the air velocity is 1 m/s. Estimate the limiting rates of continuous work for unclothed, lightly clothed and heavily clothed personnel.

Solution From Fig. 17.9 the three values of air cooling power at $t_w = 28$ °C and $u = 1$ m/s are estimated at

unclothed	313 W/m^2
lightly clothed	168 W/m^2
heavily clothed	127 W/m^2

Table 17.1 may be used to relate these cooling powers to types of work activities.

Example 2 In is predicted that in a future subsurface development the air velocity will be 2 m/s. If personnel are to sustain a metabolic rate of 220 W/m^2 while lightly clothed, estimate the upper limit of psychrometric wet bulb temperature.

Solution From the light clothing curves on Fig. 17.9, at an air velocity of 2 m/s and an air cooling power of 220 W/m^2, the limiting wet bulb temperature is estimated at 25.6 °C.

Specific cooling power (A scale), ACPA

The original concept of air cooling power as employed in this chapter arose from pioneering work conducted by the Chamber of Mines Research Organization of South Africa (Mitchell and Whillier, 1972; Wyndham, 1974; Stewart and Whillier, 1979; Stewart, 1982). This involved detailed analytical investigations as well as some thousands of tests in a large environmentally controlled and monitored wind tunnel. A thermoregulation model was developed on the basis of the physiological heat transfer equation (17.3). That model was restricted to the hot and humid conditions prevalent in the stopes of deep South African gold mines. In those conditions the influence of respiratory cooling is relatively small and was ignored. The model was further confined to fully wetted and unclothed personnel. An air cooling power chart was produced for the specific conditions of $t_d = t_r = t_w + 2\,°C$ and $P = 100$ kPa and for wet bulb temperatures in the range 25 to 35 °C (Stewart, 1982). This was termed the specific cooling power (A scale) chart and yielded curves that are comparable with those given for unclothed personnel on Fig. 17.9 within that range of wet bulb temperature.

17.5 HEAT ILLNESSES

The human thermoregulation system depends on the efficient operation of the core and skin temperature receptors, the flow of blood throughout the body but particularly between the core and skin tissues, and the production of perspiration. If any of these mechanisms loses its effectiveness then the body will progressively exhibit the symptoms of one or more of a series of disorders known collectively as the heat illnesses. These may arise as separate and recognizable ailments with identifiable causes. However, for workers in hot and humid environments they may occur in combination.

A common initial symptom is a loss of interest in the task and difficulty in remaining alert. In any adverse environment, the desire to seek more comfortable surroundings is a psychological reaction that is just as much a part of the body's defence mechanism as thermoregulatory effects. Suppression of such predilection may result in irritability or displays of anger. This may be observed even in persons who are cognizant of the effect. The physical symptoms often reveal themselves first as a loss of coordination and dexterity.

It follows, even from these initial symptoms of heat strain, that both manual and mental work productivity will suffer, morale is likely to be low, absenteeism high and standards of safety will decline in environments that are unduly hot. This is true for either a labour-intensive method of working or one that is heavily mechanized.

Heat illnesses are introduced in the following subsections and in order of increasing seriousness. However, it should be remembered that symptoms may overlap and, in cases of doubt, it should be assumed that the patient is a victim of heat stroke.

17.5.1 Heat fainting

This occurs most frequently when a person stands still for an extended period in a warm environment. Blood tends to pool in the lower parts of the body causing a temporary reduction in blood supply to the brain and, hence, a short term loss of consciousness. A common example arises when soldiers in heavy uniform are required to remain motionless for long periods of time during ceremonial parades.

The treatment is simply to restore an adequate supply of blood to the brain by allowing the patient to lie flat in a cooler area and to loosen or remove clothing. Recovery is normally rapid and complete. The probability of heat fainting is reduced by intermittent body movements.

17.5.2 Heat exhaustion

Conductance of heat within the body is facilitated primarily by the flow of blood. If the volume of blood is insufficient then heat exhaustion may ensue. A decrease in blood volume may result from dehydration caused either by an inadequate intake of fluids or by a salt-deficient diet. Alternatively, a combination of environmental heat stress and metabolic rate may cause the heartbeat to exceed some 180 bpm. The inadequate time interval between contractions of heart muscles may be insufficient to maintain an adequate supply to the heart chambers—the rate of blood flow drops. A significant increase in total volume of blood occurs during periods of heat acclimatization.

The additional symptoms of heat exhaustion are

1. tiredness, thirst, dizziness,
2. numbness or tingling in fingers and toes,
3. breathlessness, palpitations, low blood pressure,
4. blurred vision, headache, nausea and fainting, and
5. clammy skin that may be either pale or flushed.

On the spot treatment should include removal to a cool area and administration of moderate amounts of drinking water. If the patient is unconscious then heat stroke should be assumed. In any case, a medical examination should be carried out before the victim is allowed to return to work.

17.5.3 Heat cramps

If the electrolytic balance of body fluids is sufficiently perturbed then painful muscular contractions occur in the arms, legs and abdomen. This may arise from salt deficiency or drinking large amounts of water following dehydration. Medical opinion no longer favours salt tablets but recommends a diet that provides a more natural supply of salt. However, immediate treatment may include the administration of fluid containing no more than 0.1% salt.

17.5.4 Heat rash

This is sometimes known as **prickly heat** by residents of equatorial regions. It is caused by unrelieved periods of constant perspiration. The continuous presence of unevaporated sweat produces inflammation and blockage of the sweat ducts. The typical appearance of the ailment is areas of tiny red blisters causing irritation and soreness. Although not serious in itself, heat rash may lead to secondary infections of the skin and, if sweat rates are sufficiently inhibited, may produce an increased susceptibility to heat stroke. Heat rash can be prevented by frequent bathing and the provision of cool living quarters.

Skin irritations can also occur in the hot, dry environments of some evaporite mines, especially at areas of skin contact with hat bands or protective devices.

17.5.5 Heat stroke

The most serious of the heat illnesses, heat stroke, may occur when the body core temperature rises above 41 °C. Coordination of the involuntary nervous system, including body thermoregulation, is achieved by means of proteins controlled by the hypothalamus, part of the brainstem. At temperatures in excess of 41 °C, damage to those proteins results in the transmission of confused signals to the muscles and organs that control involuntary reactions. Furthermore, irreversible injury may be caused to the brain, kidneys and liver. Heat stroke carries a high risk of fatality.

The initial symptoms of impending heat stroke are similar to those of the less serious heat illnesses, i.e. headaches, dizziness, nausea, fatigue, thirst, breathlessness and palpitations. If the patient loses consciousness, then heat stroke should be assumed. The additional symptoms arise from disruption of the thermoregulatory and other nervous systems of the body:

1. perspiration ceases, the skin remains hot but is dry and may adopt a blotchy or bluish coloration;
2. disorientation may become severe involving dilated pupils, a glassy stare and irrational or aggressive behaviour;
3. shivering and other uncontrolled muscular contractions may occur;
4. there may be a loss of control of bodily functions.

Although treatment should commence as soon as possible, current medical opinion is that the patient should not be moved until core temperature has been brought under control. The administration of first aid should include the following.

1. Cooling by air movement and the application of water to the skin. The water temperature should be not less than 15 °C in order to avoid vasoconstriction of the

skin blood capillaries. Massaging the skin also assists in promoting blood flow through the surface tissues.
2. The techniques of CPR (cardiopulmonary resuscitation) in cases where perceptible heart beat or respiration are absent. These include chest massage and mouth-to-mouth breathing.
3. Frequent sips of water if the patient is conscious.

Medical assistance should be sought and treatment continued until the core temperature falls below 38.5 °C. The patient should then be covered by a single blanket and transported by stretcher to hospital. Unfortunately, death or permanent disability may still occur if irreversible damage has been incurred by vital organs. Even after a physical recovery, the heat tolerance of the victim may be impaired.

17.5.6 Precautions against heat illnesses

There is a great deal that can be done to protect a workforce against unacceptable heat strain. First, the design and engineering control of the ventilation and air quality systems of the mine or facility should be directed toward providing a safe climatic environment within all places where personnel work or travel routinely. If threshold limit values for climatic parameters have not been set by legislative action, then they should be chosen and enforced by self-imposed action of the mine or industry. Cool and potable drinking water should be available to workers in hot mines.

Clothing that is appropriate for the work environment and level of activity should be worn by personnel. In hot environments, this may mean very little clothing or lightweight garments that are loosely weaved, porous and absorbent. For unusual circumstances where personnel are occasionally required to enter hot areas that do not meet the criteria of climatic acceptability, microclimate protective garments may be used as a short-term and temporary measure. Such devices typically take the form of specialized jackets that contain distributed pockets which hold either solid carbon dioxide (dry ice), cooled gels or water ice. The microclimate garments are worn over underclothing in order to avoid ice burns. Outer clothing reduces direct heat transfer from the surroundings. It is important that this type of microclimate jacket is changed before the cooling medium is exhausted, as the garment may then become an effective insulator. More sophisticated microclimate suits, incorporating temperature-controlled fluids circulating through capillary tubing, have also been manufactured. However, these are expensive and less applicable to mining environments.

A workforce that is required to work in hot environments should be trained to recognize the initial symptoms of heat strain and to adopt sensible work habits. The latter include appropriate, but not excessive, rest periods and adequate consumption of cool drinking water. A good dietary balance should be sought that includes sufficient but, again, not excessive salt. Vitamin C supplementation is also regarded as improving heat tolerance (Visagie *et al.*, 1974).

Finally, workers should have been subjected to a process of natural or induced acclimatization prior to being employed in particularly hot conditions (section 17.7).

17.6 COLD ENVIRONMENTS

This chapter is primarily concerned with physiological reactions to heat. However, personnel at mines or other facilities that are located in cold climates or at high altitudes may be subjected to low temperatures for at least part of the year. Even in the more temperate zones, workers in main intake airways may suffer from cold discomfort during winter seasons. It is, therefore, appropriate that we include the effects of cold environments on the human body.

17.6.1 Physiological reactions to cold environments

Heat loss from the clothed body in cold surroundings is mainly by convection with lesser amounts by radiation and respiration. The most important climatic parameters in these circumstances are ambient (dry bulb) temperature and air velocity. These may be combined into an equivalent wind chill temperature (section 17.6.3).

As heat loss occurs, the initial behavioural response is to don additional clothing and to increase metabolic heat production by conscious musclar activity. Involuntary physiological reaction is initiated by reductions in either the body core temperature or mean skin temperature and consists of an increase in muscular tension. In skin tissues, this causes the familiar 'gooseflesh' and progresses into shivering within localized muscle groups. The generation of metabolic heat may be raised by up to 120 W/m^2. With further cooling, the degree of shivering increases to encompass the whole body, producing some 300 W/m^2 of metabolic heat and effectively incapacitating the person.

If the core temperature falls below 35 °C, the body thermoregulation system will be affected. Core temperatures of less than 28 °C can prove fatal, although successful recovery of individuals has been achieved from core temperatures of less than 20 °C. The subjective feeling of comfort depends on the mean skin temperature and also the surface temperature of the extremities. Commencing from the neutral mean temperature of 34 °C, the following subjective reactions have been reported for sedentary persons.

Mean skin temperature (°C)	*Subjective response*
31	Uncomfortably cold
30	Shivering
29	Extremely cold
25	Limit of tolerance without numbing

However, the increased blood flow caused by a medium or hard work rate can render these reduced values of mean skin temperature acceptable.

The large surface-to-volume ratio of fingers, toes, and ears maximizes heat loss while the influence of vasoconstriction is particularly effective in reducing blood flow to those areas. The latter is a natural reaction to protect, preferentially, the vital organs within the body core. However, it can give rise to severe discomfort and tissue (frost) damage to the extremities. The following reactions to hand surface temperatures have been reported:

1. 20 °C, uncomfortably cold;
2. 16 °C, limit of finger dexterity;
3. 15 °C, extremely cold;
4. 5 °C, painful.

Skin surface temperatures some 1.5 to 2.0 °C higher produce similar reactions at the feet.

17.6.2 Protection against cold environments

The first line of defence against any potentially adverse environment in the subsurface is the initial design and subsequent control of the ventilation and air conditioning systems. Methods of heating mine air are discussed in section 18.4. These are subject to economic limitations and can lead to problems of strata control if the mine is within permafrost. It may, therefore, be necessary to provide means of personal protection against a cold working environment.

Personnel should be instructed on physiological reactions to cold ambient surroundings and, particularly, on the use of protective clothing. The thermal insulating properties of garments depend on the entrapment of air rather than the fibres of the material. The incorporation of radiation reflective layers can also be helpful. A compromise must be sought between the body movements required to perform the work and the restraint offered by bulky insulated clothing. An ambient temperature of −35 °C appears to be the practical limit for many activities while, at −50 °C only short-term exposure should be permitted. Protection to the hands and feet is particularly important but does, of course, affect dexterity. Auxiliary heating of some 10 W to each hand and foot allows activity to continue at even lower temperatures provided that the rest of the body is well insulated.

Facial tissues do not suffer unduly from vasoconstriction. Hence, face masks are usually unnecessary. However, the effects of wind chill can result in freezing of exposed skin. Fur-lined hoods should be used to avoid direct wind on the face at low ambient temperatures. Pre-heating inspired air is not required for healthy individuals for air temperatures down to −45 °C.

A limited degree of cold acclimatization can occur in personnel who have been subjected to frequent exposure. An improved blood flow to skin tissues gives increased circulation to the extremities and reduces the risk of frostbite. Additional heat is also generated by the metabolism of body reserves of adipose (fatty) tissue.

17.6.3 Indices of cold stress

A series of experiments were carried out by Siple and Passel (1945) in Antarctica involving rates of cooling by convection and radiation from a plastic cylinder of

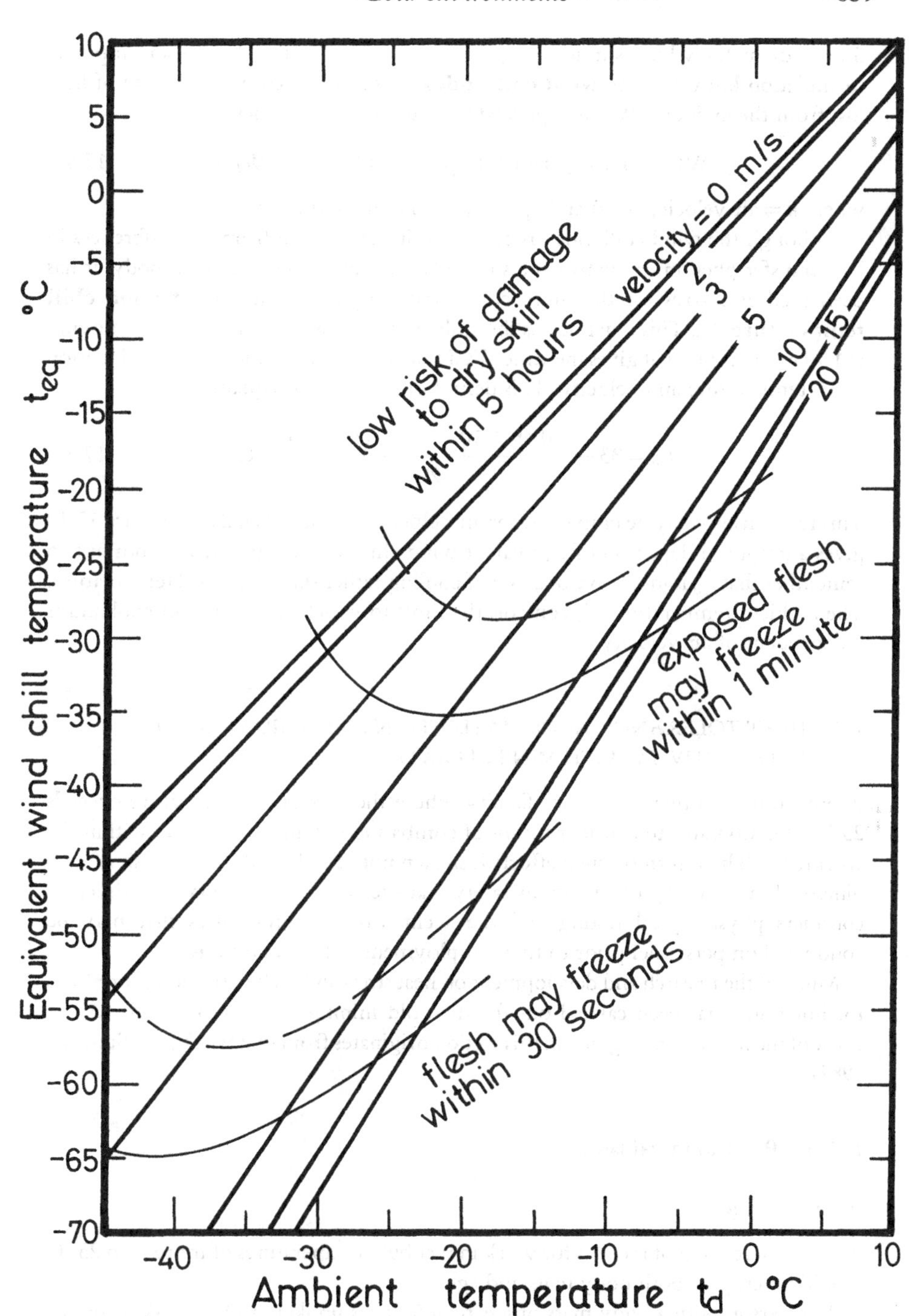

Figure 17.10 Equivalent wind chill temperature (developed from data reported by the US Army Research Institute of Environmental Medicine).

5.7 cm diameter whose surface temperature was 33 °C. This led to an empirical formulation known as the **wind chill index**, WCI, that described the rate of heat loss from the cylinder. When expressed in SI units, this becomes

$$\mathrm{WCI} = 1.163\,(10.45 + 10\sqrt{u} - u)\,(33 - t_d) \quad \mathrm{W/m^2} \qquad (17.40)$$

where u = air velocity (m/s) and t_d = dry bulb temperature (°C).

Although the wind chill index is open to criticism arising from the differences in heat transfer phenomena between a small plastic cylinder and a human body, it has been used as a basis for the more commonly employed **equivalent wind chill temperature**, t_{eq}. This is defined as the ambient temperature in an airstream moving at 1.79 m/s (4 mph) that gives the same WCI (rate of cooling) as the actual combination of air temperature and velocity. This definition leads to the equation

$$t_{eq} = 33 - \frac{(10.45 + 10\sqrt{u} - u)(33 - t_d)}{22} \quad {}^\circ\mathrm{C} \qquad (17.41)$$

This relationship may be employed for air velocities not less than 2 m/s. Figure 17.10 gives a graphical depiction of equivalent wind chill temperature. It is important to remember that t_{eq} is indicative of a rate of cooling rather than a physical temperature. The ambient temperature, t_d, controls the limit to which any body can cool and is independent of air velocity.

17.7 HEAT TOLERANCE, ACCLIMATIZATION AND VARIATION OF PRODUCTIVITY WITH MINE CLIMATE

In any mine or other subsurface facility where the wet bulb temperature exceeds 25 °C, it is advisable to consider means of combating potential heat strain within the workforce. The return on such efforts is shown not only by a decreased risk of heat illnesses but also improvements in safety statistics and productivity. This section considers physiological testing and the acclimatization procedures that may be conducted on personnel prior to their employment in hot conditions.

Much of the research and development on heat stress and acclimatization procedures for hot mines has been carried out for the gold mining industry of South Africa. Most of the numeric data given in this section originates from that work (e.g. Stewart, 1982).

17.7.1 Physiological tests

Initial screening

An initial screening of recruits for work in wet bulb temperatures of more than 25 °C should investigate both body mass and age.

Any person with a body mass of less than 50 kg is unlikely to have a skin surface area that is sufficient to ensure efficient body cooling. Conversely, individuals with a mass exceeding 100 kg have a high and constant metabolic burden in supporting their own weight. No person in either of these groups should be asked to work in hot environments.

After the age of 40 years, the probability of heat stroke increases significantly. Although considerable variation exists between individuals, it is prudent to relocate persons of more than 45 years of age to zones of reduced heat stress.

Work capacity test

A simple test of work capacity consists of a 10 minute period of constant and controlled activity that requires an oxygen consumption rate of 1.25 l/min. At the conclusion of the period, the heartbeat provides an indication of maximum rate of oxygen consumption and, hence, work capacity. Persons with test heartbeats of over 140 bpm should be prohibited from tasks requiring hard manual exertion or from working in conditions of heat stress. If the test results indicate a heartbeat of less than 120 bpm then that individual should be capable of strenuous work.

If the wet bulb temperature in the workings does not exceed 27.5 °C the initial screening and the work capacity tests are all that are required.

Heat tolerance test

This more rigorous test may be carried out on persons who have passed both the initial screening and the work capacity tests, and who may be required to work in wet bulb temperatures in excess of 27.5 °C. The test consists of requiring the participants to engage in a controlled activity corresponding to an oxygen consumption of 1.25 l/min for a period of 4 h in a climatic chamber. The wet bulb temperature should be at least equal to the highest value that will be encountered within work areas in the mine.

Oral temperatures are taken at intervals of not more than 30 min throughout the test and the following criteria applied. If oral temperature

1. remains below 37.5 °C, person eligible immediately for work in hot mine environment,
2. varies between 37.5 and 38.6 °C, eligible for short-term or microclimate acclimatization,
3. varies between 38.6 and 40.0 °C, eligible for full acclimatization,
4. exceeds 40.0 °C, test terminated and person classified as heat intolerant.

If it is suspected that heat intolerance is caused by a temporary illness or body disorder then the test may be repeated after treatment and recovery.

17.7.2 Heat acclimatization

The organized process of heat acclimatization may take place over a period of several days during which the participants are subjected to gradually increasing work rates and/or levels of heat stress. During this time, definite physiological changes occur that enable the human body to better resist the adverse effects of a hot environment.

The first attempts at organized acclimatization for work in hot mines appear to have been carried out in 1926 within the South African gold mining industry (Dreosti, 1935). Considerable effort was devoted to the improvement of acclimatization procedures and South Africa seems to have been the only country that provided formal acclimatization for the majority of recruits into deep and hot mines. However, advances in the design of ventilating and air conditioning systems have led to significant improvements in climatic conditions within those mines. The modern philosophy is to promote working environments that do not necessitate full acclimatization procedures.

Methods of acclimatization

There are essentially three methods of acclimatization for workers in hot environments. The first and, perhaps, most natural is for the individual to spend the first two weeks of underground work in tasks of increasing physical effort (or less frequent rest periods) within the actual mine environment. This may be combined with commencing in relatively cool areas of the mine and progressively moving the participants to locations of increased heat stress through the two week period.

Although such 'on-the-job' acclimatization appears to be convenient, it does have significant drawbacks. If successful acclimatization is to be achieved, then the process must be monitored and controlled by well-trained supervisors. Work rates, body temperatures and psychrometric conditions must be logged and, if necessary, modifications made to the place or rate of work. Individuals who exhibit signs of heat intolerance must be identified before they suffer unduly from heat strain. Drinking water must be administered at frequent intervals. The difficulty is that such procedures do not blend readily into the production-oriented activities of a working stope or face. It is for this reason that 'on-the-job' acclimatization has, in general, been less successful than the more easily controlled procedures.

For individuals whose oral temperature does not exceed 38.6 °C during a heat tolerance test (section 17.7.1), a shorter period of acclimatization may provide the required degree of protection. The use of microclimate jackets (section 17.5.6) during acclimatization is acceptable for candidates who are already familiar with underground operations. For the first week in hot conditions, the worker performs his/her expected rate of work while wearing a microclimate jacket. This allows the individual to be fully productive during the acclimatization period while, at the same time, offering protection against heat illness.

The use of surface or underground climatic chambers designed specifically for fully controlled acclimatization procedures was developed for the South African gold mining industry (Stewart, 1982). The success of those procedures in reducing the incidence of heat stroke and improving productivity led to the widespread use of the technique within that country. The environment within the climatic chamber is maintained at a wet bulb temperature of about 31.7 °C and an air velocity of 0.5 m/s.

The work consists of stepping on and off a bench at a rate indicated by an audible and visual metronome. The height of the bench is adjusted according to each man's body weight such that every person works at the same rate. This correlates with an oxygen consumption of some 1.4 l/min and, hence, is equivalent to hard manual work. The task is carried out for four continuous hours per day over a period of several days. Measurements are made of body temperature, and drinking water is administered every 30 min. During the first day of acclimatization, the stepping task is conducted at only half the normal pace for most of the four hours. The time spent on full work rate (oxygen consumption of 1.4 l/min) is increased gradually until the final day is entirely at the full rate.

The period of acclimatization within a climatic chamber originally encompassed eight days. However, the improvements of conditions underground, coupled with heat tolerance testing have resulted in the acclimatization period being reduced to four days with far fewer men required to undergo it. The ultimate objective is to reduce the wet bulb temperature to less than 27.5 °C in all mines at which time acclimatization will no longer be required (Burrows, 1989).

Physiological changes during acclimatization

During the process of acclimatization, several significant physiological changes take place that enhance the effectiveness of the body thermoregulation system. These are accompanied by a noticeable psychological improvement. During the first day of acclimatization, participants may become morose, irritable and refuse to complete the task. This contrasts with the final day when heavier tasks are undertaken, perhaps in more stringent psychrometric conditions, but in a much happier frame of mind.

Acclimatization produces a dramatic increase in blood volume, averaging about 25%. This, in turn, results in several beneficial reactions. First, there is an improvement in the response of blood flow conductance to increased metabolic heat production or greater heat stress. Coupled with an enhanced elasticity in the walls of blood vessels, this allows a greater flow of blood both to internal organs and to skin tissues. As a consequence, both core temperature and heartbeat decrease for any given workrate and level of environmental heat stress.

There is also a dramatic increase in sweat rate for any given core temperature during acclimatization. Perspiration commences at a lower core temperature, increasing the range and efficiency of evaporative cooling. The improved ability to produce sweat appears also to be associated with the increased blood volume as well as enhanced performance of the sweat glands.

Finally, acclimatization raises the mechanical efficiency of the human body, i.e. any given task is achieved with a lower production of metabolic heat. This, again, results in a lower core temperature and reduced thermal strain.

The combined effect of the changes obtained through acclimatization is that the worker can perform any given task in hot conditions more efficiently, with a greatly reduced risk of heat illness, and in a more stable mental state.

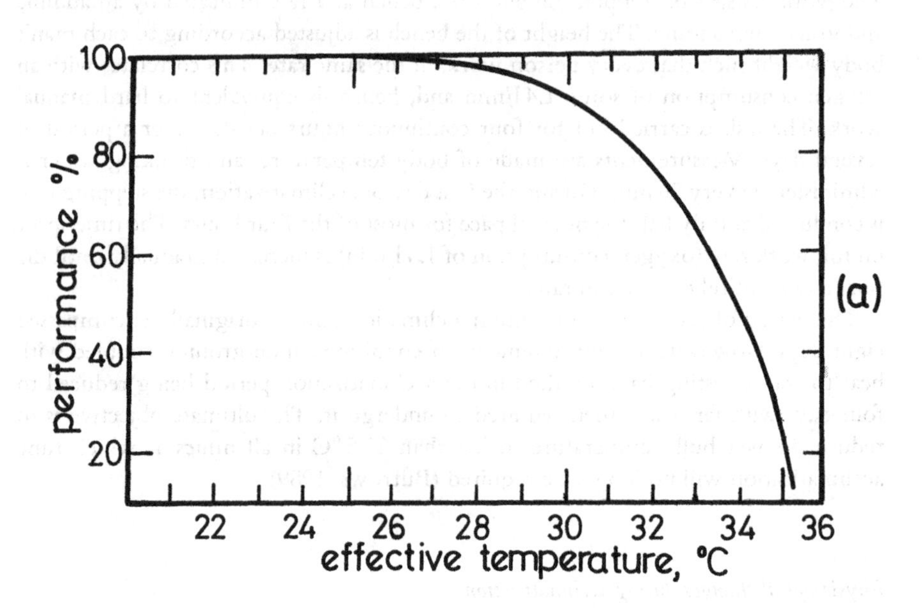

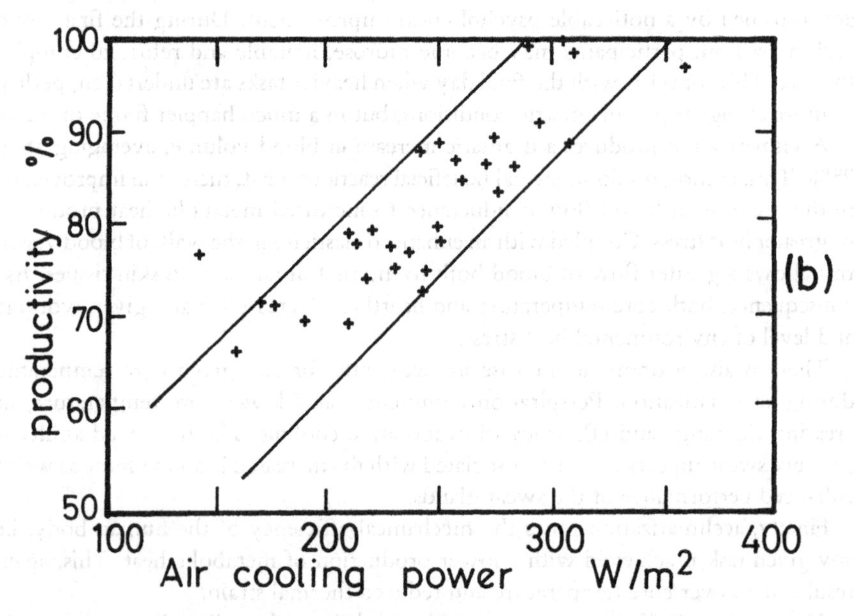

Figure 17.11 Effects of mine climate on (a) individual performance and (b) mine productivity. Both are plotted as a percentage of the largest value reported.

17.7.3 Effect of mine climate on productivity

From a consideration of the psychological as well as physiological effects of hot working conditions, it can be expected that as the cooling power of the ambient airstream decreases below some 300 W/m^2, personnel will become less effective. There will be a deterioration in individual work performance and the quality of that work. Unduly hot conditions may result in poor morale and are likely to produce increased accident rates and problems of absenteeism.

However, the same difficulties can arise from other causes including standards of supervision, management style and labour relations, geological conditions and legislative requirements. For this reason, few studies have been reported that give quantified relationships between mine environmental conditions and those factors that contribute to overall mine productivity. It is, however, relatively easy to demonstrate the debilitating influence of heat on individual work performance. Figure 17.11(a) summarizes results obtained from monitoring the effectiveness of acclimatized men engaged in manual work in a heading (Poulton, 1970). The work consisted of shovelling fragmented rock into ore cars. The temperature of the near-saturated air and the air velocity at the workplace were varied to give the range of effective temperatures (section 17.4.3) shown on the figure. The work performance at an effective temperature of 22 °C was taken to represent 100%. The results show that work efficiency begins to suffer at an effective temperature of about 27 °C and declines rapidly when the effective temperature exceeds 30 °C. The latter approximates to an air cooling power of some 270 W/m^2 for unclothed personnel.

Figure 17.11(b) depicts results reported in a comprehensive study by Howes (1978) using statistics collated by the Chamber of Mines for gold mines in South Africa. The air cooling power used in this figure was an early version of ACP (A scale).

The correlation between Fig. 17.11(a) and 17.11(b) is clear for a labour-intensive operation. However, any environment that causes physical discomfort will involve a decline in standards of care and attention. The potential for accidents and loss of production is clearly then greater in a mechanized operation where an error made by a single person can have serious repercussions. Low morale and the irritability that often appear as an early symptom of heat strain can easily lead to strained industrial relations. This can have disastrous effects on overall mine profitability. Furthermore, the costs of the additional personnel that are required to compensate for absenteeism in hot mines can be high in an expensive labour market.

Failure to create reasonably comfortable working environments underground not only puts the health and safety of the workforce at risk but also puts a brake on productivity and profitability.

REFERENCES

ASHRAE (1985) *Fundamentals Handbook*, American Society of Heating, Refrigerating and Air Conditioning Engineers.

ASHRAE (1989) *Fundamentals Handbook*, American Society of Heating, Refrigerating and Air Conditioning Engineers.

Burrows, J. H. (1989) Personal communiation.

Chamber of Mines of South Africa (1972) *Routine Mine Ventilation Measurements*, Johannesburg.

Dreosti, A. O. (1935) Problems arising out of temperature and humidity in deep mines of the Witwatersrand. *J. Chem. Metall. Min. Soc. S. Afr.* **36**, 102–29.

duBois, D. and duBois, E. F. (1916) A formula to estimate approximate surface area if height and weight are known. *Arch. Intern. Med.* **17**, 863–71.

Fanger, P. O. (1970) *Thermal Comfort*, Danish Technical Press, Copenhagen.

Gagge, J., *et al.* (1969) The prediction of thermal comfort when thermal equilibrium is maintained by sweating. *ASHRAE Trans.* **75** (Part 2), 108.

Hanoff, J. S. T. (1970) Investigation of the limit of tolerable stresses to men subjected to climatic and working conditions in coal mines. *Glückauf-Forschung.* **31** (4), 182–95.

Howes, M. J. (1978) The development of a functional relationship between productivity and the thermal environment. *J. Mine Vent. Soc. S. Afr.* **31**, 21–38.

ISO (1982) Hot environments—estimation of the heat stress on working man, based on the WBGT (wet bulb globe temperature). *ISO (International Organisation for Standardization) Standard 7243* (see also *Standard 7730*, 1984).

Mitchell, D. and Whillier, A. (1972) The cooling power of underground environments. *J. Mine Vent. Soc. S. Afr.* **25**, 140–51.

NIOSH (1986) Criteria for a recommended standard —occupational exposure to hot environments, revised criteria. *USDHHS (NIOSH) Publication 86–113*, US Department of Health and Human Services.

Onkaram, B., *et al.* (1980) Three instruments for assessment of WBGT and a comparison with WGT (Botsball). *Am. Ind. Hyg. Assoc.* **41**, 634–41.

Oohori, T., *et al.* (1984) Comparison of current two-parameter indices of vapor permeation of clothing—as factors governing thermal equilibrium and human comfort. *ASHRAE Trans.* **90** (2).

Poulton, E. C. (1970) *Environment and Human Efficiency*, C. C. Thomas, Springfield, Il.

Siple, P. A. and Passel, C. F. (1945) Measurements of dry atmospheric cooling in subfreezing temperatures. *Proc. Amn. Philos. Soc.* **89**, 177.

Stewart, J. (1982) *Environmental Engineering in South African Mines*, Chapters 20 and 21, Mine Ventilation Society of South Africa.

Stewart, J. M. and Whillier, A. (1979) A guide to the measurement and assessment of heat stress in gold mines. J. Mine Vent. Soc. S. Afr. **32** (9), 169–78.

Visagie, M. E., *et al.* (1974) Changes in vitamin A and C levels in black mine workers. *S. Afr. Med. J.* **48**, 2502–06.

Wyndham, C. H. (1974) The physiological and psychological effects of heat. *The Ventilation of South African Gold Mines*, Chapter 7, pp. 93–137, Mine Ventilation Society of South Africa.

FURTHER READING

Botsford, J. H. (1971) A wet globe thermometer for environmental heat measurement. *Am. Ind. Hyg. Assc. J.* **38**, 264.

Hill, L., Griffith, O. W. and Flack, M. (1916) The measurement of the rate of heat loss on body temperature by convection, radiation and evaporation. *Philos. Trans. R. Soc. London, Sect. B* **207**, 183–220.

Houghton, F. C. and Yaglou, C. P. (1923) Determining lines of equal comfort. *ASHVE Trans.* **29**, 163–76, 361–84.

Yaglou, C. P. and Miller, W. F. (1925) Effective temperatures with clothing. *ASHVE Trans.* **31**, 89–99.

A17 APPENDIX: LISTING OF THE THERMOREGULATION MODEL DEVELOPED IN SECTION 17.3

The program is given in the BASIC language and is suitable for running on a personal computer. The data required are indicated in lines 20 to 100 and the computed parameters identified in lines 560 to 790. The program cycles iteratively, varying the mean skin temperature and other dependent variables until the air cooling power (*M* scale) equals the input metabolic rate. Mean skin temperatures in excess of 36 °C generate a warning message.

```
10 OPEN "LPT1" FOR OUTPUT AS #1
20 INPUT "TW,TD (deg C),P(kPa) ",TW,TD,P
30 INPUT "RAD TEMP OF SURROUNDINGS (deg C) ",TR
40 P=1000*P
50 INPUT "RELATIVE AIR VELOCITY (m/s) ",UR
60 INPUT "CLOTHING RESISTANCE, RCL (degC m^2/W) ",RCL
70 INPUT "CLOTHING AREA RATIO,Fcl ",FCL
80 INPUT "CLOTH VAPOUR PERMEABILITY EFFICIENCY, Icl ",ICL
90 INPUT "METABOLIC RATE (W/m^2) ",M
100 INPUT "BODY VIEW FACTOR,Fr ",FR
110 IF UR<3 THEN 140
120 U=UR
130 GOTO 150
140 U=.8*UR+.6
150 T=TW
160 HC=.00878*P^.6*SQR(U)
170 GOSUB 930
180 SINT=S
190 GOSUB 1010
200 GOSUB 1050
210 TEX=32.6+(.066*TD)+(.0002*E)
220 EA=E
230 GOSUB 1080
240 REM
250 REM RESPIRATORY HEAT, BR
260 T=TEX
270 GOSUB 930
280 SEX=S
290 BR=.0000017*M*(SEX-SINT)
300 REM
310 REM CONVECTIVE HEAT, C
320 CNV=(TSK-TD)/(RCL+1/(FCL*HC))
330 TCL=TSK-CNV*RCL
340 REM RADIATIVE HEAT, R
350 TCL=TSK-CNV*RCL
```

```
360 TAV=(TCL+TR)/2+273.15
370 HR=4*5.67E-08*(TAV)^3
380 R=HR*FR*(TCL-TR)
390 REM
400 REM EVAPORATIVE HEAT, EV
410 GOSUB 1190
420 IF TSK>32.5 THEN GOSUB 1410
430 T=TSK
440 GOSUB 850
450 HE=.0007*HC*L/P
460 GOSUB 810
470 RECL=RCL*P/(.0007*L*ICL)
480 EV=W*(ES-EA)/(RECL+1/(FCL*HE))
490 ACPM=BR+CNV+R+EV
500 PRINT TSK
510 IF ABS(ACPM-M)<1 THEN 560
520  TSK=TSK-.01*(ACPM-M)
530 W=W1
540 IF TSK>32.5 THEN GOSUB 1410
550 GOTO 310
560 PRINT #1,"ACPM = ",:PRINT #1,USING "####";ACPM;:PRINT #1," W/m^2"
570 PRINT #1,"TCLOTH = ",:PRINT #1,USING "##.##";TCL;:PRINT #1," C
580 PRINT #1,"SKIN TEMP = ";:PRINT #1,USING "##.##";TSK;:PRINT #1," C"
590 IF TSK>=36 AND TSK<37 THEN PRINT #1,"***ONSET OF HEAT STRAIN***":GOTO 610
600 IF TSK>=37 THEN PRINT #1,"*** DANGEROUS SKIN TEMPERATURE ***"
610 PRINT #1,"TEXHAUST = ",:PRINT #1,USING "##.##";TEX;:PRINT #1," C
620 PRINT #1,"SWEAT = ",:PRINT #1,USING "##.##";W
630 PRINT #1,"HC = ",:PRINT #1,USING "###.##";HC;:PRINT #1," W/m^2 C",
640 PRINT #1,"HR = ",:PRINT #1,USING "###.##";HR;:PRINT #1," W/m^2 C"
650 PRINT #1,"HE = ",:PRINT #1,USING "###.##";HE;:PRINT #1," W/m^2 Pa",
660 PRINT #1,"FCL = ",FCL
670 PRINT #1,"RECL = ",:PRINT #1,USING "###.###";RECL;:PRINT #1," m^2 Pa/W",
680 PRINT #1,"RCL = ",RCL
690 PRINT #1,"RESP = ",:PRINT #1,USING "###.##";BR;:PRINT #1," W/m^2    ",
700 PRINT #1,"CONV = ",:PRINT #1,USING "###.##";CNV;:PRINT #1," W/m^2"
710 PRINT #1,"RAD = ",:PRINT #1,USING "###.##";R;:PRINT #1," W/M^2    ",
720 PRINT #1,"EVAP = ",:PRINT #1,USING "###.##";EV;:PRINT #1," W/M^2"
730 PRINT #1,"SINT = ",:PRINT #1,USING "####.#";SINT*.001;:PRINT #1," kJ/kg    ",
740 PRINT #1,"SEX = ",:PRINT #1,USING "####.#";SEX*.001;:PRINT #1," kJ/kg"
750 PRINT #1,"LSK = ",:PRINT #1,USING "#####.#";L*.001;:PRINT #1," kJ/kg    ",
760 PRINT #1,"EAIR = ",:PRINT #1,USING "###.##";EA*.001;:PRINT #1," kPa"
770 PRINT #1,"ESK = ",:PRINT #1,USING "###.##";ES*.001;:PRINT #1," kPa    ",
780 PRINT #1,"X = ",:PRINT #1,USING "##.####";X;:PRINT #1," kg/kg dry
790 PRINT #1,"MICRO VELOCITY = ";:PRINT #1,USING "##.##";U;:PRINT #1," m/s"
800 END
810 REM SAT VAP PRESS
820 ES=610.6*EXP(17.27*T/(237.3+T))
830 RETURN
840 REM
```

```
850 REM LATENT HEAT
860 L=(2502.5-2.386*T)*1000
870 RETURN
880 REM
890 REM MOISTURE CONTENT
900 X=.622*E/(P-E)
910 RETURN
920 REM
930 REM SIGMA HEAT
940 GOSUB 850
950 GOSUB 810
960 E=ES
970 GOSUB 890
980 S=L*X+1005*T
990 RETURN
1000 REM
1010 REM ACT. MOISTURE CONTENT
1020 X=(S-1005*TD)/(L+1884*(TD-TW))
1030 RETURN
1040 REM
1050 REM ACT VAP PRESS
1060 E=P*X/(.622+X)
1070 RETURN
1080 REM SKIN TEMP
1090 REM REF GAGGE, 1969
1100 TSK=24.85+.322*TD-.00165*TD*TD
1110 IF W<.9 THEN 1130
1120 TSK=24.49+.249*TD
1130 REM VELOCITY (HC) CORRECTION
1140 DTSK=8.999999E-03*HC*(TSK-TD)*(.15-RCL)/.15
1150 TSK=TSK-DTSK
1160 REM CLOTHING CORRECTION
1170 TSK=TSK+(RCL/.15)
1180 RETURN
1190 REM SWEAT FACTOR
1200 IF TW> 27.194*EXP(-.004*M) THEN 1230
1210 W=.06
1220 GOTO 1390
1230 IF TW<33 THEN GOTO 1260
1240 W=1
1250  GOTO 1390
1260 IF M>=200 THEN GOTO 1320
1270 A=-(1.781265E-04-2.544E-07*M+8.557743/M^2.218731)
1280 B=.012095-1.79735E-05*M+1063.187/M^2.31927
1290 C=-(.194294-3.279531E-04*M+46903!/M^2.44229)
1300 D=.998117-1.816454E-03*M+924321!/M^2.651246
1310 GOTO 1360
1320 A=-(3.97379E-05+7.075456E-08*M+17.880231#/M^2.218441)
1330 B=2.369825E-03+3.359426E-06*M+2211.1688#/M^2.316119
```

```
1340 C=-(-5.682282E-03+5.86905E-05*M+95583.18/M^2.437565)
1350 D=-.321605+7.958265E-04*M+1875636!/M^2.645912
1360 W=D+TW*(C+TW*(B+TW*A))
1370 IF W<.06 THEN W=.06
1380 IF W>1 THEN W=1
1390 W1=W
1400 RETURN
1410 REM INCREASE W FOR TSK>32.5
1420 IF TSK<38 THEN 1450
1430 W=1
1440 GOTO 1480
1450 DW=SIN((31.85*TSK-762)*.01745)*.5+.5
1460 W=W1+DW
1470 IF W>1 THEN W=1
1480 RETURN
```

18

Refrigeration plant and mine air conditioning systems

18.1 INTRODUCTION

One of the earliest methods of temperature control in underground mines was the importation of naturally produced ice from the surface. Blocks of ice were transported in ore cars to cool miners in the Comstock Lode under Virginia City in Nevada, USA, during the 1860s. The vapour compression refrigeration cycle, currently the most widespread method of artificial cooling, appears to have first been used in mining during the 1920s. Examples included the famous Morro Velho Mine in Brazil (1923) and experimental work in British coal mines (Hancock, 1926). Air cooling techniques in mining gained further recognition in the 1930s including their utilization in the gold mines of South Africa and in the Kolar Goldfields of India. However, it was the 1960s that saw the the start of a real escalation of installed mine cooling capacity.

Large centralized refrigeration plant, located underground, became popular in the South African gold mines. Limitations on the heat rejection capacity of return air, combined with the development of energy recovery devices for water pipelines in shafts and improved 'coolth' distribution systems have led to a renewed preference for surface plant. (The term 'coolth' is a decidedly unscientific but descriptive word used widely in association with chilled fluid distribution systems.)

Although the traditional role of mine cooling has been to combat geothermal heat and the effects of autocompression in deep metal mines, an additional influence has been the escalating amount of mechanized power employed underground, particularly in longwall coal mines. This has resulted in smaller scale and more localized use of air cooling units in such mines at depths where, prior to intensive mechanization, heat had not been a limiting environmental problem. Examples of these installations can be found in the United Kingdom and Germany.

In this chapter, we shall examine the essential theory of the vapour compression refrigeration cycle in addition to discussing the design of mine cooling systems and some of the methods of distributing 'coolth' to the working areas of a subsurface facility. The chapter concludes with a section on the opposite problem, that of increasing the temperature of the intake air for mines in cold climates.

18.2 THE VAPOUR COMPRESSION CYCLE

There are many transient phenomena that are known to produce a cooling effect, varying from endothermic chemical reactions to the sublimation of solid carbon dioxide (dry ice). Where a continuous cooling effect is required then a means must be employed by which a supply of mechanical, electrical or thermal energy is utilized to remove heat from some **source**, and to transport it to a **thermal sink** where it can be rejected. If the primary objective is to cool the source, then the device is known as a **refrigerator**. If, however, the desired effect is to heat the sink, then it is called a **heat pump**. In fact, both effects occur simultaneously. Hence, a domestic refrigerator cools the interior of the container but heats the air in the kitchen.

Of the several devices that have been developed to achieve a continuous refrigeration or heat pump effect, the most common is based on the vapour compression cycle. This may be used on small units such as air conditioning equipment fitted to automobiles or for very large scale cooling of mine workings where many megawatts of heat require to be transferred. In this section, we shall examine the basic principles of the vapour compression cycle, how rates of heat exchange may be calculated, and the essential components of a refrigeration plant.

18.2.1 Basic principles

When a liquid boils, it does so at constant temperature provided that the applied pressure remains fixed. The heat added is utilized in increasing the internal kinetic energy of the molecules until they can no longer remain in the liquid phase but burst free to form a vapour or gas (section 2.1.1). If, however, the applied pressure is raised to a higher value, then additional heat is required to vaporize the liquid. The boiling temperature will increase. The relationship between pressure and boiling point for any given liquid may be defined as the vapour pressure line on a pressure–temperature diagram such as that shown on Fig 18.1. The fluid is a liquid on the left of the curve and a vapour (gas) on the right of the curve. The liquid may be vaporized either by increasing the temperature or decreasing the pressure. Similarly, condensation from vapour to liquid may occur either by decreasing the temperature or increasing the pressure.

The change in physical appearance of a fluid on crossing the vapour pressure line is quite distinct, liquid to gas or vice versa. However, there is a critical pressure–temperature coordinate beyond which the change of phase is gradual rather than sudden and there is no clearly defined moment of evaporation or condensation. This is known as the critical point.

Although the vapour pressure curve appears as a single line on the pressure–temperature diagram, it takes a finite amount of time and energy exchange to cross that

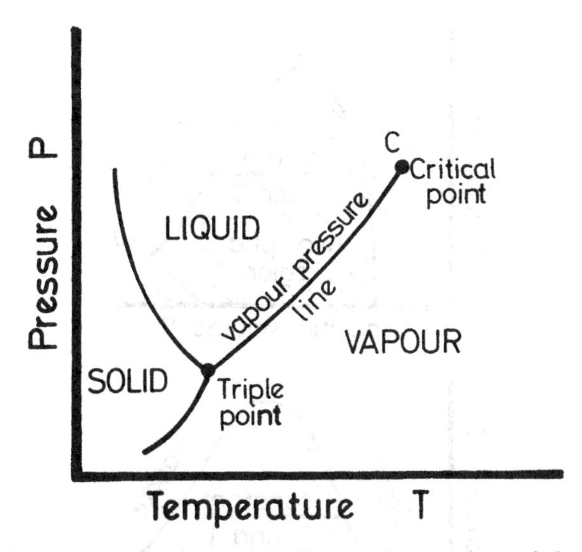

Figure 18.1 A pressure–temperature diagram showing lines of phase change.

line. During this time, part of the fluid will be liquid and the remainder will be vapour. This is the situation that exists inside a boiling kettle. The region within which the two phases coexists is more clearly shown on the *PV* (pressure against specific volume) diagram of Fig. 18.2(a). The corresponding temperature–entropy (*Ts*) and pressure–enthalpy (*PH*) diagrams are shown on Figs. 18.2(b) and 18.2(c) respectively. The latter two diagrams are particularly useful in analysing and quantifying both the work and heat transfer processes in the vapour compression cycle. In all three diagrams of Fig. 18.2, a horizontal line within the two-phase region indicates that the pressure and temperature both remain constant during any given isobaric phase change. However, the volume, entropy and enthalpy all increase significantly during vaporization. Different horizontal lines on any one of these diagrams will indicate different values of pressure and corresponding boiling (or condensation) temperature (Fig. 18.1). When a liquid boils at a given value of applied pressure, then it will extract heat from the surroundings or other available medium. If the vapour thus produced is then transported to a new location and compressed to a higher pressure, then it can be condensed at a correspondingly higher temperature, yielding up its heat of condensation to the new surroundings or any cooling medium that may be supplied. This is the basic principle underlying the vapour compression refrigeration cycle.

18.2.2 Refrigerant fluids

The differing pressure–temperature relationships of various fluids allow each of those fluids to act as a refrigerant over specified temperature ranges. Carbon dioxide and even water have been used as refrigerants. For the ranges of pressures and temperatures

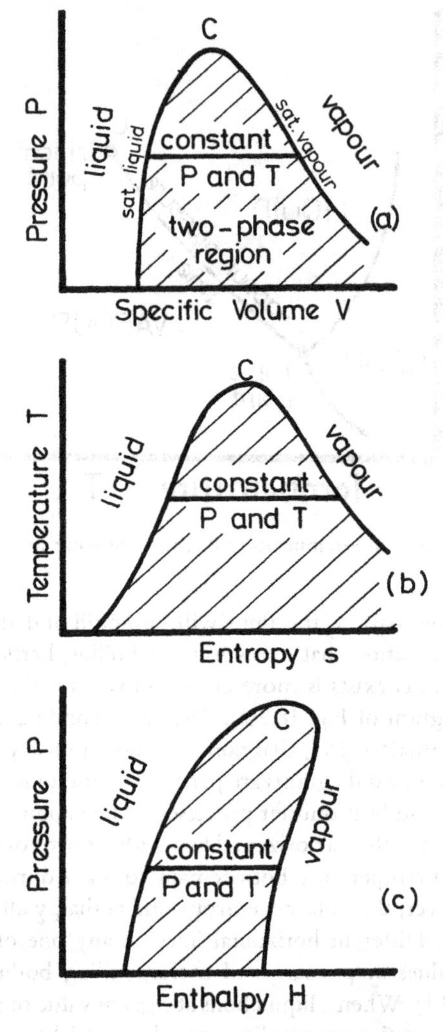

Figure 18.2 *PV*, *Ts* and *PH* diagrams for change of phase.

acceptable in refrigeration plant for air conditioning, ammonia is a particularly efficient refrigerant although its toxicity limits its use. The ideal refrigerant for mining use is one that gives a high efficiency of heat transfer, is non-corrosive to metal, non-toxic and has a boiling temperature close to 0 °C at a positive pressure with respect to the ambient atmosphere. Fluorinated hydrocarbons have been widely employed in many types of industrial plant as well as domestic refrigeration equipment. Although non-toxic, fluorinated hydrocarbons react adversely with atmospheric ozone and research is in progress to develop alternative refrigerants.

Refrigerant fluids are commonly referred to, not by their chemical names, but by *R* (refrigerant) number. Table 18.1 indicates some of those designations.

Table 18.1 Refrigerant numbers of selected fluids

Refrigerant number	*Chemical name*
R11	Trichlorofluoromethane
R12	Dichlorodifluoromethane
R22	Chlorodifluoromethane
R113	Trichlorotrifluoromethane
R717	Ammonia
R744	Carbon dioxide
R50	Methane
R290	Propane
R1150	Ethylene

18.2.3 Basic components of the vapour compression cycle

Figure 18.3 shows that there are four essential components of hardware in a vapour compression refrigeration unit. The evaporator is a heat exchanger, typically of the shell-and-tube configuration in mining refrigeration plant. In the larger units, the refrigerant liquid is on the outside of the tubes while the medium to be cooled (for example, water, brine or glycol) passes through the tubes. Smaller units employed for direct cooling of an airstream are sometimes called **direct evaporators** and contain the refrigerant within the tubes while the air passes over their outer surface.

Within the evaporator, the refrigerant pressure is maintained at a relatively low level and boils at correspondingly low temperature. For example, refrigerant R12 will boil at 4 °C if the pressure is 351 kPa (approximately 3.5 atm). The heat required to maintain the boiling is extracted from the gas or liquid passing on the other side of the tube walls. Hence, that gas or liquid is cooled. The refrigerant, now vaporized, collects at the top of the evaporator and is allowed to gain a few degrees of superheat to ensure full vaporization before it passes on the compressor. Except for direct evaporators, it may be necessary to insulate the external surface of the evaporator in order to prevent excessive heat gain from the ambient atmosphere.

The compressor is the device where mechanical work is input to the system. Reciprocating, screw or centrifugal compressors are all employed. The latter are favoured for the larger units and where the required pressure ratio remains constant. However, the development of large diameter screw compressors allows a good efficiency to be maintained under conditions of variable cooling load (Baker-Duly *et al.*, 1988). Multistage compressors are employed to give high differentials of pressure and, hence, large temperature differences between the evaporator and condenser. Electric motors are normally employed to drive the compressors on mine refrigeration units although internal combustion engines may be used on surface or as standby units. The duty of a refrigeration plant can be modified by changing the speed of the compressor. The flow of vapour through the compressor and, hence, rate of heat transfer can also be controlled by inlet guide vanes.

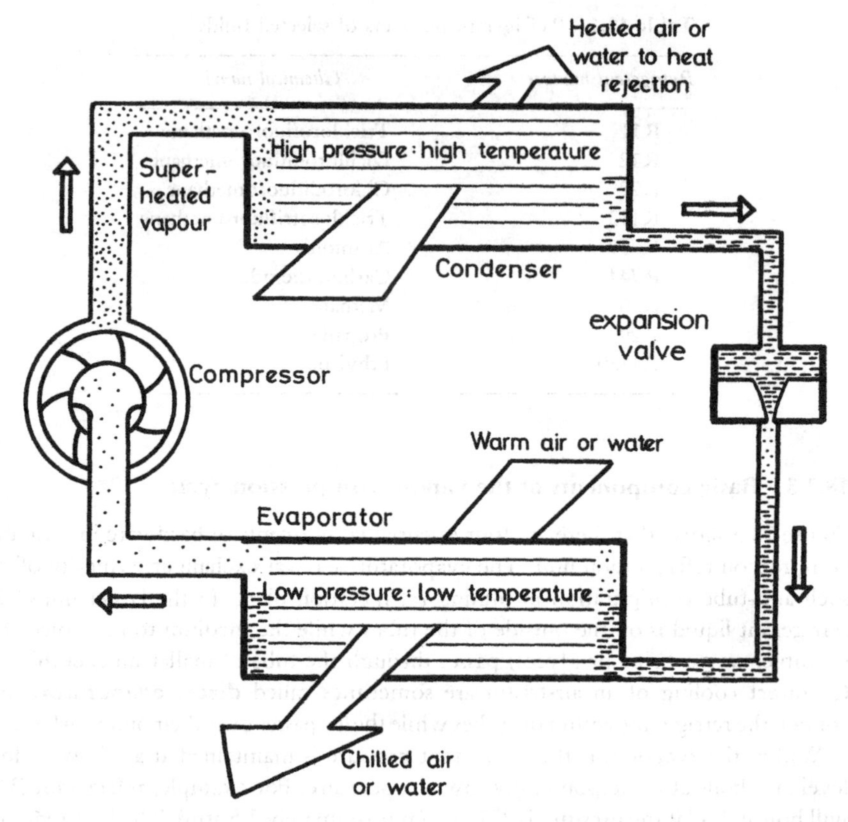

Figure 18.3 Major components of a vapour compression cycle.

The refrigerant vapour leaves the compressor and passes into the condenser at a relatively high pressure and temperature. The condenser itself may be of similar construction to the evaporator, that is, a shell-and-tube heat exchanger. Heat is removed from the refrigerant by air, water or some other fluid medium to the extent that the refrigerant cools and condenses back to a liquid. As the pressure is high, this occurs at a relatively high temperature. At a pressure of 1217 kPa, refrigerant R12 will condense at 50 °C. The latent heat of condensation is removed by the cooling fluid for subsequent rejection in a cooling tower or other type of separate heat exchanger. As the cycle is closed, the rate at which heat is removed from the refrigerant in the condenser must equal the combined rates of heat addition in the evaporator and work provided by the compressor.

The condensed refrigerant flows from the condenser to the fourth and final component of the cycle. This is the expansion valve whose purpose is simply to reduce the pressure of the refrigerant back to the evaporator conditions. An expansion valve may be a simple orifice plate or can be controlled by a float valve. At the exit from the expansion valve, the liquid is at low pressure and has a correspondingly low boiling temperature. Provided that the pipework is insulated, the latent heat for boiling can come only from the liquid refrigerant itself. Hence, the temperature of the refrigerant drops rapidly as it passes from the expansion valve to the evaporator where it enters as a mixture of liquid and vapour, thus closing the cycle.

18.2.4 Performance of a refrigeration cycle

The Carnot cycle

Electrical motors or heat engines are devices that convert one form of energy into another. Their efficiency may, therefore, be defined as an output/input ratio. However, in the case of refrigerators or heat pumps, the purpose is to remove heat from a given source and to reject it at a higher temperature to a receiving sink. A different measure of performance is required.

Figure 18.4 shows the temperature–entropy diagram for the ideal (frictionless) Carnot cycle to which an actual vapour compression cycle can aspire but never attain. We may follow the ideal cycle by commencing at position 1, the entry of refrigerant vapour to the compressor. The ideal compression process is isentropic (section 10.6). The compressed refrigerant vapour enters the condenser at position 2 and, in the ideal condenser, passes through to position 3 with neither a pressure drop nor a fall in temperature. This assumes frictionless flow and perfect heat transfer. The ideal expansion valve allows an isentropic fall in pressure and temperature to station 4, the entrance to the evaporator. The cycle closes via frictionless flow and perfect heat transfer in the evaporator.

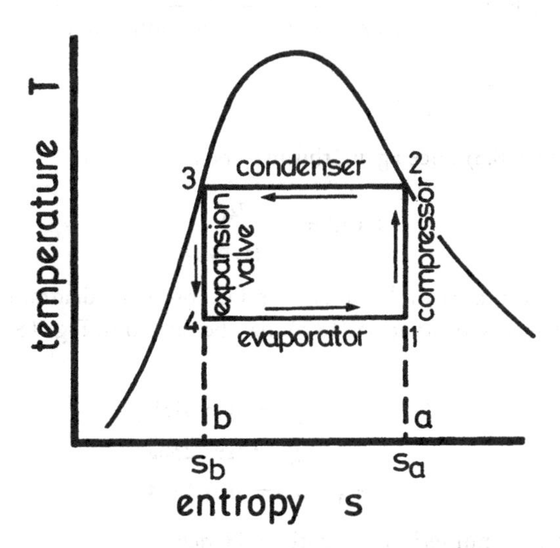

Figure 18.4 Temperature–entropy diagram for an ideal vapour compression cycle (Carnot cycle).

In order to quantify the ideal Carnot cycle, we apply the steady-flow energy equation (3.25). The kinetic and potential energy terms are small and, in any event, cancel out in a closed cycle. Using the numbered station points from Fig. 18.4, the steady-flow energy equation gives:

compressor	W_{12}	$= H_2 - H_1$ J/kg	(18.1)
condensor	q_{23}	$= H_3 - H_2$ J/kg	(18.2)
expansion valve	0	$= H_4 - H_3$ J/kg	(18.3)
evaporator	q_{41}	$= H_1 - H_4$ J/kg	(18.4)
cycle summation	$W_{12} + q_{23} + q_{41}$	$= 0$ J/kg	(18.5)

where W_{12} = mechanical energy added by the compressor, q_{23} = heat exchange in the condenser (this is numerically negative as heat is leaving the refrigerant), q_{41} = heat added to the refrigerant in the evaporator and H = enthalpy.

We can rewrite equation (18.5) as

$$W_{12} + q_{41} = -q_{23} \quad \text{J/kg} \tag{18.6}$$

thus confirming our earlier statement that the heat rejected in the condenser is numerically equal to the sum of the compressor work and the heat added to the refrigerant in the evaporator.

The measure of performance of a refrigeration cycle is known as the **coefficient of performance**, COP, and is defined as

$$\text{COP} = \frac{\text{useful cooling effect (evaporator heat transfer)}}{\text{work input from the compressor}}$$

$$= \frac{q_{41}}{W_{12}} \tag{18.7}$$

Using equation (18.5) once again, this may be rewritten as

$$\text{COP} = \frac{q_{41}}{-q_{23} - q_{41}} \tag{18.8}$$

Recalling that the area under a process line on a *Ts* diagram represents heat (section 3.5), the terms in equation (18.8) may be related to Fig. 18.4 for the Carnot cycle:

$$q_{41} = \text{area 41ab}$$

$$-q_{23} = \text{area 23ba}$$

$$-q_{23} - q_{41} = \text{area 1234}$$

(remember that q_{23} is numerically negative). Hence

$$W_{12} = \text{area 1234}$$

and equation (18.8) becomes

$$\text{Carnot COP} = \frac{\text{area 41ab}}{\text{area 1234}}$$

$$\text{Carnot COP} = \frac{T_1(s_a - s_b)}{(T_2 - T_1)(s_a - s_b)}$$

$$= \frac{T_1}{T_2 - T_1} \quad \text{or} \quad \frac{T_4}{T_3 - T_4} \tag{18.9}$$

as $T_1 = T_4$ and $T_2 = T_3$.

Hence, the ideal or Carnot coefficient of performance is given as the ratio

$$\text{Carnot COP} = \frac{\text{evaporator temperature (absolute)}}{\text{condenser temperature} - \text{evaporator temperature}} \tag{18.10}$$

Example The evaporator and condenser of a refrigeration unit have temperatures of 4 and 50 °C respectively. Determine the maximum possible coefficient of performance of this unit.

Solution The Carnot or ideal coefficient of performance is given by equation (18.10)

$$\text{Carnot COP} = \frac{273.15 + 4}{50 - 4}$$

$$= 6.025$$

The actual cycle

A real vapour compression cycle has a coefficient of performance that is necessarily lower than the corresponding Carnot COP. There are two reasons for this. First, in actual compressors and expansion valves, there is, inevitably, an increase in entropy. Furthermore, as the refrigerant passes through real condensers and evaporators there will be slight changes in pressure and temperature. In a well-designed unit, the latter are small compared with the differences in pressure or temperature between the condenser and evaporator.

Secondly, it would be impractical to design a refrigeration unit that attempted to follow the Carnot cycle. Referring again to Fig. 18.4, it can be seen that station 1, the entry to the compressor, lies within the two-phase region. The presence of liquid droplets would cause severe erosion of the compressor impeller. It is for this reason that a few degrees of superheat are imparted to the vapour before it leaves a real evaporator.

Figure 18.5 illustrates the temperature–entropy and pressure–enthalpy diagrams for a practical vapour compression cycle. The major difference between Fig. 18.5(a) and the Carnot cycle shown on Fig. 18.4 is that the compression commences slightly

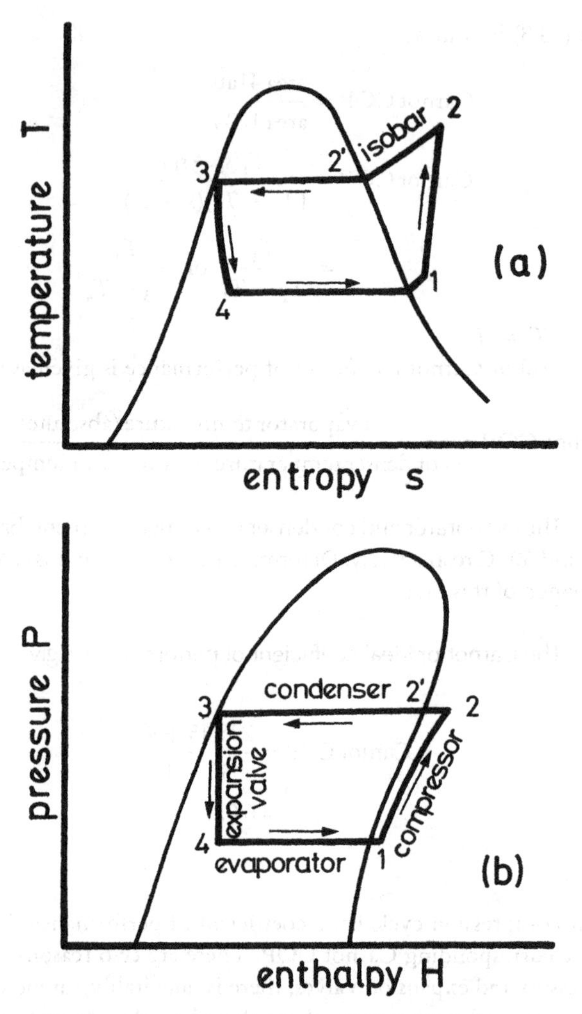

Figure 18.5 Temperature–entropy and pressure–enthalpy diagrams for an actual vapour compression cycle.

beyond the saturated vapour line (station 1) and involves an increase in entropy to station 2. On entering the condenser, the superheated vapour undergoes a near isobaric cooling process (stations 2 to 2′), losing sensible heat to the cooling medium until condensing temperature is reached at station 2′.

Although the process through the expansion valve is no longer isentropic, the fluid remains at contant enthalpy (no heat exchange takes place) and, hence, appears on the pressure–enthalpy diagram as a vertical line 3 to 4.

Equations (18.1) to (18.4) arose from those parts of the steady-flow energy equation that do not involve the friction term. They apply equally well to both the real and the ideal cases. Furthermore, the definition of coefficient of performance given in

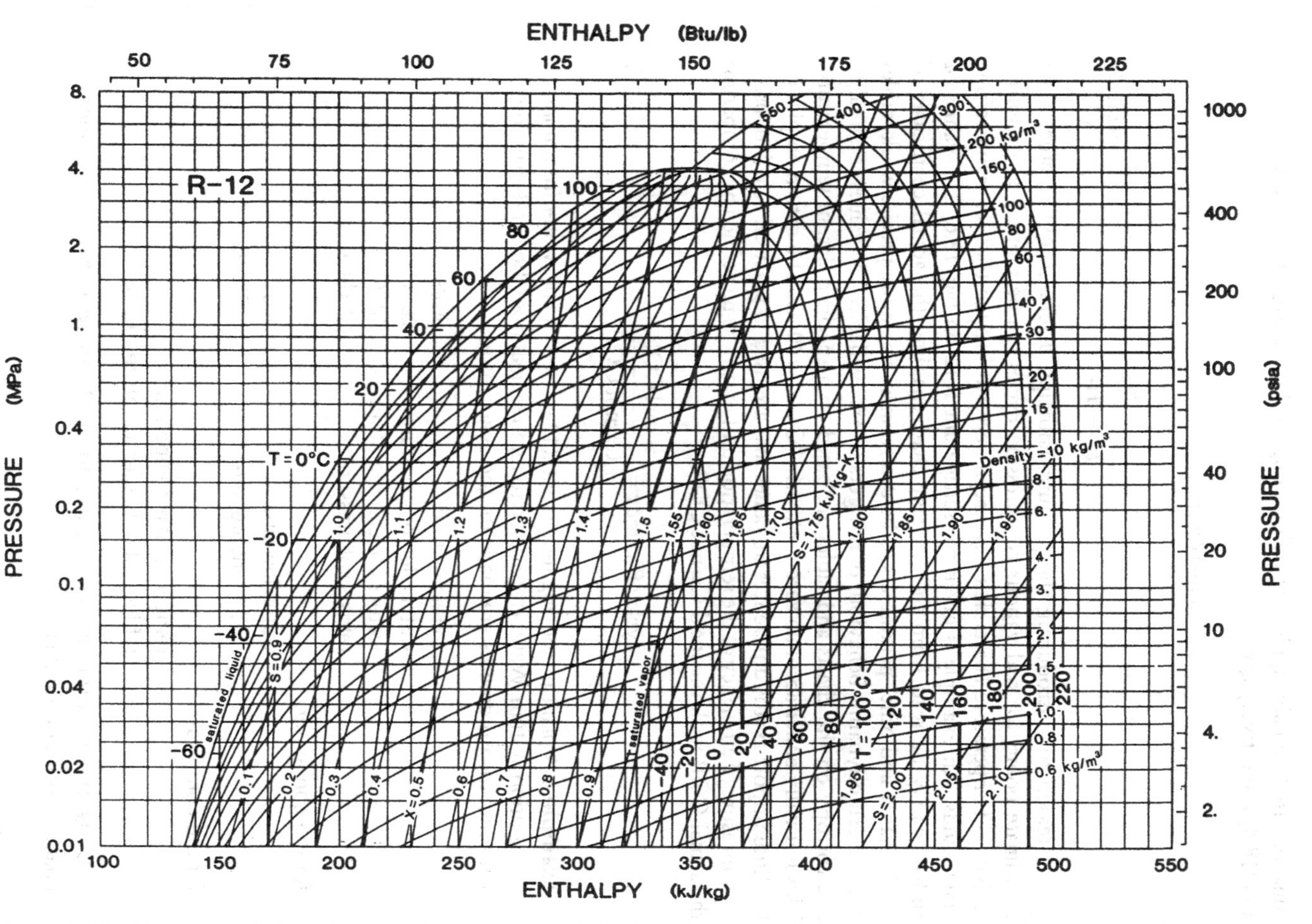

Figure 18.6 Pressure–enthalpy diagram for refrigerant R12 (based on zero kelvins). (*Reprinted with permission from the 1985 ASHRAE Fundamentals Handbook*).

equation (18.7) remains applicable. Substituting for W_{12} and q_{41} from equations (18.1) and (18.4) respectively gives

$$\text{actual COP} = \frac{q_{41}}{\dot{W}_{12}}$$

$$= \frac{H_1 - H_4}{H_2 - H_1} \tag{18.11}$$

The reason for introducing the pressure–enthalpy diagram now emerges. The coordinate points for real vapour compression cycles can be plotted on pressure–enthalpy–temperature charts that have been derived through tests on many refrigerant fluids. An example is given for refrigerant R12 on Fig. 18.6. The values of enthalpy may be read from the relevant chart enabling the actual coefficient of performance and other parameters to be determined. For more precise work, tables of the thermodynamic behaviour of refrigerants are available. The values of enthalpy and entropy on the charts and tables are based on a specified datum temperature. This is usually either absolute zero (−273.15 °C) or −40 °C. The datum employed has little import as it is differences in values that are used in practical calculations.

A measure of efficiency of the refrigeration unit may now be defined by comparing the actual COP with the ideal Carnot COP:

$$\text{cycle efficiency} = \frac{\text{actual COP}}{\text{Carnot COP}} \tag{18.12}$$

Example A plant employing refrigerant R12 is used to chill water for a mine distribution system. Water is also used to remove heat from the condenser. The following measurements are made for the evaporator,

water flowrate	50 l/s
water inlet temperature	20 °C
water outlet temperature	10 °C
refrigerant pressure	363 kPa

for the condenser,

water flowrate	140 l/s
refrigerant pressure	1083 kPa

for the compressor,

refrigerant inlet temperature	7 °C
refrigerant outlet temperature	65 °C

Using the pressure–enthalpy chart for R12, analyse the performance characteristics of the unit.

Solution *Step 1:* On the pressure–enthalpy diagram, draw horizontal lines to represent the pressures in the evaporator (363 kPa) and condenser (1083 kPa). For clarity, the relevant area of the R12 pressure–enthalpy chart has been

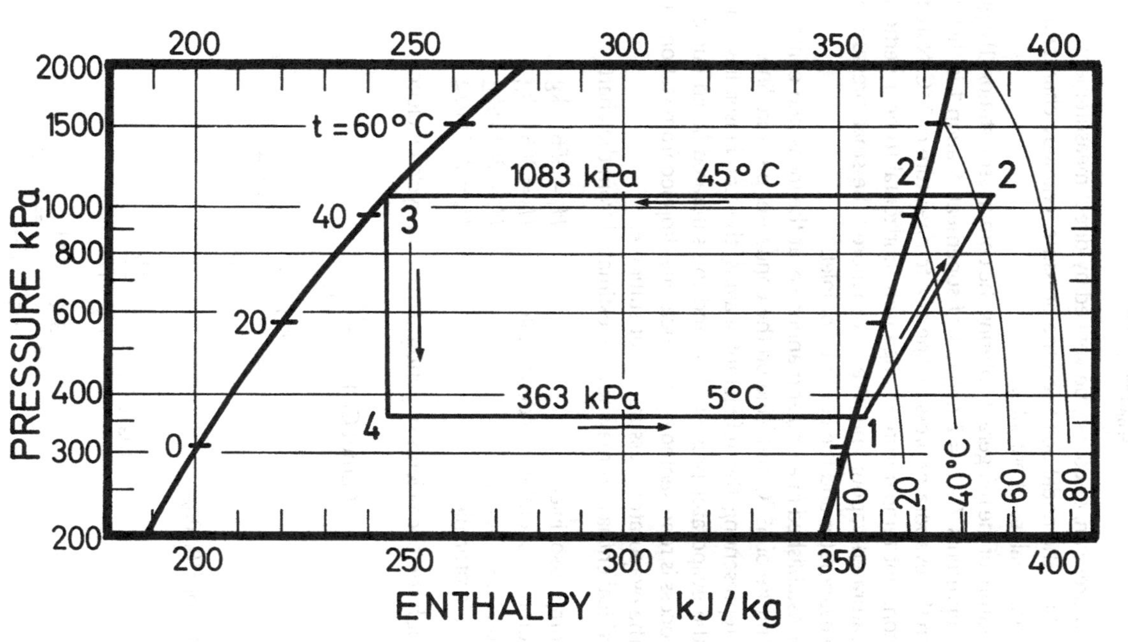

Figure 18.7 Example vapour compression cycle on the pressure–enthalpy diagram.

enlarged and reproduced as Fig. 18.7. The corresponding temperatures may be estimated directly from the chart:

$$t_{evap} = t_4 = 5\,°C$$

$$t_{cond} = t_3 = 45\,°C$$

These temperatures can, of course, be verified by direct measurement.

Step 2: Establish the nodal points on the pressure–enthalpy chart and read the corresponding values of enthalpy.

The temperature of the refrigerant entering the compressor (station 1) is given as 7 °C. This represents 7 − 5 = 2 degrees of superheat. Station 1 is therefore established on the evaporator pressure line at a temperature of 7 °C. Note that the lines of constant temperature change from horizontal to near vertical on passing the saturated vapour line. We have ignored the small pressure drop between the evaporator and the compressor inlet.

Station 2 is established in a similar manner, i.e. at the measured compressor outlet temperature of 65 °C and lying on the condenser pressure line.

Station 3, representing the refrigerant leaving the condenser lies on the junction of the evaporator pressure line and the saturated liquid curve. The expansion process is represented by a vertical line dropped from station 3 until it intersects the evaporator pressure line at station 4.

The four critical enthalpies may now be estimated from the chart:

compressor inlet	$H_1 = 356$ kJ/kg
compressor outlet–condenser inlet	$H_2 = 386$ kJ/kg
condenser outlet	$H_3 = 244$ kJ/kg
evaporator inlet	$H_4 = 244$ kJ/kg

Step 3: Establish the coefficients of performance and cycle efficiency:

$$\text{Carnot COP} = \frac{T_4}{t_3 - t_4}$$

$$= \frac{273.15 + 5}{45 - 5} = 6.954$$

(equation (18.9) or (18.10)),

$$\text{actual COP} = \frac{H_1 - H_4}{H_2 - H_1}$$

$$= \frac{356 - 244}{386 - 356} = 3.733$$

(equation (18.11)) and

$$\text{cycle efficiency} = \frac{\text{actual COP}}{\text{Carnot COP}}$$

$$= \frac{3.733}{6.954} \times 100 = 53.7\%$$

(equation (18.12)).

Step 4: Determine the useful cooling effect, or evaporator duty, q_{evap}. This can be established from the known flowrate and temperature drop of the water being cooled as it passes through the evaporator:

$$q_{evap} = m_{w,evap} C_w \Delta T_{w,evap} \quad \text{kW} \tag{18.13}$$

where water mass flowrate $m_{w,evap} = 50\,l/s$, $= 50\,kg/s$, specific heat of water, $C_w = 4.187\,kJ/(kg\,°C)$ and temperature drop of water $\Delta T_{w,evap} = 20–10 = 10\,°C$, giving

$$q_{evap} = 50 \times 4.187 \times 10\,kJ/s \text{ or kW}$$
$$= 2093.5\,kW$$

Step 5: Determine the mass flowrate of refrigerant, m_r, around the system. The evaporator duty, already found to be 2093.5 kW, can also be expressed as

$$q_{evap} = m_r(H_1 - H_4) \quad \text{kW} \tag{18.14}$$

Hence,

$$m_r = \frac{2093.5}{356 - 244} = 18.69\,kg/s$$

Step 6: Determine the compressor duty. The work input from the compressor to the refrigerant, W_{12}, can be established in two ways. First,

$$W_{12} = m_r(H_2 - H_1) \quad \text{kW} \tag{18.15}$$
$$= 18.69(386 - 356) = 560.7\,kW$$

and, secondly, from the definition of coefficient of performance given in equation (18.7):

$$W_{12} = \frac{q_{evap}}{\text{actual COP}} = \frac{2093.5}{3.733} = 560.8\,kW$$

The slight difference arises from rounding errors.

Step 7: Evaluate the condenser duty. The heat transferred from the refrigerant in the condenser is simply (equation (18.6))

$$-q_{cond} = q_{evap} + W_{12}$$
$$= 2093.5 + 560.7 = 2654.2\,kW$$

As the cooling water flowrate through the condenser is known to be $m_{w,cond} = 140\,kg/s$, the temperature rise of this water, $\Delta T_{w,cond}$, may be determined from

$$-q_{cond} = m_{w,cond} C_w \Delta T_{w,cond}$$

giving

$$\Delta T_{w,cond} = \frac{2654.2}{140 \times 4.187}$$

$$= 4.53\,^{\circ}\mathrm{C}$$

The performance of a refrigeration unit may be enhanced beyond that illustrated in the example. For large units, it may be necessary to employ two or more stages of compression in order to achieve the desired difference in temperature between evaporator and condenser. In that case, it is usual to employ the same number of expansion valves as stages of compression. A fraction of the refrigerant is vaporized on passing through an expansion valve. That fraction can be read from the corresponding point on the pressure–enthalpy diagram or calculated from the isobaric enthalpies at the saturated liquid and saturated vapour curves. This 'flash gas' may be separated out and piped to the corresponding intermediate point between compressor stages. This is known as **interstage cooling** or, simply, **intercooling** and reduces the total mass of refrigerant that passes through all stages of compression. The result is to reduce the required compressor power and/or increase the useful refrigerating effect.

In some units, the hot liquid refrigerant leaving the condenser is partially cooled by a secondary cooling circuit. This **subcooling** reduces the fraction of flash gas produced at the downstream side of expansion valves and gives a small increase in the performance of the plant.

The coefficient of performance as defined by equation (18.7) takes into account the isentropic efficiency of the compressor. However, it does not cater for the efficiency of the motor or other device that drives the compressor. Hence, the W_{12} in equation (18.7) may be replaced by the power fed to the compressor drive unit to give a more practical coefficient of performance. Furthermore, the compressor is not the only device that consumes energy in a refrigeration unit. There will also be water pumps or fans to promote the flow of the cooled medium at the evaporator and the cooling medium at the condenser. An **overall** coefficient of performance may be defined as

$$\text{overall COP} = \frac{\text{heat transfer in the evaporator}}{\text{total energy consumption of the refrigeration unit}} \qquad (18.16)$$

Care should, therefore, be taken that any quoted values of COP are interpreted in the correct manner.

18.3 COMPONENTS AND DESIGN OF MINE COOLING SYSTEMS

18.3.1 Overview of mine cooling systems

In the majority of large-scale mine cooling systems, there are usually three sets of heat transfer involved:

1. transfer of heat from the work areas to evaporators of the refrigeration units;
2. transfer of heat from the evaporators to the condensers in the refrigeration units (the vapour compression cycle);
3. transfer of heat from the condensers to the free atmosphere on surface.

There exist tremendous variations in the duty, complexity and efficiency of the hardware involved in these three phases, dependent on the severity and dispersion of the heat problem in the mine. In this section we shall discuss a range of systems varying from a simple spot cooler for very localized application, to a large integrated system that may be required for a deep and hot mine. This will serve both as an overview and an introduction to a more detailed examination of individual components and overall system design.

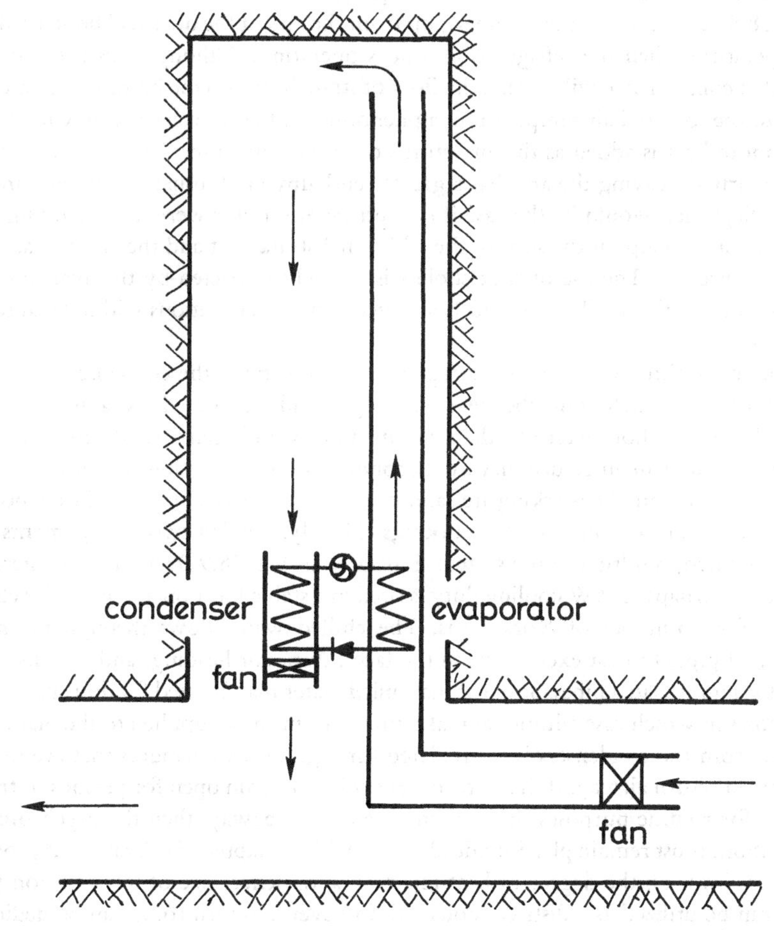

Figure 18.8 The simplest application of a direct evaporator or spot cooler with local heat rejection in a mine heading.

Figure 18.8 shows a simple application of a direct evaporator or 'spot cooler' to a specific working area in a mine. The evaporator of the refrigeration unit takes the form of copper or cupronickel tubular coils located within an air duct. The refrigerant passes through the inside of the evaporator tubes and cools the air flowing along the duct. The heat from the condenser is rejected directly to the return air. This system has the advantage that the cooling effect produced by the refrigeration unit is utilized directly and immediately at the place where cooling is required. There is no loss of efficiency caused by an intervening water reticulation system between the evaporator and working area. Similarly, heat rejection from the condenser is direct and immediate. The spot cooler is merely an industrial version of a domestic air conditioning unit where heat is rejected directly to the outside atmosphere.

Unfortunately, the spot cooler has a major disadvantage that limits its application in subsurface ventilation systems. Glancing, again, at the blind heading illustrated in Fig. 18.8, it is clear that the air emerging from the end of the duct will be at a reduced temperature when the refrigeration unit is operating. Although this is the desired effect, a consequence will be that the flow of strata heat into the heading will increase due to the lowered air temperature, a phenomenon that is examined in Chapter 15. Yet more heat is added as the air returns over the condenser. The net effect is that the return air leaving the area has a greater enthalpy (and, usually, temperature and humidity) than would be the case if the refrigeration unit were not operating. The increase in enthalpy is the sum of the additional strata heat and the energy taken by the compressor. The use of spot coolers is severely restricted by the availability of return air and the debilitating effect on psychrometric conditions within local return airways.

There are three ways of alleviating the situation. First, the condenser could take the form of a shell-and-tube heat exchanger and be cooled by a water circuit (Fig. 18.3). The hot water can then be piped away and recirculated through a heat exchanger in a main return airway. Secondly, the refrigeration unit itself may be located away from the working area and water that has been chilled by the evaporator piped to a heat exchanger in the workings. Thirdly, the first two arrangements may be combined, resulting in the system illustrated in Fig. 18.9. Here, the refrigeration plant of, perhaps, 2 MW cooling duty is sited in a stable location and provides chilled water for a number of work areas. The chilled water flows through thermally insulated pipes to heat exchangers in the faces, stopes or headings and returns to the plant via uninsulated pipes. Some of the chilled water may be used for dust suppression purposes in which case additional make-up water must be supplied to the plant. Hot water from the condensers is recirculated through heat exchangers that are located in a main return airway. If that return airway is to remain open for persons to travel, either for routine purposes or as an emergency escapeway, then the psychrometric conditions must remain physiologically acceptable (Chapter 17). Again, this provides a severe limit on the degree of heat rejection and, hence, size of refrigeration plant that can be utilized as a district cooler. If, however, a return route can be dedicated fully to heat rejection, then physiological acceptability limits may be exceeded. In this case, the dedicated return or 'dirty pipe', as it is sometimes called, must be made

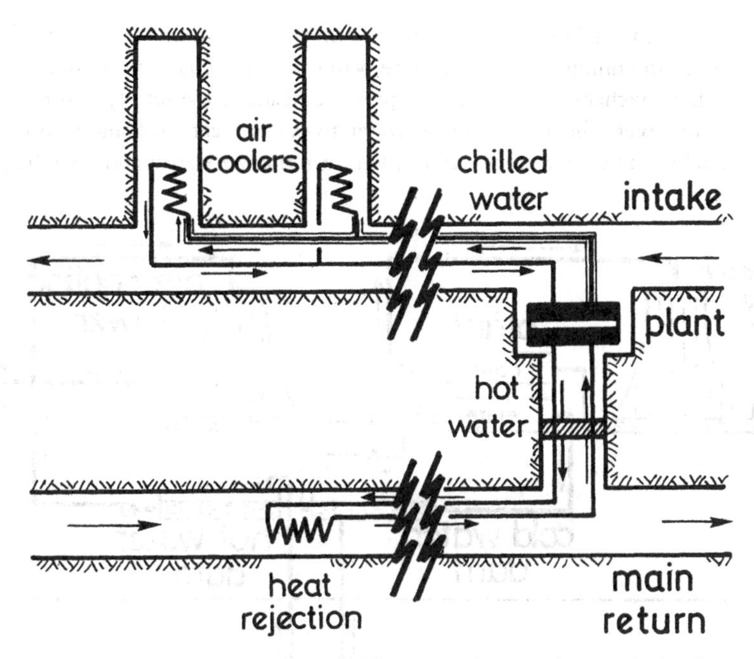

Figure 18.9 A district chiller supplying cold water to heat exchangers in headings, stopes or faces and rejecting heat into a main return.

inaccessible to inadvertent entry by personnel. Inspection or maintenance can be carried out either when the plant is shut down or by persons wearing protective clothing (section 17.5.6). It may be necessary to seek exemption from national or state legislation in order to utilize a 'dirty pipe' arrangement.

For mines that have a widespread heat problem, the economies of scale and the need for flexibility indicate a requirement for large centralized cooling facilities. Banks of individual refrigeration units, each producing a typical 3.5 MW of cooling capacity, may be assembled to give a total duty which may exceed 100 MW for a large and deep metal mine. The 'coolth' is normally distributed via chilled water lines to provide any required combination of bulk air cooling, face or stope air cooling and chilled service water.

Until the mid-1970s, centralized plant tended to be located underground in excavated refrigeration chambers close to shaft bottoms. All of the mine return air could then be utilized for heat rejection. In South Africa, it was common for large cooling towers to be situated in, or adjacent to, the upcast shaft bottoms. Chilled water from the centralized underground plant could be transmitted to other levels. However, at elevations greater than some 500 m below the plant, water pressures in the pipe ranges become excessive. This can be counteracted either by pressure-reducing valves (adjustable orifices) or by employing water-to-water heat exchangers and secondary, low pressure, cold water circuits. Unfortunately, there are difficulties associated with both of these types of devices. Pressure-reducing valves necessitate the employment of high duty pumps to raise the heated water back to the plant elevation. Maintaining the high pressure water in a closed circuit by means of water to water heat exchangers balances the pressure heads in the supply and return shaft ranges. However, the intermediate water-to-water heat exchangers produce an additional loss of heat transfer efficiency between the work areas and the refrigeration

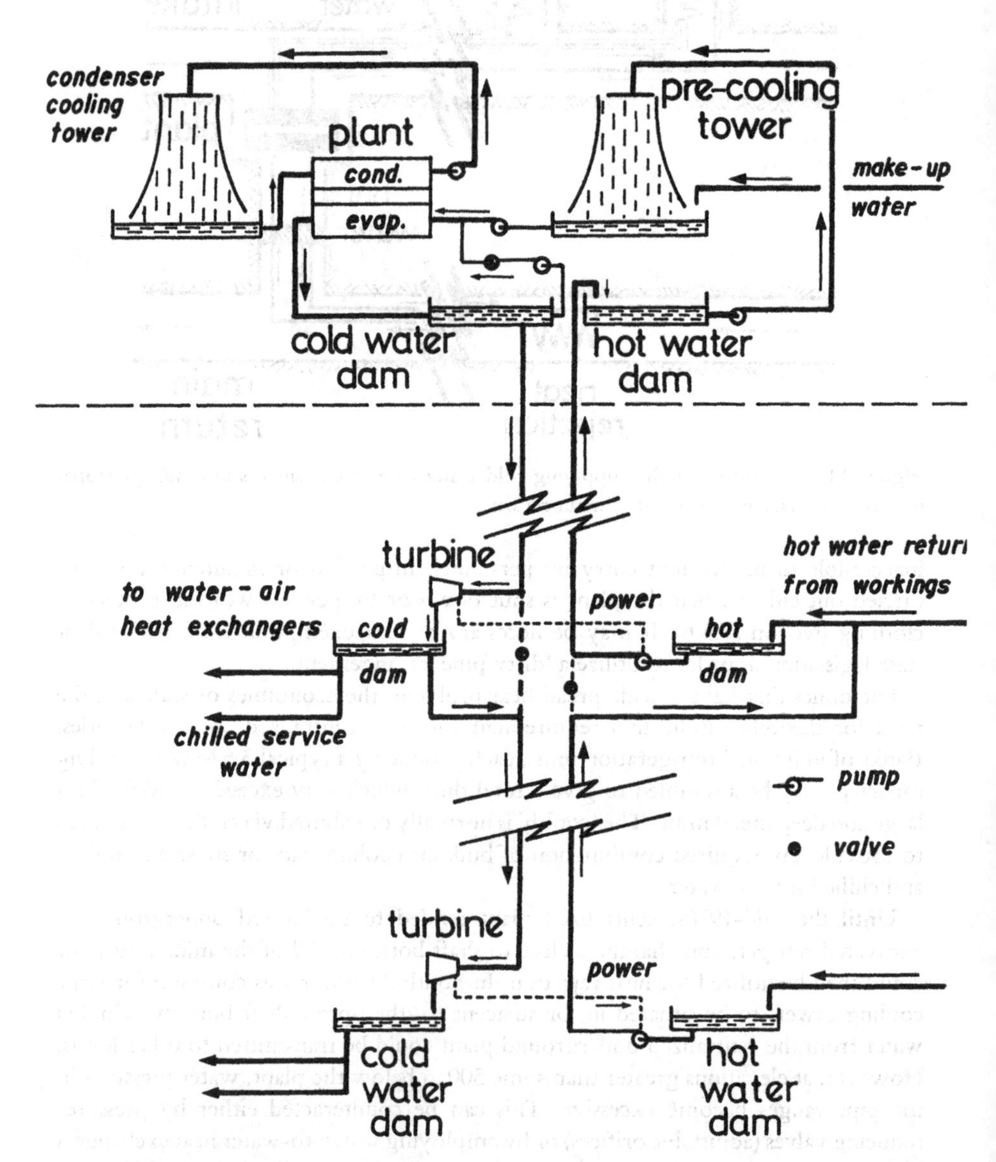

Figure 18.10 Schematic of centralized plant on surface supplying chilled water for underground heat exchangers and mine service water.

plant. Furthermore, they are a further source of potential corrosion and fouling of internal tubes.

In the mid-1970s a number of factors coincided to promote a trend towards the location of centralized refrigeration plant at the surface of deep and/or hot mines. First, there was a significant drop in the wet bulb temperatures at which mining workforces could be expected to work. Cut-off wet bulb temperatures have been reduced from 32 °C to 28 °C with an expectation of further improvements to 27 °C or less. Secondly, the combination of greater depths of workings and more intensive mechanization produced higher heat loads to be handled by the mine environmental control system. These factors combined to give very significant increases in the required cooling capacities of mine refrigeration plant, to the extent that untenable and uneconomic quantities of return air would be required for heat rejection. There was no choice but to locate the larger new plant on surface where heat rejection is relatively straightforward. Thirdly, the problem of high pressures developing in the shaft pipes was combated by the development of energy recovery devices including water turbines (section 18.3.4).

A simplified schematic of one arrangement involving surface refrigeration plant is indicated on Fig. 18.10. Let us commence with the supply of cold water at the surface and follow the circulation of that water around the mine. The figure illustrates water being cooled by a surface refrigeration plant and collected temporarily in a cold water dam. The hot water from the condensers is cycled around a cooling tower for heat rejection to the atmosphere. If a natural supply of sufficiently cool water is available from a stream or river, then there may be no need for the capital and operating expense of refrigeration plant. Alternatively, in dry climates, the required degree of cooling may be achievable simply by spraying the water supply through a surface cooling tower. Furthermore, it may be necessary to operate the refrigeration plant during the summer months only.

The water passes from the surface cold water dam to an insulated shaft pipe through which it falls to the working levels. The water may then be passed through a turbine at one or more subsurface levels before being stored temporarily in underground cold water dams. The turbine(s) achieve three beneficial results. First, the pressure of the water is reduced to that of the mine atmosphere at the corresponding level. Hence, no extensive high pressure water systems need exist in the workings. Secondly, the mechanical output power produced by the turbines may be employed directly to assist in driving the pumps that raise the return hot water to surface. However, because the demand for cold water may be out of phase with the availability of hot return water, it is preferable to use the turbines to drive generators for the production of electrical power. Third, the removal of energy from the water by the turbines results in a reduced temperature rise of that water, improving the cooling efficiency of the system.

From the cold water dams, the chilled water may be utilized for a variety of purposes including water-to-air heat exchangers for cooling the air at the entrance to a stope or face, or for recooling at intermediate points along the stope or face. Water-to-water heat exchangers may be employed to cool secondary water circuits such as the supply of potable drinking water or dust suppression service water. However, the modern trend is to supply chilled service water directly from the cold water dam.

Warm water returning from the heat exchangers or via drainage channels is directed into hot water dams adjacent to a shaft. The pumps that return this water to surface may be powered partly by the energy recovered by the turbines. Where refrigeration plant is in use, the hot water from the mine may be sprayed through pre-cooling towers prior to its return to the plant evaporators. This gives an additional low cost supplement to the cooling capacity of the system. Any required make-up water and anticorrosion chemicals are added at this stage.

The purpose of the dams on both the cold and the hot water sides of the layout is to provide capacitance to the system. This permits short-term fluctuations in demand for chilled water while using smaller refrigeration plant than would otherwise be necessary. At times of low demand, the temperature of the water in the surface dams can be further reduced by recycling that stored water through the plant. This smooths the variations in cooling load required of the refrigeration units.

Having introduced the broad concepts of subsurface cooling systems, we are now in a better position to examine the component parts in greater detail.

18.3.2 Heat exchangers

In general, a heat exchanger may be defined as a device that promotes the transfer of thermal energy from one solid or fluid system to another. In subsurface air conditioning systems, there are two classifications of heat exchanger in common use, both involving heat transfer between fluids.

An **indirect heat exchanger** promotes heat transfer across a solid medium that separates the two fluids. There is no direct contact between the fluids. Examples include

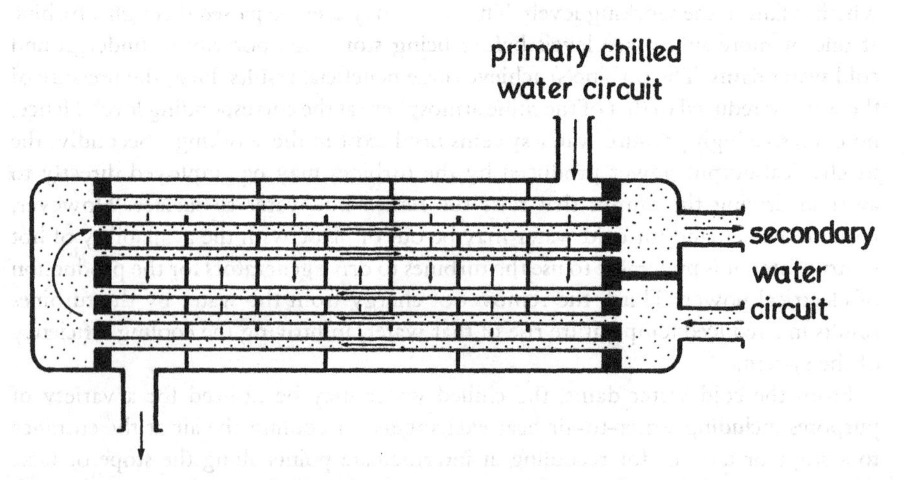

Figure 18.11 A water-to-water heat exchanger.

1. the shell-and-tube evaporators and condensers of refrigeration units (Fig. 18.3),
2. shell-and-tube water-to-water heat exchangers employed to transfer heat between water systems of differing pressure and/or water quality (Fig. 18.11), and
3. tubular coil heat exchangers to transfer heat from air to water (Fig. 18.12).

As the term implies, **direct heat exchangers** involve direct contact between the two fluids. Cooling towers and other types of spray chambers fall into this category. The objective may be to reject heat from the water to the air as in a conventional cooling tower. Conversely, a spray chamber supplied with chilled water provides a means of cooling an airstream.

The higher efficiency of heat transfer associated with direct heat exchangers caused a distinct trend away from tube coil air coolers during the 1980s. Good performance of large horizontal spray chambers for bulk air cooling promoted the further development of compact and enclosed spray chambers for more localized use.

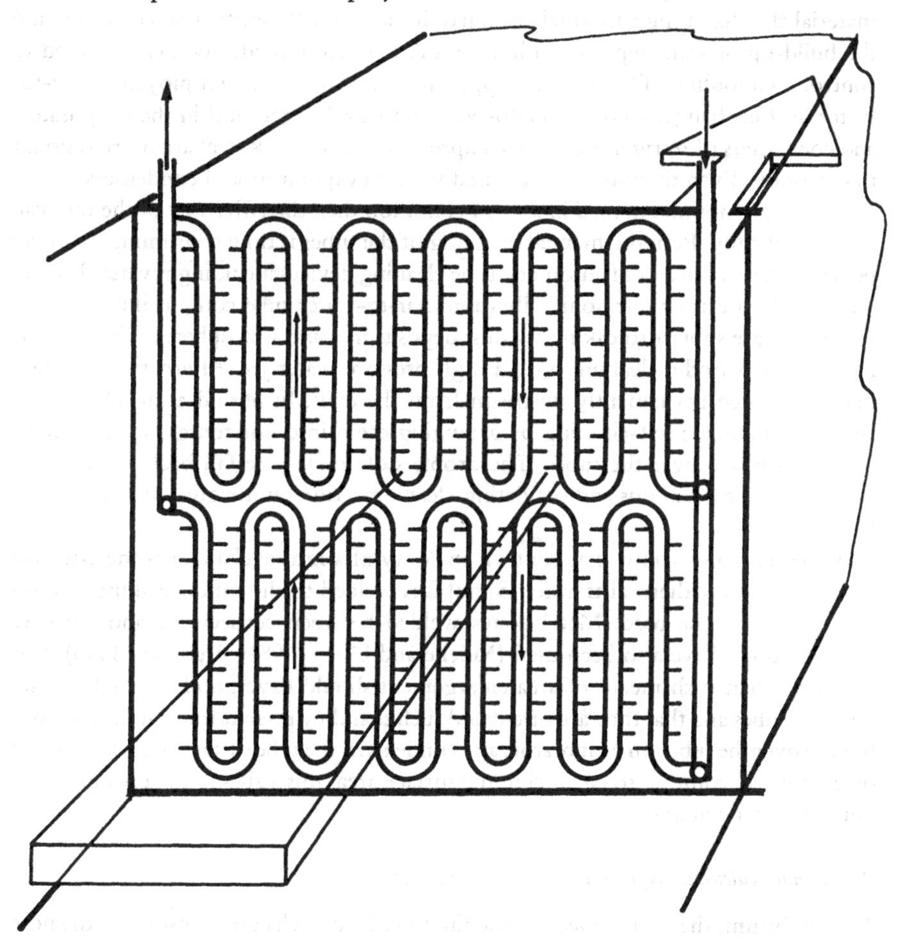

Figure 18.12 An air-to-water heat exchanger showing one layer of tubes.

In this section we shall examine the essential structure and operating principles of both indirect and direct forms of heat exchanger.

Indirect heat exchangers

A shell and tube heat exchanger may contain over 200 tubes. In a counterflow heat exchanger, the fluids inside and outside the tubes move in opposite directions, maximizing the total heat transfer. However, the water-to-water heat exchanger illustrated in Fig. 18.11 shows that this may be sacrificed in the interest of compactness.

In an air-to-water indirect heat exchanger, a bank of tubes carrying chilled water is located within an air duct. Figure 18.12 shows one layer of the tube bank.

The two important features of a heat exchanger are that a good efficiency of heat transfer is obtained and that the pressure drops suffered by both fluid streams should remain within acceptable limits. Heat transfer is facilitated by choosing a tube material that has a high thermal conductivity and is also resistant to corrosion and the build-up of scale deposits within the tubes. Chemical additives can be used to control such fouling of the tubes. Copper tubes are commonly employed for air-to-water heat exchangers. For water-to-water heat exchangers and in the evaporators and condensers of refrigeration units, cupronickel or stainless steel are more resistant to corrosion. Even titanium may be used within evaporators and condensers.

If tubes are not maintained in a clean condition then the efficiency of the unit can fall dramatically. Periodic manual brushing of the tubes is time consuming and may be replaced by a variety of automatic tube cleaning devices including reverse flushing with brush inserts or ultrasonic vibration. Air-to-water tube coils, as illustrated on Fig. 18.12, are subject to caking by dust deposits in mine atmospheres. This may be handled by periodic cleaning with a high pressure water jet. Alternatively, sprays may be located permanently upstream from the coils. In humid atmospheres, the coils may be made self-cleaning by arranging them into horizontal banks with air passing upwards over the tubes. The droplets of condensate fall back through the coils giving a continuous and oscillating cleaning action on the outer surfaces of the tubes.

A second major factor that controls the duty of a heat exchanger is the effective area available for thermal transfer. Spiral fins welded to the tubes, as illustrated on Fig. 18.12, may be used. Metal plates welded between consecutive and partially flattened tubes have also been used (Mücke and Uhlig, 1984; Weuthen, 1975). It is important that such means of area enlargement should have good thermal contact with the tubes and that they are orientated such that they present minimum resistance to flow over the tubes. In evaporators and condensers, the tube surfaces may be knurled or sintered. In addition to enhancing the surface area, this assists in the promotion of boiling or condensation.

Performance calculations for indirect heat exchangers

At equilibrium, the heat gained by one fluid in a heat exchanger must equal the heat lost by the other fluid. Hence, taking the example of an air to water heat exchanger, we can express the rate of heat transfer to be

$$q = m_w C_w \Delta t_w = m_a \Delta S \quad \text{W} \tag{18.17}$$

where m_w = mass flow of water (kg/s), C_w = specific heat of water (4187 J/(kg °C)), Δt_w = rise in temperature of the water (°C), m_a = mass flow of air (kg/s) and ΔS = fall in sigma heat of the air (J/(kg °C)) (section 14.6).

As each of the factors in equation (18.17) can be calculated easily from measured temperatures and flowrates, either the water or the air may be used to determine the rate of heat transfer, q. As measurements on the water circuit can normally be made more accurately than those on the airflow, the former are preferred for a determination of heat transfer. If condensation occurs on the outside of the coils then the third part of equation (18.17) is slightly in error as it does not take into account the heat removed from the system by the condensate. However, this is usually small and may be neglected.

Another way of expressing the heat transfer is in terms of an overall heat transfer coefficient, U (W/(m^2 °C)), for the coils and adjacent fluid boundary layers, and the difference between the mean temperatures of the air and the water, $t_{ma} - t_{mw}$, giving

$$q = UA(t_{ma} - t_{mw}) \quad \text{W} \tag{18.18}$$

where A = area available for heat transfer. This is analogous to equation (15.16) for a rock surface.

As the temperatures of both the air and water streams are likely to vary in a logarithmic rather than a linear fashion through the heat exchanger, it is more accurate to employ logarithmic mean temperature difference in equation (18.18), giving

$$q = \frac{UA(\Delta t_1 - \Delta t_2)}{\ln(\Delta t_1 / \Delta t_2)} \quad \text{W} \tag{18.19}$$

where ln indicates natural logarithm, and Δt_1 and Δt_2 are the temperature differences between the fluids at each end of the heat exchanger.

The UA product is usually quoted as a measure of the **effectiveness** of an indirect heat exchanger.

A performance check should be carried out at monthly or three-monthly intervals in order to determine any deterioration in UA caused by corrosion, scaling, or other forms of deposition on the tube surfaces.

Example Measurements made on a counterflow air to water cooling coil give the following results:

air:	inlet wet bulb and dry bulb temperatures	28 °C, 34 °C
	outlet wet bulb/dry bulb temperature	22.7 °C, 22.7 °C
	mass flowrate	5.1 kg/s
	barometric pressure	105 kPa

water	inlet temperature	17 °C
	outlet temperature	23 °C
	mass flowrate	4 kg/s.

Determine the operating duty of the coil and the *UA* value.

Solution From the psychorometric equations given in section 14.6, the sigma heats of the air at inlet and outlet conditions of 28 °C, 34 °C and 22.7 °C, 22.7 °C, respectively, are found to be

$$S_{in} = 84\,696 \text{ J/kg}$$

and

$$S_{out} = 63\,893 \text{ J/kg}$$

From equation (18.17) the heat lost from the air is

$$\begin{aligned} q_a &= m_a(S_{in} - S_{out}) \\ &= 5.1(84\,696 - 63\,893) \\ &= 106.1 \times 10^3 \text{ W} \end{aligned}$$

Also, from equation (18.17), the heat gained by the water is

$$\begin{aligned} q_w &= m_w C_w \Delta t_w \\ &= 4 \times 4187 \times (23 - 17) \\ &= 100.5 \times 10^3 \text{ W or } 100.5 \text{ kW} \end{aligned}$$

At equilibrium, q_a and q_w must be equal, showing that the errors in measurements have caused a deviation of some 5.5%. Much larger discrepancies often occur due mainly to difficulties in making measurements in highly turbulent and, often, saturated airflows. We shall continue the calculation using the rate of heat transfer given by the water circuit, 100.5 kW.

In order to calculate the *UA* value from equation (18.19), we must first evaluate the temperature differences between the water and air at the air inlet and air outlet ends (subscripts 1 and 2, respectively):

$$\Delta t_1 = t_{a,in} - t_{w,out} = 34 - 23 = 11\ °C$$

$$\Delta t_2 = t_{a,out} - t_{w,in} = 22.7 - 17 = 5.7\ °C$$

Notice that we use dry bulb temperature at the air inlet and before saturation conditions are attained for the determination of Δt_1. This is equivalent to taking the latent heat of condensation as a factor contributing towards the *UA* value of the heat exchanger.

Equation (18.19) now gives

$$UA = q_w \frac{\ln(\Delta t_1/\Delta t_2)}{\Delta t_1 - \Delta t_2}$$

$$= 100.5 \frac{\ln(11/5.7)}{11 - 5.7}$$

$$= 12.47 \text{ kW/°C}$$

The UA value of a clean coil may lie between 10 and 25 kW/°C depending on the design of the heat exchanger and the configuration of fluid flows. Records should be kept of the periodic performance tests on each heat exchanger. Significant reductions in UA values indicate the need for cleaning or replacement of the tubes.

Although UA values are normally determined by measurement as illustrated in the example, they may also be defined by the following equation:

$$\frac{1}{UA} = \frac{1}{h_i A_i} + \frac{1}{h_{fi} A_i} + \frac{x}{k_t A_m} + \frac{1}{h_o A_o} + \frac{1}{h_{fo} A_o} \quad \text{°C/W} \qquad (18.20)$$

where h = heat transfer coefficients (W/(m^2 °C)), A = area available for heat transfer (m^2), x = thickness of tube walls (m) and k_t = thermal conductivity of tube material (W/(m °C)), and the subscripts are i for the inside surface of tubes, o for the outside surface of tubes and m for the mean of inner and outer surfaces.

h_{fi} and h_{fo} are the heat transfer coefficients associated with fouling (deposits) on the inside and outside surfaces of the tubes respectively. On a clean tube there are no such deposits; h_{fi} and h_{fo} are both then infinite.

Radiation terms have been left out of equation (18.20) as these are normally small in a heat exchanger. They may, however, become significant in situations such as an uninsulated pipe suspended in an airway.

Equation (18.20) can be further simplified for practical application. First, the term $x/k_t A_m$ is very small compared with the others. Secondly, the terms involving fouling of the tubes are often combined and attributed to the inside surface only, giving

$$\frac{1}{UA} = \frac{1}{h_i A_i} + \frac{1}{h_o A_o} + \frac{1}{h_f A_i} \quad \text{°C/W} \qquad (18.21)$$

As fouling of the tubes occurs, h_f will decrease causing the UA value also to fall. Whillier (1982) quotes a typical value of some 3000 W/(m °C) for h_f in a mine refrigeration plant.

For turbulent flow inside smooth tubes, the heat transfer coefficient can be determined from the Colburn equation (15A.15):

$$h_i = 0.023 \frac{k}{d} \mathrm{Re}^{0.8} \mathrm{Pr}^{0.4} \quad \text{W/(m}^2\text{ °C)} \qquad (18.22)$$

where k = thermal conductivity of fluid (W/(m °C)), d = internal diameter of tube, Re = $\rho u d/\mu$ (dimensionless) is Reynolds' number, ρ = fluid density (kg/m^3), u = fluid velocity (m/s), μ = dynamic viscosity (N s/m^2), Pr = $C_p \mu/k$ (dimensionless, may be taken as 0.7 for air) is the Prandtl number and C_p = Specific heat at constant pressure (J/(kg °C)).

For air and water, the values of viscosity and thermal conductivity within the temperature range 0 to 60 °C are given in sections 2.3.3 and 15.2.4 respectively. Other expressions for the heat transfer coefficient at rough surfaces are given in Table 15A.1.

The values of h_o for the outer surfaces of the tubes vary widely according to geometry and the configuration of tubes. For turbulent cross flow of air over tubular surfaces, McAdams (1954) gives, for a single tube,

$$h_o = 0.24 \frac{k_a}{d} \mathrm{Re}^{0.6} \quad \mathrm{W/(m^2\,°C)} \tag{18.23}$$

and, for a bank of staggered tubes,

$$h_o = 0.29 \frac{k_a}{d} \mathrm{Re}^{0.6} \quad \mathrm{W/(m^2\,°C)} \tag{18.24}$$

In these relationships, d is the outer diameter of the tubes and Reynolds' numbers are determined on the basis of the maximum velocity of the air as it flows between the tubes. The selection of cooling coils is facilitated greatly by tables and graphs provided by manufacturers.

Direct heat exchangers

In direct heat exchangers, air is brought into contact with water surfaces. The purpose may be to cool the water used in removing heat from the condensers of a refrigeration plant. In this case the hot water is sprayed into a **cooling tower** and descends as a shower of droplets through an ascending airstream. Heat is transferred from the water droplets to the air by a combination of convection (sensible heat) and evaporation (latent heat). The cooled water that collects at the base of the cooling tower is then returned to the condenser (see the condenser cooling tower on Fig. 18.10).

Alternatively, the objective may be to cool the air. In this case, chilled water is sprayed into a vertical or horizontal **spray chamber**. Provided that the airflow enters with a wet bulb temperature that is higher than the temperature of the water, then heat will be transferred from the air to the water by a combination of convection and condensation.

Although there are significant differences between the designs of cooling towers and spray coolers, there are several common factors that influence the amount and efficiency of heat exchange:

1. water mass flowrate;
2. supply temperature of water;
3. air mass flowrate;
4. psychrometric condition of the air at inlet;
5. duration and intimacy of contact between the air and the water droplets.

The last of these factors depends on the physical design of the heat exchanger, in particular,

1. the relative velocity between the air and water droplets, and
2. the size and concentration of water droplets—governed by the flow and pressure of the supply water, the type and configuration of spray nozzles, and the presence of packing within the heat exchanger.

The traditional use of direct heat exchangers in mine air conditioning systems has been for cooling towers either underground or on surface. Tube coil heat exchangers with closed circuit water systems have been used extensively for air cooling in or close to mine workings. However, the greater efficiency of direct heat exchangers led to the development of large permanent spray chambers for bulk air cooling. Furthermore, through the 1980s, smaller portable spray chambers began to be employed for local cooling.

Cooling towers

Let us examine, first, the essential features of cooling towers. Figure 18.13 illustrates a vertical forced draught cooling tower of the type that may be used on the surface of a mine. Hot water from refrigeration plant condensers is sprayed into the cooling tower and moves downward in counterflow to the rising air current. The purpose of the packing is to distribute the water and air flows uniformly over the cross-section and to maximize both the time and total area of contact between the air and water surfaces. The packing may take the form of simple splash bars or riffles arranged in staggered rows, egg-crate geometries or wavy (film-type) surfaces located in vertical configurations. The materials used for packing may be treated fir or redwood timber, galvanized steel, metals with plastic coatings and injection-moulded PVC or polypropylene. Concrete is used primarily for casings, structural reinforcements and water sumps or dams. Air velocities through counterflow packed cooling towers lie typically in the range 1.5 to 3.6 m/s.

In underground cooling towers, the packing is often eliminated completely or takes the form of one or two screens arranged horizontally across the tower. Such heat exchangers are essentially vertical spray towers. Other means of maximizing contact time are then employed. These include designs that introduce a swirl into the air at entry, and directing the sprays upward rather than downward. Again, in underground installations, airflow is induced through the cooling towers by the mine ventilating pressure or by booster fans in return airways rather than by fans connected directly to the tower. The pressure drop through underground cooling towers is further reduced by replacing the mist eliminator screens with an enlarged cross-sectional area; the lower air velocity in this zone reduces the carry-over of water droplets. The optimum air velocity in open spray towers lies in the range 4 to 6 m/s with a maximum of some 8 m/s. Water loadings should not be greater than 16 l/(s m^2) of cross-sectional area (Stroh, 1982).

Cooling towers of the type used for mine air conditioning are typically 10 to 20 m in height and some 3 to 8 m in diameter, depending on the rate at which heat is to be exchanged. Heat loads may be as high as 30 MW.

Natural draught cooling towers do not have fans but rely upon air flow induction

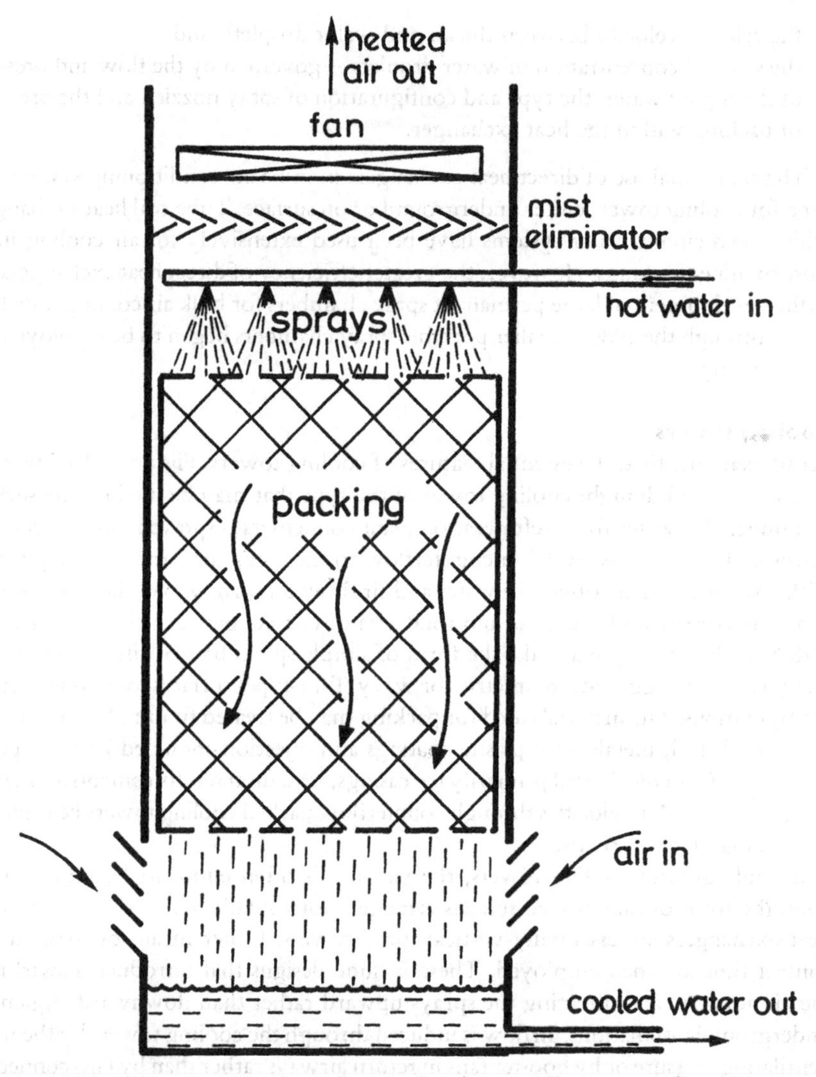

Figure 18.13 Direct contact counterflow cooling tower. (The same configuration may be used to cool air by supplying the heat exchanger with chilled water.)

by the action of the sprays, or by density difference between the outside atmosphere and the hot moist air within the tower. The hyperbola-shaped cooling towers employed commonly for the high heat loads of power stations are of this type and may be over 150 m high in order to accentuate the chimney effect.

The air leaving a cooling tower is usually saturated. This often results in the formation of a fog plume in the free atmosphere. The surface environmental effects of such a plume should be considered carefully and may influence the siting of a surface cooling tower. In large installations with unacceptable fog plumes, part of the hot condenser water may be cooled within a finned tube indirect heat exchanger. This involves sensible heat exchange only and reduces the humidity of the air leaving the cooling tower. However, such an arrangement detracts from the overall efficiency of heat transfer.

In a direct exchange cooling tower, some water is lost continuously from the circuit both by evaporation and by drift (or carry-over) of small droplets. The evaporation loss approximates some 1% for each 7 °C of water cooling and drift loss is usually less than 0.2% of the circulation rate (ASHRAE, 1988). However, evaporation can result in a rapid escalation in the concentration of dissolved solids and other impurities leading to scaling, corrosion and sedimentation within the system. In order to limit the build-up of such impurities, some water is continuously removed from the system (bleed-off or blow-down). The bleed-off rate is controlled by monitoring the quality of the water but may be of the order of 1% of the circulation rate. The combined losses from evaporation, drift and bleed-off are made up by adding clean water to the circuit.

Further protection of metal components and, particularly, the tubes of condensers is obtained by administering anticorrosion compounds. These generally take the form of chromates, phosphates or polyphosphonates of zinc, and promote the formation of a protective film on metal surfaces. Some of these compounds are toxic and precautions may be necessary against accidental release into natural drainage systems. Biocides such as chlorine are also added on a periodic basis to control the growth of algae and other organic slimes.

Chilled water spray chambers

If the water supplied to a direct contact heat exchanger is at a temperature below that of the wet bulb temperature of the air, then cooling and dehumidification of the air is achieved. Chilled water spray chambers fall into two categories with respect to size. First, the larger installations are constructed at fixed sites for bulk cooling of main airflows. Secondly, portable spray chambers for localized use within working areas have many advantages over the more traditional tube coil (indirect) stope or face coolers. We shall consider each of these two applications in turn.

For bulk air cooling, spray chambers may be designed in vertical or horizontal configuration. Indeed, if the cooling tower shown on Fig. 18.13 were supplied with chilled water, then it would act equally well as a vertical spray air cooler. Such designs may be employed either on surface or underground for bulk cooling of intake air and may have heat transfer duties up to 20 MW.

Horizontal spray chambers have more limited maximum capacities of some 3.5 MW. They are, however, more convenient for underground use in that existing airways may be utilized without additional excavation. Figure 18.14 illustrates a single-stage horizontal spray chamber. The sprays may be directed against or across the airflow. The nozzles can be distributed over the cross-section as shown in the sketch or, alternatively, located at the sides or near the base of the chamber. Although the position of the nozzles appears not to be critical, it is important that both the sprays and the airflow are distributed uniformly over the cross-section. The spray

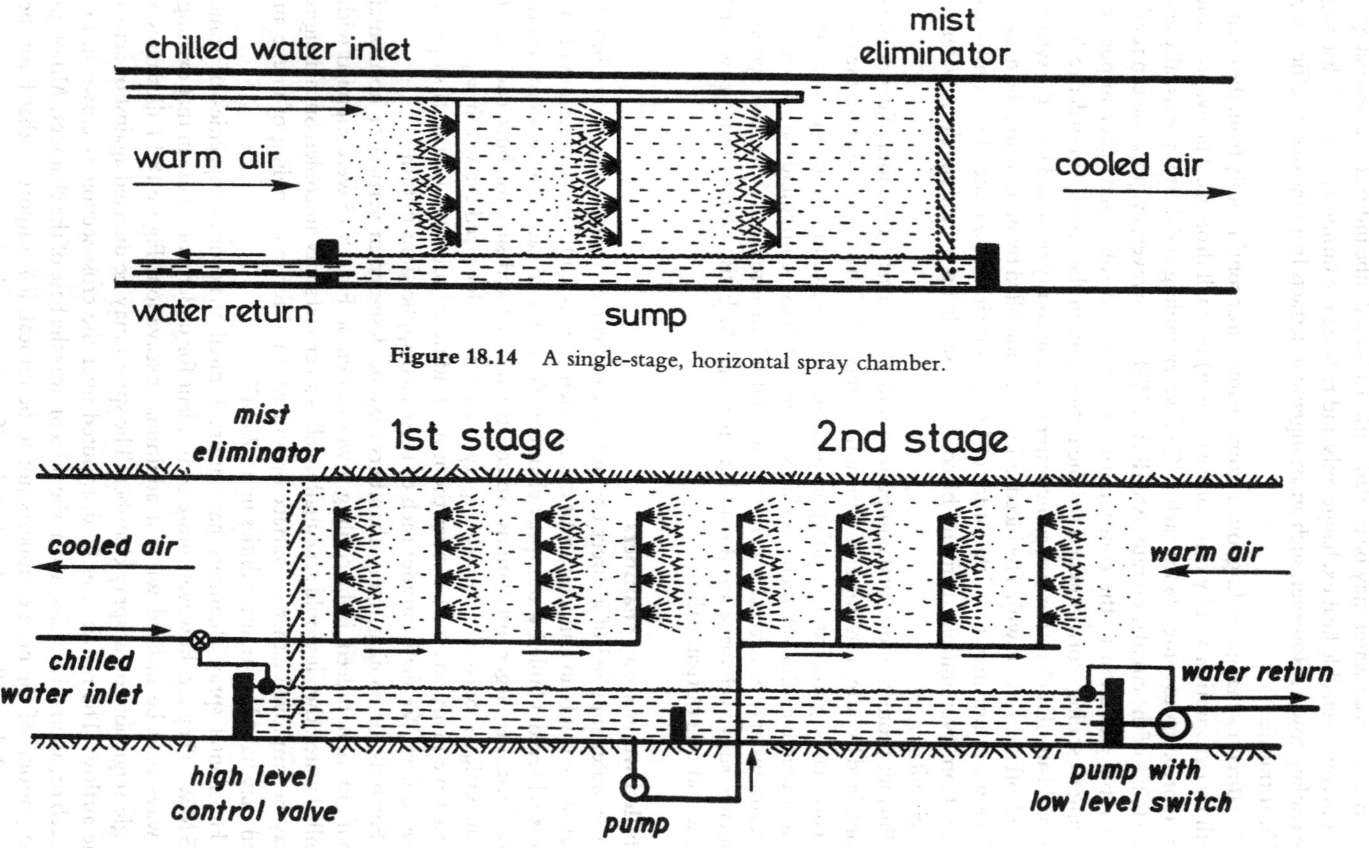

Figure 18.14 A single-stage, horizontal spray chamber.

Figure 18.15 A two-stage horizontal spray chamber.

density should lie within the range 2 to 5 l/s for each square metre of cross-sectional area (Bluhm, 1983).

The area of liquid–air interface and efficiency of heat exchange increase as the size of droplets falls. However, very small droplets result in excessive carry-over or necessitate highly constrictive mist eliminators. Furthermore, higher water pressures and, therefore, pumping costs are required to produce fine sprays. In practice, droplet diameters of some 0.5 mm and water pressures in the range 150 to 300 kPa give satisfactory results in horizontal spray chambers (Reuther *et al.*, 1988).

At positions fairly close to the nozzles, the relative directions of the air and water droplets may be counterflow or cross-flow, dependent upon the orientation of the nozzles. However, aerodynamic drag rapidly converts the spray to parallel flow, particularly for the smaller droplets. In order to regain the higher efficiency of counterflow heat exchange, the spray chamber may be divided into two or three stages so that the air leaves the chamber at the position of the coldest sprays. Chilled water should be supplied at as low a temperature as practicable but, in any event, not higher than 12 °C. Figure 18.15 illustrates a two-stage spray chamber. An additional sump at each end assists in balancing the pumping duties.

If the surrounding strata are unfractured and unaffected by water, then the spray chamber may utilize the full cross-section of a bypass airway. If, however, the rock must be protected against the effects of water, then concrete lining or prefabricated sections may be employed to contain the spray chamber. In cases where the strata are very sensitive to water (such as evaporite strata) then it is advisable to protect the rock surfaces for 50 to 100 m downstream from the spray chamber.

The cross-sectional area of the spray chamber should be chosen to give a preferred air velocity of some 4 to 6 m/s, but not more than 7 m/s. Higher air velocities will reduce the efficiency of heat exchange and can result in excessive pressure drops in the airflow.

In addition to cooling and dehumidifying the air, spray coolers can also reduce dust concentrations. However, the build-up of dust particles in the recirculating water system may cause fouling of the pipes and other heat exchangers. This may require filters or sedimentation zones to be included in the design.

Internal packing can also be employed to improve the efficiency of horizontal spray chambers (Stroh, 1980). This may increase the air frictional pressure drop across the cooler. However, the water supply pressure (gauge) may be as low as 30 kPa since the spray nozzles can be replaced by low resistance dribbler bars.

In order to extend the applicability of direct spray coolers, enclosed and portable units have been developed (Thimons *et al.*, 1980; Ramsden and Bluhm, 1984; Reuther *et al.*, 1988). These may be mounted on wheels or sleds and are often referred to as **spray mesh coolers**. Figure 18.16 illustrates the principle of operation. In order to maintain the design to acceptable dimensions for portable application, the water loading may be much higher than open spray chambers and it becomes even more important to maximize the area and time of contact between the air and water. This occurs in three stages. First, the air passes through the lower mesh of plastic or knitted stainless steel. Secondly, the airflow is directed through the upward pointing and

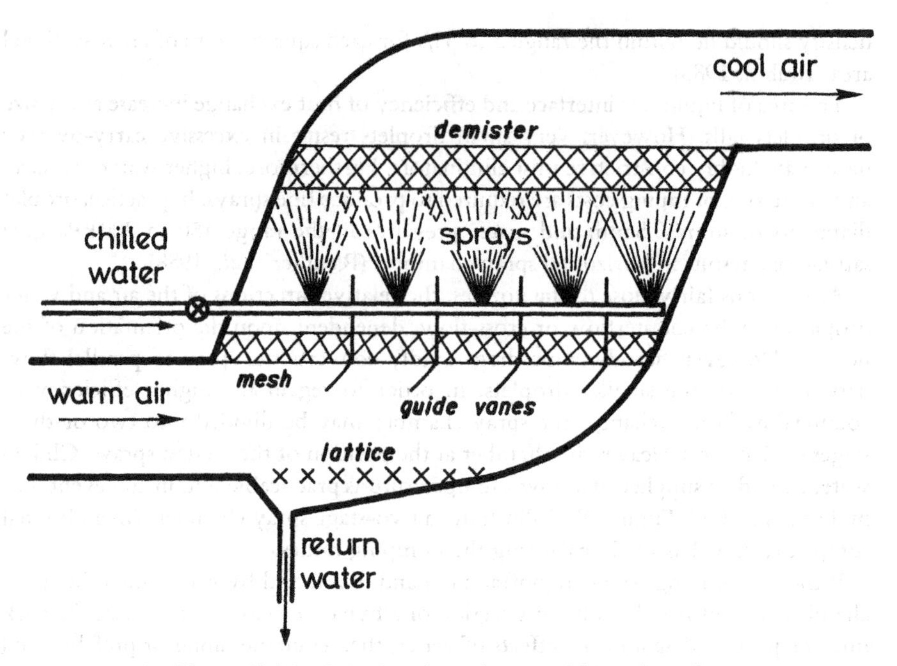

Figure 18.16 A section through a portable mesh cooler.

finely divided sprays. The final stage of cooling occurs within the upper mesh which also acts as a demister. The distribution of droplets of water falling from the upper to the lower mesh may be assisted by installing drip fins. These also help to prevent water running down the walls (Heller *et al.*, 1982). Between the two meshes, heat exchange occurs by a combination of counterflow, cross-flow and parallel flow. Another type of portable spray cooler employs film packing and cross-flow heat exchange (Reuther *et al.*, 1988).

The dry bulb temperature of air can be reduced by passing it through a spray chamber supplied with unchilled water. The device is then known as an **evaporative cooler**. No heat is removed from the air. Hence, if the water is supplied at wet bulb temperature, then the exhaust air may attain that same wet bulb temperature while the moisture content will have increased. Evaporative coolers have an application for surface structures in hot dry climates but are seldom used in underground mines. They have been employed for very localized effects in hot evaporite mines. However, the reduced dry bulb temperature encourages enhanced heat flow from the strata (section 15.2.2) which, when combined with the raised moisture content, reduces the cooling power of the air downstream from the cooled area.

Performance calculations for direct heat exchangers

A common theoretical analysis may be applied to direct heat exchangers irrespective of the direction of heat transfer. The results of such analyses apply equally well to cooling towers and chilled water spray chambers. In order to avoid unnecessary repetition, we will conduct the following analysis on the assumption of a cooling tower (Fig. 18.13). Water recycles continuously through the condenser of a refrigeration plant (where it gains heat) and the cooling tower where it rejects that heat to the atmosphere.

The first observation is that if we ignore the small heat losses from interconnecting pipes and the equally small effects of make-up water, then, at steady state, the heat rejected in the cooling tower must be equal to the heat gained in the condenser. This leads to the initially surprising statement that the rate of heat rejection in the cooling tower depends only on the heat load imposed by the condenser and not at all on the design of the cooling tower. However, if the cooling tower is inefficient in transferring heat from the water to the air, then the temperature of the water throughout the complete circuit will rise until balance is attained between heat gain in the condenser and heat rejection in the cooling tower. This would be unfortunate as the coefficient of performance of the refrigeration plant deteriorates as the condenser temperature increases (section 18.2.4).

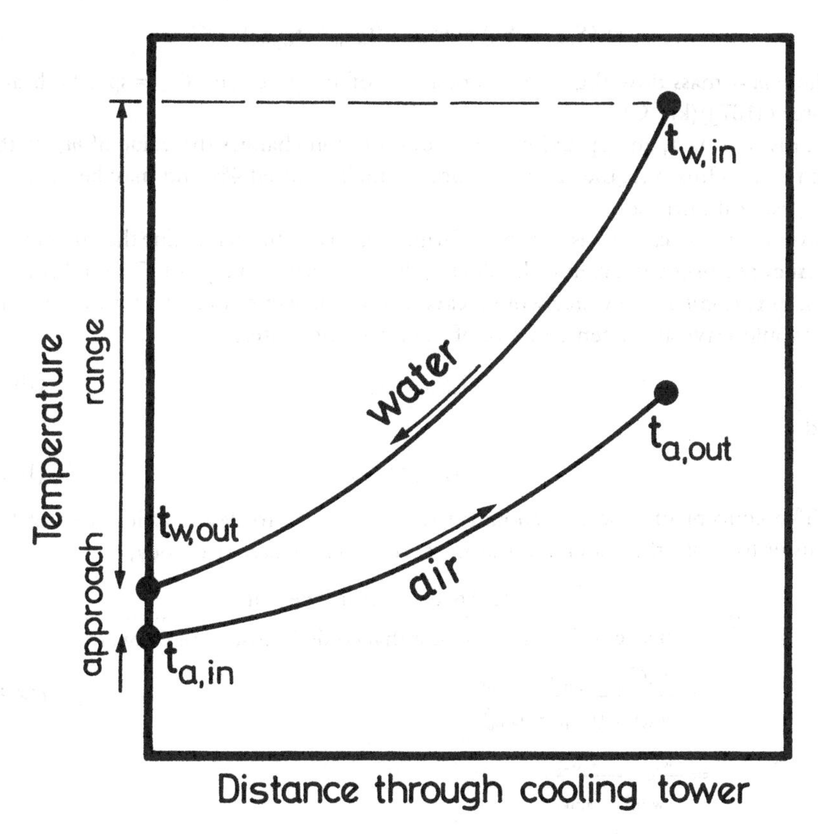

Figure 18.17 Variation of water temperature, t_w, and wet bulb temperature of the air, t_a, through a cooling tower.

Figure 18.17 illustrates the decrease in temperature of the water as it falls through the cooling tower and the corresponding increase in the wet bulb temperature of the ascending air. This figure also defines the meaning of two terms that are commonly employed: the **range** is the change in temperature of the water

$$\text{range} = t_{w,in} - t_{w,out} \quad °C \tag{18.25}$$

while the **approach** is the difference between the temperatures of the water outflow and the wet bulb temperature of the air inflow:

$$\text{approach} = t_{w,out} - t_{a,in} \quad °C \tag{18.26}$$

where t = temperature (°C), and the subscripts are w for liquid water, a for air (wet bulb), in for inflow and out for outflow. Manufacturers will usually accept requests for approach values down to 2 °C.

The quantitative analysis of the cooling tower commences by writing down the balance that must exist between rates of heat gained by the air and heat lost from the water:

$$m_a(S_{out} - S_{in}) = m_w C_w(t_{w,in} - t_{w,out}) \quad W \tag{18.27}$$

where m = mass flow (kg/s), S = sigma heat of air (J/kg) and C_w = specific heat of water (4187 J/(kg °C)).

This is, in fact, an approximation as evaporation changes the value of m_w within the tower. However, the error does not normally exceed 4% and may be neglected for practical purposes.

As in so many cases of assessing performance, it is useful to imagine the unattainable **perfect cooling tower**. In such a device, the two curves on Fig. 18.17 would coincide and, in particular, the water would leave at inlet air wet bulb temperature while the air would leave at the temperature of the incoming water, that is

$$t_{w,out} = t_{a,in} \tag{18.28}$$

and

$$t_{a,out} = t_{w,in} \tag{18.29}$$

The concept of a perfect cooling tower allows us to devise efficiencies of heat transfer for both the water and the air streams. The water efficiency,

$$\begin{aligned}\eta_w &= \frac{\text{actual heat lost from the water}}{\text{theoretical maximum heat that could be lost from the water}} \\ &= \frac{m_w C_w(t_{w,in} - t_{w,out})}{m_w C_w(t_{w,in} - t_{a,in})} \\ &= \frac{t_{w,in} - t_{w,out}}{t_{w,in} - t_{a,in}} \\ &= \frac{\text{range}}{\text{range} + \text{approach}}\end{aligned} \tag{18.30}$$

(see Fig. 18.17). Similarly, the air efficiency,

$$\eta_a = \frac{\text{actual heat gained by the air}}{\text{theoretical maximum heat that could be gained by the air}}$$

$$= \frac{m_a(S_{out} - S_{in})}{m_a(S_{w,in} - S_{in})} \tag{18.31}$$

$$= \frac{S_{out} - S_{in}}{S_{w,in} - S_{in}}$$

where $S_{w,in}$ = sigma heat of saturated air at a temperature equal to that of the inlet water (J/kg). (The term 'thermal capacity' is sometimes employed where we have used 'theoretical maximum heat'.)

Although the water and air efficiencies are useful indicators of the efficacy of heat transfer for each of the two fluids, neither gives any clue to the overall quality of design for the complete cooling tower. For example, if the water flow were low giving relatively few small droplets falling through a large airflow, then the water efficiency would be very high. However, the air would be used to only a small fraction of its thermal capacity and, hence, the air efficiency and overall heat transfer would be low.

To examine this problem let us consider, again, perfect heat transfer in a cooling tower. Then,

$$\text{maximum thermal capacity of water} = m_w C_w (t_{w,in} - t_{a,in}) \quad \text{J}$$

and

$$\text{maximum thermal capacity of airstream} = m_a(S_{w,in} - S_{in}) \quad \text{J}$$

The ratio of these maximum thermal capacities defines the **tower capacity factor**, R:

$$R = \frac{m_w}{m_a} C_w \frac{t_{w,in} - t_{a,in}}{S_{w,in} - S_{in}} \tag{18.32}$$

The liquid-to-gas ratio, m_w/m_a, normally lies within the range 0.4 to 2.0.

The theoretical concept of tower capacity factor ignores the practical reality that in a real cooling tower the rate of heat loss from the water must equal the rate of heat gain by the air. It may be defined in words as follows:

> If both of the given fluid streams were used to their maximum thermal capacity, then R would be the number of watts lost by the water for each watt gained by the air.

An important observation here is that R depends only on the flowrates and the inlet conditions of the water and air. It is completely independent of the construction of the cooling tower. Notice also that we have been quite arbitrary in choosing to define R as the ratio of water thermal capacity divided by air thermal capacity, rather than the other way round. The values of tower capacity factor vary from 0.5 to 2.

The tower capacity factor is also related to the ratio of air and water efficiencies. From equations (18.30) and (18.31)

$$\frac{\eta_a}{\eta_w} = \frac{S_{out} - S_{in}}{S_{w,in} - S_{in}} \frac{t_{w,in} - t_{a,in}}{t_{w,in} - t_{w,out}}$$

and substituting for

$$S_{out} - S_{in} = \frac{m_w}{m_a} C_w(t_{w,in} - t_{w,out})$$

from equation (18.27) gives

$$\frac{\eta_a}{\eta_w} = \frac{m_w}{m_a} C_w \frac{t_{w,in} - t_{a,in}}{S_{w,in} - S_{in}} \tag{18.33}$$

As this is identical to equation (18.32), then

$$\frac{\eta_a}{\eta_w} = R \tag{18.34}$$

indicating that the ratio of air and water efficiencies is also independent of the design of the cooling tower or spray chamber.

The overall effectiveness of heat transfer within the cooling tower must be dependent on the fluid having the smaller thermal capacity, i.e. the fluid having the larger efficiency. To quantify this concept, we can say that the effectiveness, E, is the larger of the values of η_a and η_w. Another way of expressing this, using equation (18.34), is

$$\begin{aligned} E &= \eta_a \text{ if } R \geqslant 1 \\ E &= \eta_w \text{ if } R \leqslant 1 \end{aligned} \tag{18.35}$$

η_a and η_w are, of course, equal when $R = 1$.

Prior to 1977, there seems to have been no single parameter that could be used to describe the quality of design for any given cooling tower. The rate of heat transfer, air and water efficiencies, and the tower capacity factor are all dependent on inlet temperatures. It is, therefore, somewhat misleading to quote a single value of cooling duty to characterize a heat exchanger.

In the mid-1970s, Whillier of the South African Chamber of Mines conducted an analysis on results published earlier by Hemp (1967, 1972) relating to a series of tests during which a spray tower was successively supplied with a range of inlet water temperatures. Whillier (1977) noticed that two distinct curves were produced when (a) water efficiency and (b) air efficiency were plotted against tower capacity factor. Furthermore, these two curves appeared to be mirror images of each other. In a moment of inspiration he recalled that the order of the ratio that we choose to define R is quite arbitrary (equation (18.32)). Whillier took the values of tower effectiveness,

E, as defined by equations (18.35) and plotted them against R^*, where

$$R^* = R \text{ when } R \leqslant 1$$
$$R^* = 1/R \text{ when } R \geqslant 1 \qquad (18.36)$$

(i.e. inverting the ratio when R exceeds unity).

To his delight, all the points for the spray tower then lay on a single curve within an acceptable experimental scatter, and no matter whether the tower was used to cool water or air. The shape of that curve is illustrated on Fig. 18.18. The fact that all test results lay on a single curve held promise that an equation or characteristic number for that curve could be used as a means of quantifying the overall merit of the design and construction of the cooling tower, and independent of the inlet temperatures or flowrates of the air and water. Whillier attempted several curve-fitting exercises on his results and found that a reasonable correlation was obtained by the exponential relationship

$$E = F^{R^*} \qquad (18.37)$$

where F was a constant for that particular spray tower. Different values of F in the range 0 to 1 would be obtained for other cooling towers or spray chambers. Whillier termed this parameter the **factor of merit** for the cooling tower. A means had been

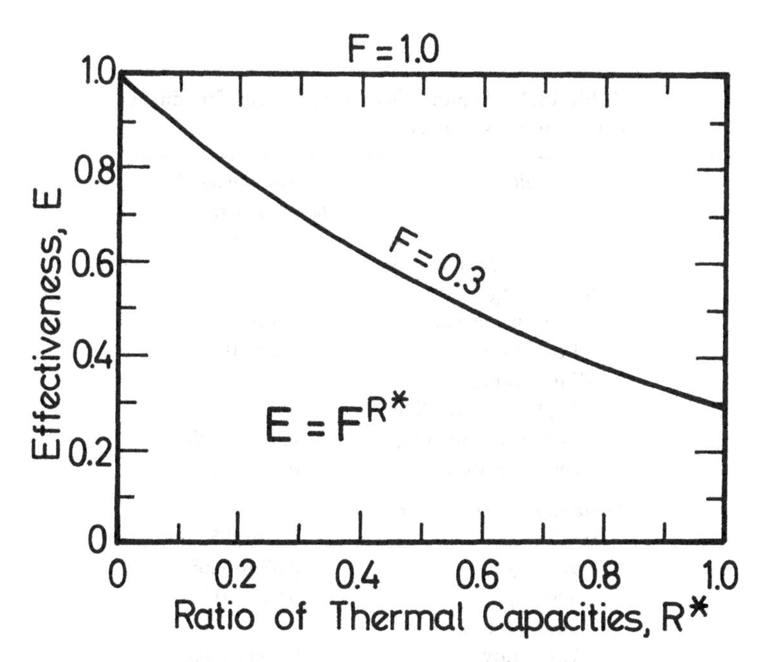

Figure 18.18 Form of the curve of E against R^* for a direct-contact heat exchanger that has a factor of merit $F = 0.3$.

found by which the performance characteristic of a direct heat exchanger could be quantified in a single number. A factor of merit of 1.0 would indicate a perfect counterflow direct-contact heat exchanger. Practical values lie within the range 0.3 to 0.8.

Further work by Bluhm (1980, 1983) investigated the factor of merit for multistage spray chambers, and produced an improved formula for spray coolers:

$$\eta_w = \frac{1 - \exp[-N(1-R)]}{1 - R\exp[-N(1-R)]} \tag{18.38}$$

where

$$N = \frac{F}{R^{0.4}(1-F)} \tag{18.39}$$

It is recommended that this formula is used for factors of merit below 0.4 or above 0.7.

Typical factors of merit suggested by Whillier (1977) and Bluhm (1980) are shown on Table 18.2. Once the factor of merit has been established for any direct-contact heat exchanger, either by testing or by estimation from Table 18.2, then the performance of the unit can be established for any given flow rates and inlet conditions for the air and water streams.

Table 18.2 Typical factors of merit for direct contact heat exchangers

Type of unit	*Typical values of factor of merit, F*
Vertical spray filled towers	
No packing	
High water loading	0.5 to 0.6
Low water loading	0.6 to 0.7
With packing	
High water loading	0.55 to 0.65
Low water loading	0.65 to 0.75
Industrial packed tower	0.7 to 0.8
Horizontal spray chambers	
Single stage	0.4 to 0.5
Two stage	0.58 to 0.68
Three stage	0.65 to 0.75
Packed air coolers	
Cross-flow	0.55 to 0.65
Counterflow	0.68 to 0.78
Mesh coolers	0.55 to 0.65

Example 1 A cooling tower operates at a barometric pressure of 95 kPa. The following temperature measurements are obtained:

air inlet wet bulb temperature, $t_{a,in}$	18.0 °C
air outlet wet bulb temperature, $t_{a,out}$	28.0 °C
water inlet, $t_{w,in}$	32.0 °C
water outlet, $t_{w,out}$	22.5 °C

Determine the factor of merit for the cooling tower.

Solution Using the psychrometric equations given in section 14.6 (or from a 95 kPa psychrometric chart), sigma heats may be determined to be

at air inlet (18.0 °C), S_{in}	52.05 kJ/kg
at air outlet (28.0 °C), S_{out}	90.90 kJ/kg
at water inlet temperature (32.0 °C), $S_{w,in}$	111.64 kJ/kg

Air efficiency. From equation (18.31),

$$\eta_a = \frac{S_{out} - S_{in}}{S_{w,in} - S_{in}}$$

$$= \frac{90.90 - 52.05}{111.64 - 52.05} = 0.652$$

Water efficiency. From equation (18.30),

$$\eta_w = \frac{t_{w,in} - t_{w,out}}{t_{w,in} - t_{a,in}}$$

$$= \frac{32.0 - 22.5}{32.0 - 18.0} = 0.679$$

Tower capacity factor. From equation (18.34),

$$R = \frac{\eta_a}{\eta_w} = \frac{0.652}{0.679} = 0.961$$

Tower effectiveness. From equations (18.35), as $R < 1$ then $E = \eta_w = 0.679$. Also, from equation (18.36),

$$R^* = R = 0.961$$

Factor of merit. From equation (18.37),

$$E = F^{R^*}$$

$$\ln(E) = R^* \ln(F)$$

or

$$F = \exp\left[\frac{\ln(E)}{R^*}\right]$$

$$= \exp\left[\frac{\ln(0.679)}{0.961}\right]$$

$$= 0.668$$

Note that we have been able to calculate the factor of merit using measured temperatures and the barometric pressure only. We have not required any fluid flowrates.

(Bluhm's equations (18.38) and (18.39) give the factor of merit to be 0.659, a difference of some 1.3% in this example.)

Example 2 Suppose the cooling tower of the previous example were to be converted into a vertical spray cooler by supplying it with chilled water. Calculate the operating characteristics of the cooler given

water inlet temperature, $t_{w,in}$	5 °C
air inlet wet bulb temperature, $t_{a,in}$	18 °C
water flowrate, m_w	100 l/s or kg/s
air flowrate, (at an air density of 1.1 kg/m^3),	120 m^3/s
	or $m_a = 120 \times 1.1 = 132$ kg/s
barometric pressure, P	95 kPa

Solution The wet bulb temperature of the inlet air is the same as in the previous example. Hence, the corresponding sigma heat remains at

$$S_{in} = 52.05 \text{ kJ/kg}$$

However, at the new inlet water temperature of 5 °C, the psychrometric equations of section 14.6 give $S_{w,in} = 19.38$ kJ/kg.

Tower capacity factor. From equation (18.32),

$$R = \frac{m_w}{m_a} C_w \frac{t_{w,in} - t_{a,in}}{S_{w,in} - S_{in}}$$

$$= \frac{100}{132} \times 4187 \frac{5 - 18}{(19.38 - 52.05)1000} = 1.262$$

Tower effectiveness. As $R > 1$ then $R^* = 1/R$ (equation (18.36)):

$$R^* = \frac{1}{1.262} = 0.7923$$

Now the factor of merit for the tower, F, was calculated in the previous example as 0.668. This remains the same despite the new inlet conditions. Equation (18.37) gives

$$E = F^{R^*} = 0.668^{0.7923}$$

$$= 0.7264$$

Fluid efficiencies. AS $R > 1$, $E = \eta_a$ (equation (18.35)). Hence $\eta_a = 0.7264$ and from equation (18.34)

$$\eta_w = \frac{\eta_a}{R}$$

$$= \frac{0.7264}{1.262} = 0.5755$$

Now, equation (18.30) gives

$$\eta_w = \frac{t_{w,in} - t_{w,out}}{t_{w,in} - t_{a,in}}$$

or

$$t_{w,out} = t_{w,in} - \eta_w(t_{w,in} - t_{a,in})$$
$$= 5 - 0.5755(5 - 18)$$
$$= 12.48\ ^\circ C$$

and equation (18.31) gives

$$\eta_a = \frac{S_{out} - S_{in}}{S_{w,in} - S_{in}}$$

from which

$$S_{out} = \eta_a(S_{w,in} - S_{in}) + S_{in}$$
$$= 0.7264(19\,380 - 52\,050) + 52\,050\ \text{J/kg}$$
$$= 28\,319\ \text{J/kg or } 28.3\ \text{kJ/kg}$$

The air wet bulb temperature giving this sigma heat may be read from the 95 kPa psychrometric chart or calculated by iterating equations (14.44) to (14.47), giving $t_{a,out} = 9.18\ ^\circ C$.

Rate of heat transfer, q. Using the water circuit

$$q = m_w C_w(t_{w,out} - t_{w,in})$$
$$= 100 \times 4187(12.48 - 5)$$
$$= 3.132 \times 10^6\ \text{W or } 3.132\ \text{MW}$$

As a cross-check, we may use the air circuit to give

$$q = m_a(S_{out} - S_{in})$$
$$= 132(28\,319 - 52\,050)$$
$$= -3.132 \times 10^6\ \text{W or } -3.132\ \text{MW}$$

The heat gained by the water is shown to balance the heat lost by the air.

Example 3 A mesh spray cooler is to be supplied with water at a temperature of 10 °C. The unit is required to cool 12 kg/s of air from a wet bulb temperature of 29 °C to 23 °C and at a barometric pressure of 100 kPa. Determine the rate of water flow that must be sprayed through the cooler, the temperature of the return water and the cooling duty of the unit.

Solution In the absence of further data concerning the device, we take a mean value of 0.6 as the factor of merit for a mesh cooler (from Table 18.2).

The data may then be summarized as follows:

$$t_{a,in} = 29\,°C \qquad t_{a,out} = 23\,°C \qquad m_a = 12\,kg/s$$
$$t_{w,in} = 10\,°C \qquad t_{w,out} = \text{unknown} \qquad m_w = \text{unknown}$$
$$F = 0.6$$

The sigma heats can be calculated from the psychrometric equations given in section 14.6 or approximated from the 100 kPa psychrometric chart:

$$S_{in}(\text{at 29 °C wet bulb temperature}) = 92.28\,kJ/kg$$
$$S_{out}(\text{at 23 °C}) = 67.11\,kJ/kg$$
$$S_{w,in}(\text{at 10 °C}) = 29.21\,kJ/kg$$

Air efficiency. From equation (18.31),

$$\eta_a = \frac{S_{out} - S_{in}}{S_{w,in} - S_{in}} = \frac{67.11 - 92.28}{29.21 - 92.28} = 0.3991$$

Water efficiency. From equation (18.30),

$$\eta_w = \frac{t_{w,in} - t_{w,out}}{t_{w,in} - t_{a,in}} = \frac{10 - t_{w,out}}{10 - 29} = \frac{t_{w,out} - 10}{19} \qquad (18.40)$$

Capacity factor. From equation (18.34),

$$R = \frac{\eta_a}{\eta_w} = \frac{0.3991 \times 19}{t_{w,out} - 10} = \frac{7.583}{t_{w,out} - 10} \qquad (18.41)$$

We now have a decision to make—whether to assume that R is greater or less than unity, allowing us to ascribe values for E and R^*. Let us assume that $R > 1$ (we shall see in a moment that this was the wrong decision). Using equations (18.35) and (18.36), if $R > 1$, then $E = \eta_a$ and $R^* = 1/R$.

Equation (18.37) gives

$$E = F^{R^*}$$

or

$$0.3991 = 0.6^{(t_{w,out} - 10)/7.583}$$

giving

$$t_{w,out} - 10 = 7.583 \frac{\ln(0.3991)}{\ln(0.6)} = 13.64$$

or

$$t_{w,out} = 23.64\,°C$$

In order to check the consistency of the value of R, we can return to equation (18.41):

$$R = \frac{7.583}{23.64 - 10} = 0.56$$

showing that our assumption of $R > 1$ was incorrect.

Let us, therefore, repeat the analysis assuming that $R < 1$. Then equations (18.35) and (18.36) give $E = \eta_w$ and $R^* = R$.

Equation (18.37) gives

$$E = F^{R^*}$$

or, using $\eta_w = (t_{w,out} - 10)/19$ from equation (18.40) and $R = 7.583/(t_{w,out} - 10)$ from equation (18.41),

$$\frac{t_{w,out} - 10}{19} = 0.6^{7.583/(t_{w,out} - 10)}$$

To simplify the algebra, we may substitute

$$t_{w,out} - 10 = a$$

giving

$$a = 19 \times 0.6^{7.583/a}$$

This can easily be solved iteratively on a pocket calculator to give

$$a = 14.56$$

and

$$t_{w,out} = 24.56\,°C$$

Using equation (18.41), again, to check the value of R gives

$$R = 7.583/14.56 = 0.521$$

In this case, our assumption that $R < 1$ is shown to be consistent.

Water flowrate. From equation (18.32),

$$m_w = R \frac{m_a(S_{w,in} - S_{in})}{C_w(t_{w,in} - t_{a,in})}$$

$$= 0.521 \times \frac{12(29\,210 - 92\,280)}{4187(10 - 29)}$$

$$= 4.955\,\text{kg/s}$$

Cooling duty. This can be calculated directly from the air data:

$$q = m_a(S_{out} - S_{in})$$
$$= 12(67.11 - 92.28) = -302 \text{ kW}$$

To check the heat balance, the heat gain by the water is given as

$$q = m_w C_w (t_{w,out} - t_{w,in})$$
$$= 4.955 \times 4.187 \times (24.56 - 10) = 302 \text{ kW}$$

Heat exchange across the walls of pipes and ducts

When a fluid flowing through a pipe or duct is at a temperature different to that of the ambient air, then heat exchange will occur across the walls of the pipe and any insulating material that surrounds it.

The analysis of indirect heat exchangers given in earlier in this section applies equally well for pipes and ducts.

Consider the insulated pipe shown on Fig. 18.19 carrying a fluid of temperature t_{in} when the temperature of the outside air is t_{out}.

The rate of heat transfer from a short length of pipe over which neither t_{in} nor t_{out}

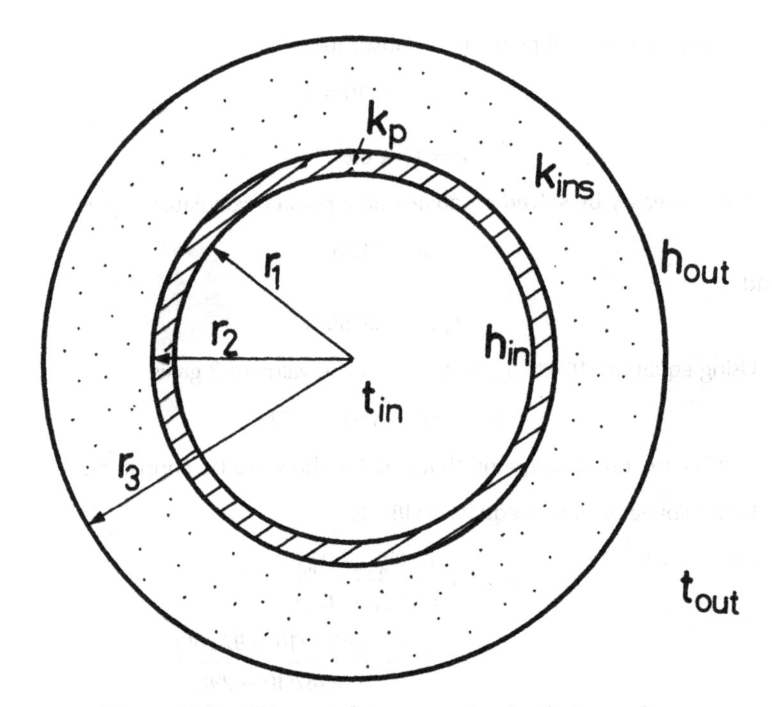

Figure 18.19 Heat transfer across an insulated pipe or duct.

changes significantly is given as (see equation (18.18)):

$$q = UA(t_{in} - t_{out}) \quad \text{W}$$

The UA value is given by

$$\frac{1}{UA} = \frac{1}{h_{in}A_{in}} + \frac{r_2 - r_1}{k_p A_{12}} + \frac{r_3 - r_1}{k_{ins}A_{23}} + \frac{1}{(h_{out} + 0.95\, h_r)A_{out}} \quad \text{°C/W} \qquad (18.42)$$

where h_{in} = heat transfer coefficient at the inner surface (W/(m^2 °C)), h_{out} = heat transfer coefficient at the outer surface (W/(m^2 °C)), h_r = radiative heat transfer coefficient at the outer surface (W/(m^2 °C)), r = radius (m), k_p = thermal conductivity of pipe (W/(m °C)), k_{ins} = thermal conductivity of insulation (W/(m °C)), A = area (or mean area) across which heat transfer occurs (m) and 0.95 = assumed value for the product of emissivity and view factor (Whillier, 1982). Values of thermal conductivity for pipe materials can be obtained from Table 18.3.

The heat transfer coefficients may be estimated from the following relationships which take into account the variations of fluid viscosity and thermal conductivity with respect to temperature.

Air inside a duct

The Colburn equation (18.22) gives the heat transfer coefficient for turbulent flow inside smooth tubes. For air, the Prandtl number, P_r, remains essentially constant at 0.7 over a wide range of temperatures and pressures (section 15A.3). Further analysis of the Colburn equation gives the approximation

$$h_{in}(\text{air}) = 3.169 \frac{(\rho_a u_a)^{0.8}}{d^{0.2}} (1 + 9.4 \times 10^{-4} t_{in} - 6.49 \times 10^{-6} t_{in}^2) \quad \text{W/(m}^2\,\text{°C)} \qquad (18.43)$$

where ρ_a = air density (kg/m^3), u_a = air velocity in duct (m/s) and d = internal diameter of duct (m). The effect of temperature is fairly small over the range of 0 to

Table 18.3 Thermal conductivities of pipe materials

Material	*Thermal conductivity (W/(m °C))*
Still air	0.028
Expanded polystyrene	0.038
Polyurethane and polystyrene	0.034
Polyisocyanurate	0.023
Phenol formaldehyde	0.032
Glass fibre	0.029
Unplasticized polyvinylchloride (UPVC)	0.14
Still water	0.62
Wood	0.17
Steel	45

60 °C. Hence, taking a value of 20 °C gives the simpler approximation

$$h_{in}(\text{air}) = 3.22 \frac{(\rho_a u_a)^{0.8}}{d^{0.2}} \quad \text{W/(m}^2\,^\circ\text{C)} \tag{18.44}$$

The incorporation of air density allows this relationship to be employed for compressed air pipes as well as ventilation ducts.

Water inside a pipe

A similar analysis for water inside a pipe leads to the approximation

$$h_{in}(\text{water}) = (1430 + 20.9\,t_{in}) \frac{u_w^{0.8}}{d^{0.2}} \quad \text{W/(m}^2\,^\circ\text{C)} \tag{18.45}$$

where u_w = water velocity (m/s).

In this case, the near-constant value of water density allows it to be eliminated as a variable. However, the viscosity of water varies significantly in the range 0 to 60 °C. Hence, the effect of temperature should not be ignored in equation (18.45).

Air outside a pipe

This time, the relationship given by McAdams for the outside of a single tube (equation (18.23)) provides the basis for further analysis. This leads to an approximation that may be used for both longitudinal flow and cross-flow in subsurface airways:

$$h_{out}(\text{air}) = 4.24 \frac{(\rho_a u_a)^{0.6}}{d^{0.4}} (1 + 0.0015\,t_{out}) \quad \text{W/(m}^2\,^\circ\text{C)} \tag{18.46}$$

Radiative heat transfer coefficient

This can be calculated from the Stefan–Boltzmann equation (15.27) using the dry bulb temperature of the surrounding air, giving

$$h_r = 22.68 \times 10^{-8} (273.15 + t_{out})^3 \quad \text{W/(m}^2\,^\circ\text{C)} \tag{18.47}$$

However, an adequate accuracy for most purposes is obtained simply by assuming a value of $h_r = 6$ W/(m^2 °C).

Over long lengths of pipe or ducting the inner and outer temperatures, t_{in} and t_{out}, are each likely to change in a non-linear manner. In this case the logarithmic mean temperature difference should be employed giving

$$q = UA \frac{\Delta t_1 - \Delta t_2}{\ln(\Delta t_1/\Delta t_2)} \quad \text{W} \tag{18.48}$$

where Δt_1 and Δt_2 are the values of the temperature difference $t_{in} - t_{out}$ at the two ends of the pipe and ln = natural logarithm.

Example A chilled water pipeline is 1000 m long and comprises a 200 mm internal diameter steel pipe of thickness 4 mm, inside a 30 mm sheath of phenol formaldehyde insulation. The water has a velocity of 2 m/s and a mean temperature of 8 °C while the surrounding air has a velocity of 4 m/s, a mean dry bulb temperature of 25.5 °C and a density of 1.12 kg/m^3.

Determine

1. the *UA* value per metre length of pipeline,
2. the actual heat gain and temperature rise of the water, and
3. the temperature of the outer surface of the insulation.

Solution

1. Let us first collate the data required by equation (18.42).

Thermal conductivities (from Table 18.3).

$$\text{steel} \qquad k_p = 45\ \text{W/(m °C)}$$
$$\text{phenol formaldehyde} \qquad k_{ins} = 0.032\ \text{W/(m °C)}$$

Radii (with reference to Fig. 18.19).

$$r_1 = 0.1\ \text{m}; \quad r_2 = 0.104\ \text{m}; \quad r_3 = 0.134\ \text{m}$$

Areas of heat transfer (per metre length).

$$A_{in} = 2\pi r_1 = 2\pi \times 0.1 = 0.6283\ \text{m}^2$$
$$A_{out} = 2\pi r_3 = 2\pi \times 0.134 = 0.8419\ \text{m}^2$$
$$A_{12} = 2\pi \frac{r_1 + r_2}{2} = 2\pi \frac{0.1 + 0.104}{2} = 0.6409\ \text{m}^2$$
$$A_{23} = 2\pi \frac{r_2 + r_3}{2} = 2\pi \frac{0.104 + 0.134}{2} = 0.7477\ \text{m}^2$$

Heat transfer coefficients. For water in the pipe (from equation (18.45)):

$$h_{in}(\text{water}) = (1430 + 20.9 t_{in}) \frac{u_w^{0.8}}{d_1^{0.2}}$$
$$= (1430 + 20.9 \times 8) \frac{2^{0.8}}{(0.2)^{0.2}}$$
$$= 3837\ \text{W/(m}^2\,°\text{C)}$$

For air outside the pipe (from equation (18.46)):

$$h_{out}(\text{air}) = 4.24 \frac{(\rho_a u_a)^{0.6}}{d_3^{0.4}} (1 + 0.0015 t_{out})$$
$$= 4.24 \frac{(1.12 \times 4)^{0.6}}{0.268^{0.4}} (1 + 0.0015 \times 25.5)$$
$$= 18.33\ \text{W/(m}^2\,°\text{C)}$$

For radiative heat transfer at the outside of the pipe (from equation (18.47))

$$h_r = 22.68 \times 10^{-8}(273.15 + 25.5)^3$$
$$= 6.04\ \mathrm{W/(m^2\,°C)}$$

The large value of h_{in}(water) shows the relative ease with which heat is transferred from the water to the steel pipe.

The UA value per metre length can now be calculated from equation (18.42). Inserting the numerical value gives

$$\frac{1}{UA} = \underset{\text{inside pipe}}{\frac{1}{3837 \times 0.6283}} + \underset{\text{steel}}{\frac{0.004}{45 \times 0.6409}} + \underset{\text{insulation}}{\frac{0.03}{0.032 \times 0.7477}} + \underset{\text{outside pipe}}{\frac{1}{(18.33 + 0.95 \times 6.04)0.8419}}$$

$$= 0.000\,41 \quad + 0.000\,14 \quad + 1.2538 \quad + 0.0493$$

$$= 1.3037\ \mathrm{°C/W}$$

or

$$UA = 0.7670\ \mathrm{W/°C}$$

The dominant effect of the insulation is shown here. Indeed, the terms involving the water inside the pipe and the steel pipe itself may be neglected without significant loss of accuracy.

2. The average heat loss per metre length of pipe is then simply

$$q = UA(t_{in} - t_{out})$$
$$= 0.7670(8 - 25.2)$$
$$= -13.42\ \mathrm{W/m}$$

(negative as heat is transferred from the air to the water). Over the 1000 m pipeline, the total heat transferred to the water is then

$$q_{tot} = 13.42 \times 1000 = 13\,420\ \mathrm{W}$$

In order to determine the actual increase in temperature of the water, Δt_w, it is first necessary to calculate the mass flow of water:

$$m_w = \rho_w \times u_w \times \text{area of flow}$$
$$= 1000 \times 2 \times \frac{\pi \times 0.2^2}{4}$$
$$= 62.83\ \mathrm{kg/s}$$

where ρ_w = density of water (1000 kg/m^3). Then

$$\Delta t_w = \frac{q_{tot}}{C_w \times m_w}$$

where C_w = specific heat of water (4187 J/(kg °C)), giving

$$\Delta t_w = \frac{13\,420}{4187 \times 62.83} = 0.051\,°\text{C}$$

This small rise in temperature justifies the assumption of a 'constant' water temperature at a mean value. In practice, the actual temperature rise may be significantly greater because of damaged or imperfectly applied insulation, particularly at flange points.

3. The heat flux remains the same at each boundary of the pipe and insulation. This allows the temperatures at those boundaries to be determined. In particular, at the outer surface of the insulation,

$$q = h_{out} A_{out} (t_{surf} - t_{out})$$

where t_{surf} = surface temperature of the insulation. Inserting the known values gives

$$t_{surf} = \frac{-13.42}{18.33 \times 0.8419} + 25.5$$
$$= -0.87 + 25.5$$
$$= 24.63\,°\text{C}$$

No condensation will occur on the outside of the insulation provided that the wet bulb temperature of the air remains below 24.63 °C. However, any imperfections in the insulation may allow the ingress of water vapour and condensation on the steel tube. This can result in problems of corrosion.

The techniques described in this section can be incorporated into mine climate simulation programs (Chapter 16) in order to take into account heat transfers between pipes or ducts and the surrounding airflow. If the fluid inside the pipe moves in the same direction as the ventilating airstream, then the heat transfers and corresponding temperature changes can be computed sequentially along each incremental length of simulated airway (section 16.2.2). In the case of the piped fluid and the ventilating airstream moving in opposite directions, then an iterative procedure becomes necessary to compute the equilibrium condition.

18.3.3 Water distribution systems

A significant item in the design and economic analysis of mine air conditioning systems is the choice of piping that carries water between the refrigeration plant and heat exchangers or mining equipment. There are, essentially, two matters of importance. First, the size of the pipe should be selected carefully. If it is too small for the given flowrate, then the frictional pressure drops and pumping costs will be high. On the other hand, if too large a pipe is chosen, then the costs of purchase and insulation will be high while the larger surface area and low velocity will exacerbate unwanted heat transfer with the surroundings.

This brings us to the second consideration, the thermal properties of the pipe material and the need for additional insulation. We shall discuss each of these matters in turn.

Pipe sizing

The first step in determining the diameters of the various pipes that constitute a water network is to assess the flowrates that are to be carried in each branch. The cooling duty, q, of a heat exchanger is given as

$$q = m_w C_w (t_{w,out} - t_{w,in}) \quad \text{W} \tag{18.49}$$

where m_w = water flowrate through the heat exchanger (l/s or kg/s), C_w = specific heat of liquid water (4187 J/(kg °C)), $t_{w,in}$ = inlet water temperature (°C) and $t_{w,out}$ = outlet water temperature (°C).

The heat transferred is dependent equally on the water flowrate and the temperature rise of the water through the unit. The latter, in turn, is governed by the construction of the heat exchanger and the supply temperature of the water. Calculations of the type included in section 18.3.2 allow flowrates to be assessed. For service water lines, the flowrates must also satisfy the machine requirements for dust suppression or hydraulic power.

Having established the water flowrate, the actual sizing of a pipe becomes a combination of simple analysis and practical considerations. Taking into account capital, operating and installation costs, van Vuuren (1975) suggested optimum water velocities of 1.8 to 2.0 m/s for 4000 kPa pipes and 2.0 to 2.2 m/s for 7000 kPa pipes. A common rule of thumb is to employ 2 m/s for an initial exercise in pipe sizing. The size of pipe is then given as

$$\frac{\pi d^2}{4} = A = \frac{Q}{u} \quad \text{m}^2$$

or

$$d = \sqrt{\frac{4Q}{\pi u}} \quad \text{m} \tag{18.50}$$

$$= 1.13 \sqrt{\frac{Q}{u}} \quad \text{m}$$

where d = pipe diameter (m), A = pipe cross-sectional area (m^2), Q = flowrate (m^3/s) and u = velocity (m/s).

The frictional pressure drop, p, along the pipe may then be determined from the Chézy–Darcy relationship (equation (2.45)):

$$p = \frac{4fL\rho_w u^2}{2d} \quad \text{Pa} \tag{18.51}$$

For practical calculations, water density, ρ_w, may be taken as constant at 1000 kg/m^3. The length L is often set at unity to give the frictional pressure loss in pascals per metre.

New steel pipes will have a coefficient of friction, f, approximating 0.0039. However, the ageing processes of corrosion and scaling may cause this to increase by 20% to 50%. Plastic pipes have a coefficient of friction averaging about 0.0032. Manufacturers usually provide charts relating frictional pressure losses to pipe diameter and flowrate for their products. The pressure loss that can be tolerated depends on the pumping cost function which is proportional to the product pQ (equation (9.17)), and the pressure required at the end of the pipeline. If the pumping costs are shown to be high, then consideration should be given to reducing the velocity and increasing the pipe diameter.

Pressure losses are also incurred at bends or pipe fittings. Again, manufacturer-specific data can be employed to determine such losses. However, for mine water systems, it is common simply to cater for bends and fittings by adding 10% to the computed pressure drop.

Pipe diameters determined through this procedure should be rounded up to the next highest standard size. Consideration should be given to increases in water flow that may be required in the future. Furthermore, any pipe ranges that might be used for fire fighting should be sized to give adequate flow and pressure at the hydrants. The sizes of pipes that are to be installed within existing airways or shafts may also be constrained by the space available.

Pipe insulation

The performance of heat exchangers is dependent on the temperature of the supply water. It is, therefore, important that heat gain along chilled water pipelines is minimized. This is normally accomplished by providing thermal insulation on those pipes. One exception is the special case where the pipeline is located within an intake airway which, itself, requires air cooling. In such circumstances, the insulation may be partially, or totally, omitted.

In addition to having a low thermal conductivity, the insulating materials for chilled water pipes in mines should conform with several other criteria.

1. The insulation should provide a barrier against the ingress of water that condenses on the outer surfaces. A porous and permeable insulation that becomes waterlogged will be much more conductive to heat. Furthermore, water that reaches the surface of metal pipes will cause problems of corrosion. The required vapour barrier may be achieved either by covering the insulation with an impermeable sheath or by using an insulating material that is, itself, non-permeable. An example is cellular glass in which the small air pockets are isolated from each other.
2. The insulation or outside sheath should have sufficient mechanical strength to withstand the rigours of transport and utilization in mine environments. Insulation on shaft pipelines must be able to withstand impact from falling objects.
3. The flammability of insulating materials and the toxicity of gases that they may

produce when heated in a mine fire is of great importance. Polyurethane foams have been employed widely in the past. However, they emit carbon monoxide and, perhaps, cyanide vapours at temperatures that may be reached in mine fires. Phenolic foams in which phenols are bound into a resin base are safer in that they have a higher ignition temperature, and a much lower rate of flame spread (Rose, 1989).

4. The cost of purchase and ease of installation of insulating materials or pre-insulated piping should also be taken into account.

A common practice has been to apply insulating materials to pipes on the mine site—either on the surface or after the pipework has been installed in its underground location. The application of foam insulators may be carried out by injecting the foam within a PVC vapour barrier sheath that has been pulled over spacers on the outside of the pipe. Pre-cast semicircular sections of cellular glass or other rigid insulators may be bound on to the pipes with adhesives or an outer sealing material. Expanded polystyrene granules and fibreglass have also been used widely (Arkle *et al.*, 1985; Ramsden, 1985). The thickness of applied insulators varies from 25 to 40 mm.

The continued development of thermoplastics has allowed the manufacture of pipes that combine mechanical strength with good thermal insulating properties. These may consist of concentric and impermeable cylinders with the annulus containing a cellular infill. Typical values of thermal conductivity are given in Table 18.3.

Energy and temperature changes within water systems

In section 3.4.1 we saw that fluid frictional effects have no influence on temperature changes within a steady-state airflow. This is not the case for water or other near-incompressible fluids. Furthermore, in section 15.3.1, the autocompression of air in a mine shaft was shown to produce an increase in air temperature that depended on the depth of the shaft. Again, this is not true for water in a shaft pipe.

In order to reassess the situation for water, let us return to the steady flow energy equation (3.25):

$$\frac{u_1^2 - u_2^2}{2} + (Z_1 - Z_2)g + W_{12} = \int_1^2 V\,dP + F_{12} = H_2 - H_1 - q_{12} \quad \frac{\text{J}}{\text{kg}} \qquad (18.52)$$

where u = velocity (m/s), Z = height above datum (m), g = gravitational acceleration (m/s^2), W = work input (J/kg), V = specific volume of fluid (m^3/kg), $= 1/\rho$ where ρ = density (kg/m^3), F = work done against friction (J/kg), H = enthalpy (J/kg) and q = heat added from surrounding (J/kg).

In applying this equation to water, we recall the basic definition of enthalpy from equation (3.19):

$$H = PV + U \quad \text{J/kg} \qquad (18.53)$$

where U = internal energy (J/kg) and, by assuming the water to be incompressible, V = constant.

The right-hand sides of equation (18.52) then give

$$V(P_2 - P_1) + F_{12} = P_2V - P_1V + U_2 - U_1 - q_{12}$$

or

$$F_{12} = U_2 - U_1 - q_{12} \quad \text{J/kg} \tag{18.54}$$

However, we may take a single value for the specific heat of water, $C_w = 4187$ J/(kg °C). Equation (3.28) then gives

$$U_2 - U_1 = C_w(T_2 - T_1) \quad \text{J/kg} \tag{18.55}$$

Substituting in equation (18.54),

$$F_{12} = C_w(T_2 - T_1) - q_{12} \quad \text{J/kg} \tag{18.56}$$

Furthermore, the frictional pressure drop, p, is given by equation (2.46):

$$p = \rho_w F_{12} = \rho_w C_w(T_2 - T_1) - \rho_w q_{12} \quad \text{Pa} \tag{18.57}$$

where ρ_w = density of water (kg/m³).

In the particular case of adiabatic flow ($q_{12} = 0$), the increase in water temperature along a pipeline becomes

$$T_2 - T_1 = \frac{p}{\rho_w C_w} \quad °\text{C} \tag{18.58}$$

The term $1/\rho_w C_w$ is sometimes known as the Joule–Thompson coefficient and, for water, has a standard value of $1/(1000 \times 4187) = 2.39 \times 10^{-7}$ °C/Pa.

The adiabatic rise in temperature along a water pipe is shown by equation (18.58) to be directly proportional to the frictional pressure drop and is independent of changes in elevation. The frictional pressure drop, p, may occur by viscous or turbulent action within the length of the pipe or by shock loss across a valve or other obstruction.

For flow through a water pump or turbine, we can apply the left- and right-hand sides of equation (18.52). However, the change in kinetic energy and the difference in elevation across the device are both negligible, giving

$$\begin{aligned} W_{12} &= H_2 - H_1 - q_{12} \\ &= V(P_2 - P_1) + U_2 - U_1 - q_{12} \\ &= \frac{P_2 - P_1}{\rho_w} + C_w(T_2 - T_1) - q_{12} \quad \frac{\text{J}}{\text{kg}} \end{aligned} \tag{18.59}$$

If there is negligible heat exchange across the casing of the pump or turbine, then we may assume adiabatic flow ($q_{12} = 0$) giving the temperature rise as

$$T_2 - T_1 = \frac{1}{C_w}\left(W_{12} - \frac{P_2 - P_1}{\rho_w}\right) \quad °\text{C} \tag{18.60}$$

While W_{12} is the total work applied to a pump impeller, the term $(P_2 - P_1)/\rho_w$ represents the output of **useful mechanical energy**. The difference between the two indicates the degradation of some of the input power to thermal energy and produces a measurable temperature rise. In a turbine, it is the other way round with $(P_1 - P_2)/\rho_w$ representing the work input and $-W_{12}$ being the output work.

For a pump, we can also write that the **actual work output** is given by

$$W_{12}\eta = \frac{P_2 - P_1}{\rho_w} \quad \frac{\text{J}}{\text{kg}} \tag{18.61}$$

and, for a turbine, the actual work output is

$$W_{12} = \frac{(P_2 - P_1)\eta}{\rho_w} \quad \frac{\text{J}}{\text{kg}} \tag{18.62}$$

where η = efficiency of the device.

Substituting for W_{12} in equation (18.60) gives the temperature rise across a pump

$$T_2 - T_1 = \frac{P_2 - P_1}{C_w \rho_w}\left(\frac{1}{\eta} - 1\right) \quad {}^\circ\text{C} \tag{18.63}$$

and across a water turbine

$$T_2 - T_1 = \frac{P_2 - P_1}{C_w \rho_w}(\eta - 1)$$

$$= \frac{P_1 - P_2}{C_w \rho_w}(1 - \eta) \quad {}^\circ\text{C} \tag{18.64}$$

In these analyses, it has been assumed that water is incompressible. In fact, the slight compressibility of water does give it a small isentropic temperature change with respect to pressure (Whillier, 1967). This is given approximately as

$$\frac{\Delta t}{\Delta P} = (0.759t - 0.2) \times 10^{-6} \quad {}^\circ\text{C/kPa} \tag{18.65}$$

where t = initial temperature (°C). This effect may be neglected for most practical purposes. However, if equation (18.63) or (18.64) is transposed in order to determine the efficiency of the device from temperature observations, then the measured temperature rise should be corrected according to equation (18.65) (negatively for a pump and positively for a turbine).

Example 1 The barometric pressures at the top and bottom of a 1500 m deep shaft are 90 and 107 kPa respectively. Chilled water at 3 °C and ambient atmospheric pressure enters the top of a 150 mm diameter steel pipe that is suspended in the shaft. A pressure-reducing valve at the bottom of the pipe controls the water flowrate to 35 l/s and reduces the water pressure to that of the ambient surroundings. Determine the pressures and temperatures of the water at the inlet and outlet of the pressure reducing valve. Assume that no heat exchange occurs across the pipe wall.

Solution The frictional pressure drop in the pipe is given by equation (18.51):

$$p = \frac{4fL\rho_w u^2}{2d} \quad \text{Pa}$$

The known values are $L = 1500$ m, $\rho_w = 1000$ kg/m^3, $d = 0.15$ m and $u = Q/A$ where Q = flowrate = 35/1000 = 0.035 m^3/s and $A = \pi \times (0.15)^2/4 = 0.017\,67$ m^2, giving $u = 1.981$ m/s.

Allowing for ageing effects, we shall assume the coefficient of friction for the steel pipe to be $f = 0.005$. We shall also allow an additional pressure loss of 10% to allow for bends and fittings. Then

$$p = \frac{4 \times 0.005 \times 1500 \times 1000 \times (1.981)^2}{2 \times 0.15}(1 + 0.1)$$

$$= 431.68 \times 10^3 \text{ Pa}$$

The water pressure at the shaft top is

$$90 \times 10^3 \quad \text{Pa}$$

The water pressure at shaft bottom (inlet to pressure reducing valve) is

$$\begin{aligned} &90 + \rho_w g L - p \\ &= 90 \times 10^3 + 1000 \times 9.81 \times 1500 - 431.68 \times 10^3 \\ &= 14\,373 \times 10^3 \text{ Pa or } 14\,373 \text{ kPa} \end{aligned}$$

At outlet from the valve, the water pressure is reduced to the shaft bottom barometric pressure of 107 kPa. Hence,

$$\begin{aligned} \text{pressure drop across the valve} &= \text{gauge pressure at valve inlet} \\ &= 14\,373 - 107 = 14\,266 \text{ kPa} \end{aligned}$$

The temperature rise in the shaft pipe is given by equation (18.58) as

$$\Delta t_{sh} = \frac{p}{\rho_w C_w} = \frac{431.68 \times 10^3}{1000 \times 4187} = 0.103\ ^\circ\text{C}$$

The temperature at the valve inlet is

$$\begin{aligned} &\text{water temperature at shaft top} + 0.103 \\ &= 3.103\ ^\circ\text{C} \end{aligned}$$

As no energy is added or extracted by the pressure-reducing valve, the flow is purely frictional. Hence,

$$\Delta t_{valve} = \frac{\text{pressure loss across valve}}{\rho_w C_w}$$

$$= \frac{14\,266 \times 10^3}{1000 \times 4187} = 3.407\,°\text{C}$$

giving the water temperature at the valve outlet to be

$$3.103 + 3.407 = 6.510\,°\text{C}$$

(The same total increase in water temperature of 3.51 °C from shaft top to outlet of the valve at the shaft bottom would occur whatever the cause of the restriction on flow—a valve, rougher lining of the pipe or a smaller diameter of pipe—provided that no energy of any kind is added or extracted from the water and that the water is emitted at the shaft bottom barometric pressure.)

As a check, the overall temperature rise of the water between the shaft top and the outlet of the shaft bottom valve can be calculated directly. From the steady flow energy equation (18.52) with no work input ($W_{12} = 0$), for an incompressible fluid $V = 1/\rho$ = constant, and ignoring changes in kinetic energy

$$(Z_1 - Z_2)g = V(P_2 - P_1) + F_{12}$$

or

$$\begin{aligned} p_{tot} = \rho_w F_{12} &= (Z_1 - Z_2)g\rho_w - (P_2 - P_1) \\ &= 1500 \times 9.81 \times 1000 - (107 - 90)1000 \\ &= 14.715 \times 10^6 - 0.017 \times 10^6 \\ &= 14.698 \times 10^6\ \text{Pa} \end{aligned}$$

(The difference in barometric pressure is shown to have very little effect on p_{tot}.)
Then (from equation (18.58))

$$\Delta t_{tot} = \frac{p_{tot}}{\rho_w C_w}$$

$$= \frac{14.698 \times 10^6}{1000 \times 4187} = 3.51\,°\text{C}$$

This check calculation illustrates that when the water is emitted from the pipe at the underground barometric pressure, the temperature rise is given approximately as

$$\frac{(Z_1 - Z_2)g}{4187} \text{ or } 2.34\,°\text{C per } 1000\,\text{m depth}$$

In the past, this has mistakenly been referred to as an 'autocompression temperature rise'. The misconception arose because, in the special case of a pipe open at the bottom, the frictional pressure drop in the pipe approximates the static pressure caused by a column of fluid in the shaft pipe.

Example 2 If, in example 1, the pressure-reducing valve at the shaft bottom is replaced by a water turbine of efficiency 75%, what would be the temperature of the water leaving the turbine? Calculate, also, the power produced by the turbine. Assume that all other parameters, including the flowrate of 35 l/s, remain unchanged.

Solution As the condition and flow of water in the pipe have not been altered, the increase in water temperature down the pipe remains at $\Delta t_{sh} = 0.103\,°C$ and the water enters the turbine at a pressure of $P_{in} = 14\,373$ kPa and a temperature of 3.103 °C.

At outlet from the turbine, the water pressure has fallen to the shaft bottom barometric pressure of $P_{out} = 107$ kPa. The temperature rise across the turbine is given by equation (18.64):

$$\Delta t_{turb} = \frac{P_{in} - P_{out}}{C_w \rho_w}(1 - \eta)$$

$$= \frac{(14\,373 - 107)10^3}{4187 \times 1000}(1 - 0.75)$$

$$= 0.852\,°C$$

Then the temperature of the water leaving the turbine is

$$3.103 + 0.852 = 3.955\,°C$$

Comparing this with the temperature of water leaving the pressure reducing valve of example 1, i.e. 6.510 °C, shows that replacing the valve by a water turbine not only produces useful power but reduces the temperature of the chilled water leaving the shaft station by 6.510 − 3.955 = 2.555 °C.

This represents an equivalent refrigeration capacity of

$$m_w C_w \Delta t = 35 \times 4187 \times 2.555$$

$$= 374.5 \times 10^3\ \text{W or } 374.5\ \text{kW}$$

The output work from the turbine is given by equation (18.62):

$$W_{12} = \frac{P_{out} - P_{in}}{\rho_w}\eta$$

$$= \frac{(107 - 14\,373)10^3}{1000} \times 0.75$$

$$= -10\,700\ \text{J/kg}$$

(negative, as work is leaving the water). To transform this to an output power, we simply multiply by the mass flowrate, i.e.

$$10\,700 \times 35 = 374.5 \times 10^3\ \text{W or } 374.5\ \text{kW}$$

We have, therefore, demonstrated that the power output from the turbine is equal to the equivalent refrigeration capacity that has been added to the system. This suggests a rapid means of approximating the overall temperature rise of the water from the shaft top to the turbine outlet at shaft bottom barometric pressure:

$$\Delta t_{tot} = \frac{(Z_1 - Z_2)g}{C_w} - \frac{\text{turbine output power}}{m_w C_w}$$

$$= \frac{1500 \times 9.81}{4187} - \frac{374\,500}{35 \times 4187}$$

$$= 3.514 - 2.555$$

$$= 0.959\ ^\circ\text{C}$$

This method remains acceptable provided that the frictional pressure losses are small compared with the static pressure caused by the column of water.

18.3.4 Energy recovery devices

The trend towards siting main refrigeration plant on the surface of mines accelerated during the 1980s. This was caused in large part by the utilization of energy recovery devices, particularly water turbines. Prior to that time, the provision of chilled water at surface had three major drawbacks. First, it caused high water pressures at shaft bottoms which necessitated a choice between high pressure heat exchangers and associated pipework or, alternatively, to pass the water through pressure reducing valves. The latter choice not only resulted in high operating costs of pumping to return the water to surface, but also converted the potential energy of the water to thermal energy, giving a rise in temperature and eroding the cooling capacity of the chilled water.

Energy recovery devices address each of these problems and divide into two major groups. First, water turbines or hydraulic motors employ the potential energy of shaft water lines to drive electrical generators or other devices. Secondly, hydrolift systems utilize the weight of descending chilled water to help raise heated return water back to surface. We shall discuss each of these systems in turn.

Water turbines

As illustrated by example 2 in the previous section, shaft bottom turbines provide the following:

1. work output that may be used to drive pumps, generators or any other mechanical device; this may provide part of the energy required to raise return water back to surface;
2. reduced temperature of the subsurface supply of chilled water, and hence, savings in refrigeration costs;
3. a low pressure water system underground.

Two types of turbines are in use. Figure 18.20 illustrates an impulse turbine, commonly known as a Pelton wheel. In this device, the pressure of the supply water

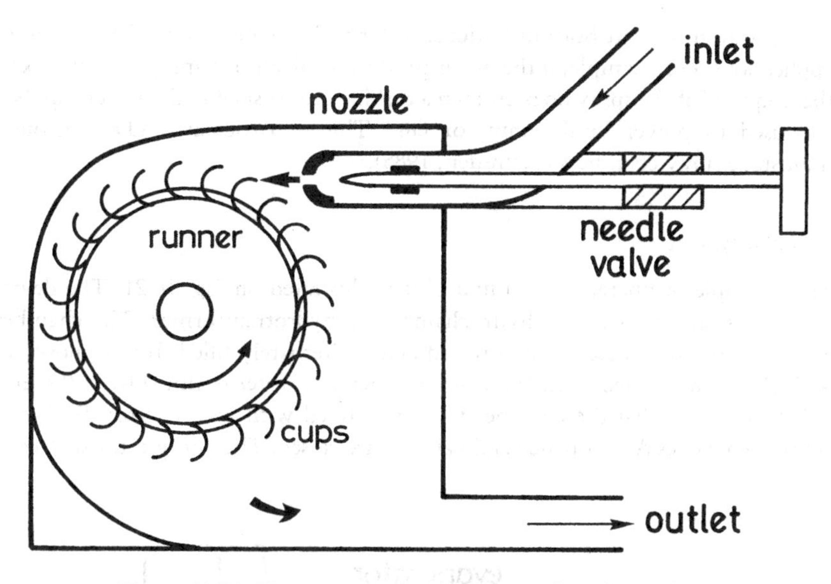

Figure 18.20 Diagrammatic representation of a Pelton wheel.

is used to produce kinetic energy by means of one or more nozzles. The high velocity jets are directed at cups attached to the periphery of a rotating wheel. Inside the casing of the device, the jets operate at the ambient atmospheric pressure. Hence, the water leaves at zero gauge pressure. Control of the mechanical output may be achieved by a needle valve in the nozzle which changes the water flowrate. Alternatively, if the flow is to be maintained, deflector plates may be used to direct part of the jets away from the cups.

The advantages of a Pelton wheel are that it is simple and rugged in construction, reliable and may reach efficiencies of over 80%. The speed control is effective and no high pressure seals are required on the casing.

The second type of water turbine is the reaction device and is often referred to as a Francis or Kaplan turbine. This is essentially a centrifugal runner (Fig. 10.2) with the impeller designed to operate in reverse and incorporating guide vanes to minimize shock losses. Indeed, a reversed centrifugal pump may be used as a reaction turbine although at a considerably reduced efficiency. Control of the device is by valves located upstream and/or downstream. Unlike the Pelton wheel, a reaction turbine operates in a 'flooded' mode giving rise to its primary advantage, i.e. the water pressure at exit can be maintained above the ambient air pressure and, hence, may be utilized further downstream to overcome pipe losses, to provide the required pressure at spray chambers or to activate further hydraulic devices.

Pelton wheels give better efficiencies at the higher duties, typically 2 MW. The Francis turbines are better suited for lower duties. As a guide, if the pressure differential available across the turbine (in kPa) is less than $70m_w^{0.67}$ (where m_w is the flowrate in l/s), then a Francis turbine will give a better performance.

In addition to shaft bottom stations, water turbines may be used for more local applications. For example, if the water pressure available at a spray chamber exceeds that required at the spray nozzles, then a turbine can be sited in the water supply line and used to power local pumps or fans. This is particularly advantageous for multistage spray chambers (Ramsden, 1985).

Hydrolift systems

The principle of operation of a hydrolift is illustrated on Fig. 18.21. The device is sometimes also known as a hydrochamber or hydrotransformer. The chamber is excavated or constructed underground and is alternately filled from opposite ends with chilled water delivered from surface, then hot water returned from the mine.

Let us assume that the chamber is initially filled with warm water. Both of the automated valves A are opened and valves B are closed. The pump at the surface cold

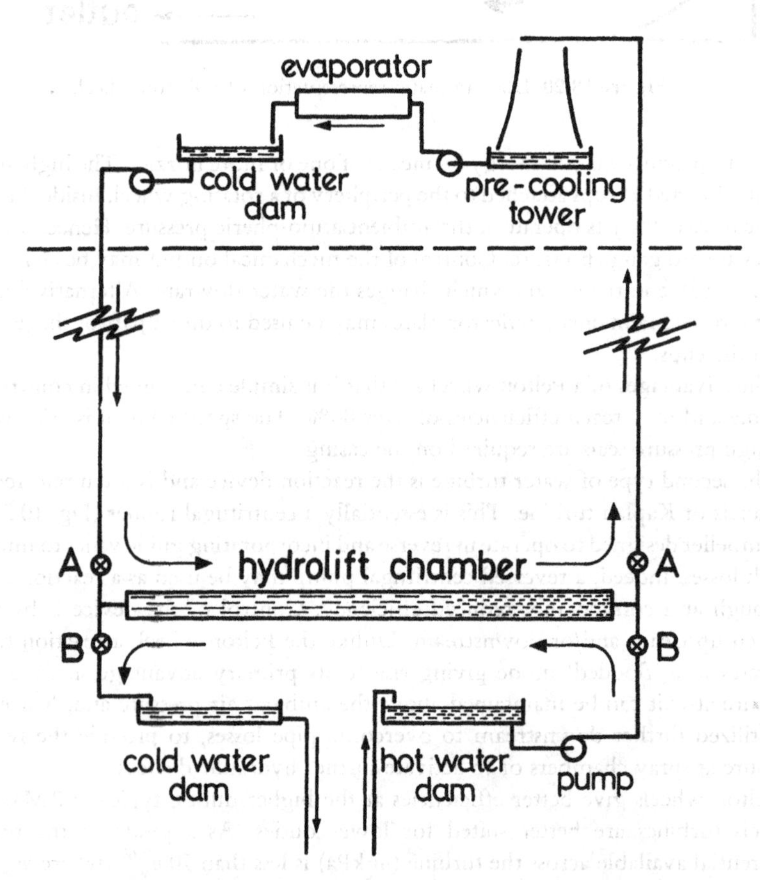

Figure 18.21 Operating principle of a hydrolift system.

water dam causes chilled water to flow down the shaft pipe and into the chamber. This, in turn, forces the hot water to exit the chamber and to ascend the return shaft pipe. As the two columns of water are nearly balanced, the duty of the pump is simply to overcome pipe friction. When the chamber is filled with chilled water valves A are closed and valves B are opened. The underground hot water dam pump then forces the return water into the chamber, expelling the chilled water to the underground cold water dam.

The cyclic nature of the system can be negated by having three hydrolift chambers with the sets of valves phased to give a continuous flow into and out of the water dams. The size of chambers and water dams should be determined on the basis of the cooling requirements of the mine. An inherent disadvantage of the hydrolift system is that the rate of chilled water flow must be equal to that of the heated return water. It is, therefore, advisable that the water dams should be sized to accommodate several hours' supply.

A loss of efficiency occurs because of heat exchange with the walls of the chamber and due to mixing and thermal transfer at the hot and cold interface within the chamber. The latter may be minimized by arranging for the hot water entry to be at a higher elevation.

18.3.5 Design of mine cooling systems

The demand on a subsurface cooling system is dictated by the magnitude and distribution of the heat load and the surface climate. The availability of cold natural water at the surface can result in large savings in cooling costs. Similarly, passing the water supply through a surface cooling tower either prior to or instead of a refrigeration plant can produce essentially free cooling.

Section 18.3.1 gave an overview of mine cooling systems and introduced spot coolers for isolated heat loads (Fig. 18.8), water chiller plant for sections of a mine (Fig 18.9) and a large centralized system (Fig. 18.10). In this section, we shall concentrate on the latter and discuss the alternative arrangements by which the hardware components may be integrated into a unified mine cooling system.

Location of main plant

Figures 18.22 to 18.24 give examples of system configurations using main refrigeration plant located on surface, underground and a combination of the two. In these figures, the ancillary components of pumps, valves and retention dams have been omitted for clarity.

Surface plant

The ease of maintenance, construction and heat rejection, coupled with the utilization of energy recovery devices have led to a preference for the main refrigeration units to be located on surface. Figure 18.22(a) illustrates the simplest system that employs surface plant. Chilled water from the evaporator passes down an insulated pipe in

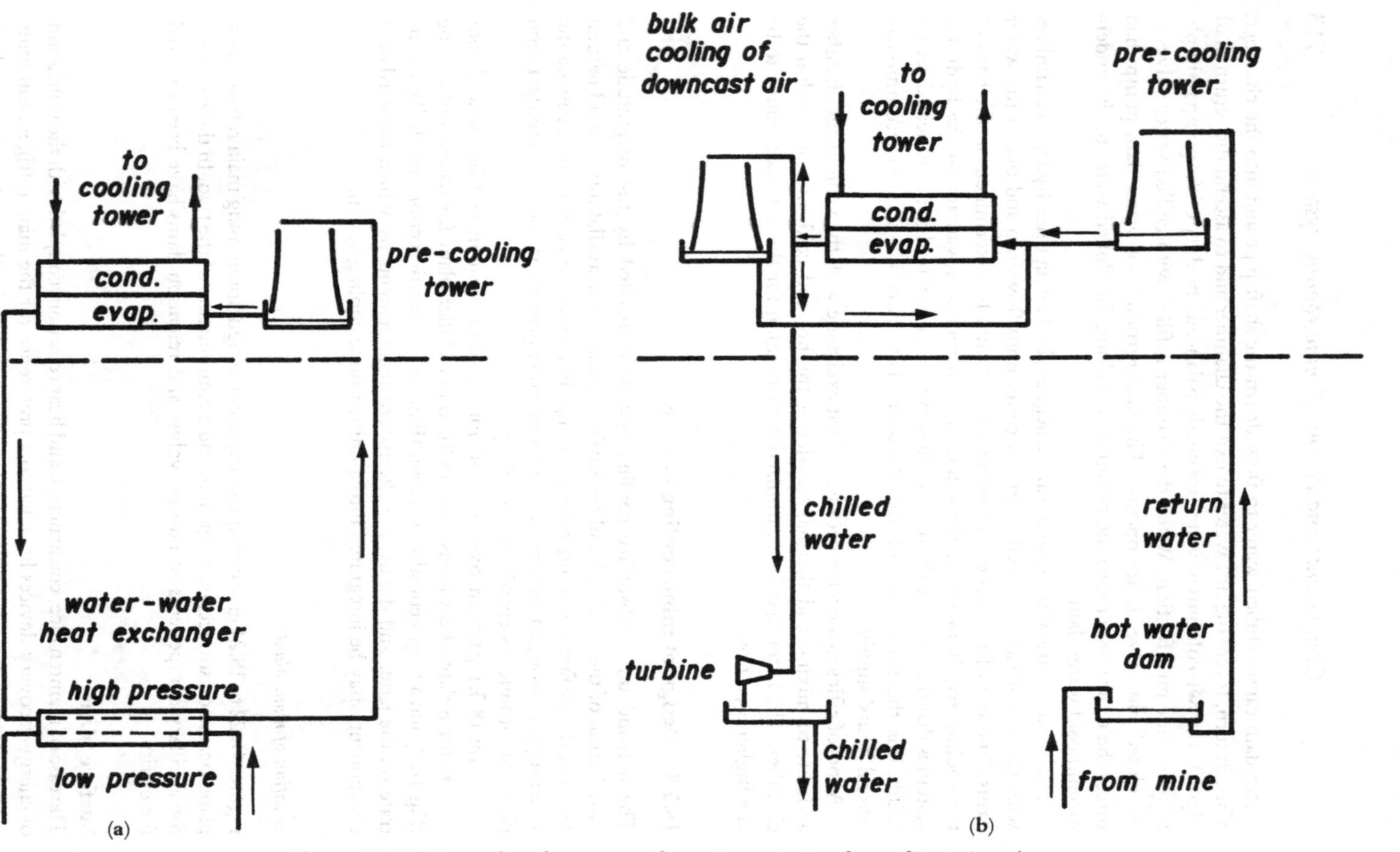

Figure 18.22 Examples of system configurations using surface refrigeration plant.

the shaft to a high pressure water-to-water heat exchanger at one or more shaft stations. This heat exchanger is a source of potential loss of system efficiency and care should be taken with the design and choice of tube materials to minimize problems of corrosion, scaling and erosion. Figure 18.22(b) shows the water-to-water heat exchanger replaced by a turbine. This configuration requires the water flow in the shafts to approximate that cycling through the mine, with capacitance provided by water dams.

In both cases, the heated water ascending to surface may be sprayed through a pre-cooling tower in order to lower its temperature before returning to the refrigeration plant. This can be advantageous in reducing the required capacity and operating costs of the refrigeration plant. However, the effectiveness of a pre-cooling tower is dependent on the difference between the temperature of the return water and the wet bulb temperature of the surface atmosphere. An 'approach' (see Fig. 18.18) of 2 °C is typical. For this reason, a pre-cooling tower may be effective only during the colder seasons.

Figure 18.22(b) also shows part of the chilled water being used for bulk cooling of the air before it enters the downcast shaft. The vertical packed spray chambers used for this purpose are normally of the induced-draught type with the fan located between the spray chamber and the shaft. This helps to prevent fogged air from entering the shaft. If the experience of any given mine is that a heat problem exists during the summer months only, then all that may be necessary is to install a surface bulk air cooling system and to utilize it during the warm season. However, if the workings are distant from the shaft bottoms and the heat production is concentrated heavily within those working zones, then bulk air cooling of surface will have greatly reduced effectiveness.

Underground plant

Figure 18.23 illustrates the principles of the system favoured by the South African gold mines prior to the 1980s. The refrigeration plant is located entirely underground. Hot water from the condensers is cycled through open spray cooling towers situated at or near the base of upcast shafts, and supplied by air returning from the mine. The advantages of this system are that it eliminates the need for surface-connecting pipe ranges and the associated pumping costs. It also avoids any environmental problems that may arise from surface plant. The major disadvantage of the system is that its duty is limited by the capacity of the return air to accept heat rejected in the underground cooling towers. Furthermore, saturated air subjected to decompression in the upcast shaft can cause problems of heavy fogging, water blanketing (section 9.3.6) and reduced life of main exhaust fans.

Combination of surface and underground plant

As the depth of mining increases, so also does the severity of geothermal and autocompressive heat problems. Unfortunately, increasing depth also exacerbates the costs of installing, maintaining and operating pipelines in shafts and, hence, eroding the advantages of a surface plant. Figure 18.24 illustrate alternative means of com-

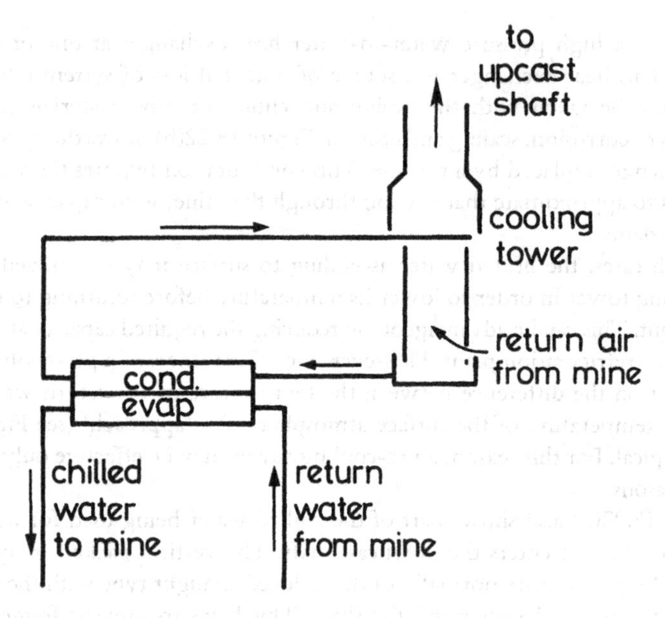

Figure 18.23 System configuration for centralized underground plant.

bining refrigeration plant on surface and underground into an integrated system. In each of these configurations, the essential role of the underground plant is to concentrate the heat in the return mine water into a lower flow, higher temperature system for transmission to surface. This enables smaller shaft pipes to be selected and also reduces the pumping costs. Furthermore, the increased temperature of the water returning to surface improves the effectiveness of the pre-cooling tower, allowing it to be utilized throughout the year. Another advantage of the combined system is that it adds considerably to the flexibility and future upgrading of the cooling duty. Surface plant alone is ultimately limited by the flow capacity of the shaft pipes. Similarly, underground plant alone is limited by the thermal capacity of the return air. However, with a combined system the cooling duty can be more easily upgraded by adding to the capacity of both surface and underground plant and, hence, increasing the temperature difference between the descending and ascending shaft pipes.

The simplest combined system is shown on Fig. 18.24(a) where the surface plant is used to cycle water through the shaft pipes and to remove heat from the subsurface condensers. Figure 18.24(b) shows a system of potentially improved efficiency in which the chilled water from surface first passes through water-to-water heat exchangers before entering the condensers. The warm return water from the mine is pre-cooled in the water-to-water heat exchangers prior to passing through the evaporators.

Figure 18.24(c) shows the water-to-water heat exchanger replaced by a turbine. In this case, the chilled water emerging from the turbine passes on to the mine via a

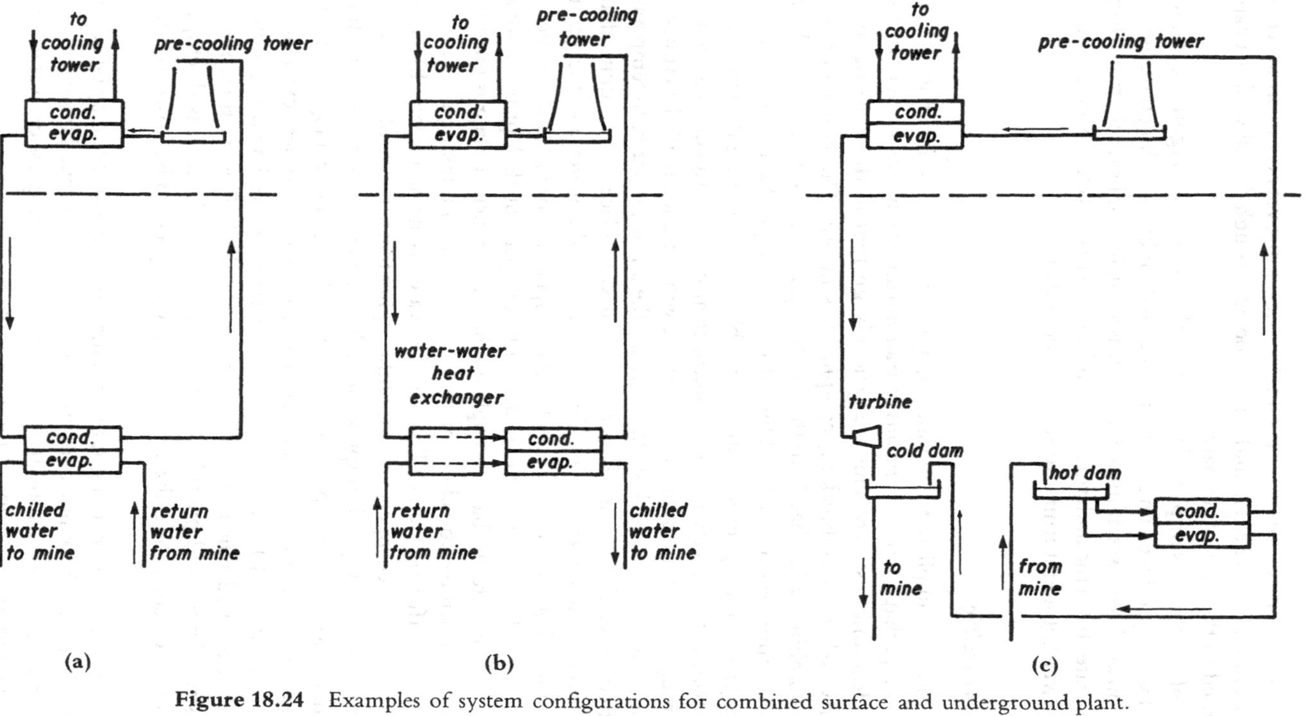

Figure 18.24 Examples of system configurations for combined surface and underground plant.

cold water dam. The return water is split. Part of it removes heat from the condenser and returns to surface. The remainder is re-cooled through the subsurface evaporator and returned to the cold water dam.

In each of the three examples of combined plant shown on Figures 18.24, some of the chilled water produced on surface may be used for bulk cooling the downcast air. Similarly, in all three cases, if mine return air is available for heat rejection, then some of the hot water from the underground condenser may be cycled through a subsurface cooling tower before returning to the hot water dam.

Service water cooling

Service water is supplied to mining equipment primarily for dust suppression purposes and perhaps, also, as a means of transmitting hydraulic power (see the next section). Heat and dust problems are generally greatest in the working areas and, particularly, where rock-breaking equipment is in use. The heat emitted by the machines combines with that from the newly exposed strata to cause rapid increases in air temperature (section 15.2.2). Uncooled water supplied at dry bulb temperature to dust suppression sprays exacerbates the problem by increasing the humidity of the air. However, if the service water is chilled, then it will both cool and dehumidify the air through which it is sprayed. This effect will continue until the effluent water running into drains or on rock surfaces attains the ambient wet bulb temperature.

The effect of chilled service water in a local workplace can be quite dramatic, depending on the temperature and flowrate of both the water and the ventilating airstream. Indeed, it is recommended that the temperature of the service water should not be less than 12 °C to prevent cold discomfort of personnel. A primary advantage of chilled service water is that it is applied where and when cooling is most needed, i.e. where and when rockbreaking or rock transportation is taking place.

The cooling efficiency of chilled service water is greater than that normally obtained from heat exchangers because of the higher temperature of the return water. For example, water supplied to sprays at 12 °C may eventually leave the district at the prevailing wet bulb temperature of, say, 27 °C, giving a range of 15 °C. For comparison, typical temperature ranges of water across local heat exchangers may be of the order of 6 to 10 °C. It follows that, for the same cooling duty, lower water flowrates can be used when the service water is chilled. This, in turn, reduces the required size of pipes and pumping costs. Furthermore, the increased temperatures of return water enhance the utilization of the lowcost pre-cooling towers on the mine surface.

The effectiveness of air cooling at the entrance to a stope or face may be diminished by the large heat flux from newly exposed or fragmented strata. However, such heat can be partially counteracted by the use of chilled dust suppression water, enabling intake air cooling to be employed more effectively.

The benefits to be gained from chilled service water clearly depend on how much water is required by the equipment, and any adverse effects on the strata or travelling conditions that may arise from too copious a supply of water. The practical approach is to determine the rate of water flow required for dust suppression or other machine purposes and the expected temperature range. The fraction of the total heat load that can be removed by service water can then be estimated.

Example Dust suppression water is supplied to a workplace at a rate of 100 l per tonne of rock mined. If the water is supplied to the machine at 12 °C and leaves the district at 26 °C, determine the cooling provided by a service water at a mining rate of 5 t/min.

Solution Water flowrate,

$$m_w = \frac{100 \times 5}{60} \text{ l or kg/s}$$
$$= 8.333 \text{ kg/s}$$

Rate of heat removal

$$m_w C_w \Delta t$$
$$= 8.333 \times 4187 \times (26 - 12)$$
$$= 488.5 \times 10^3 \text{ W or } 488.5 \text{ kW}$$

The amount of service water used may vary from 20 to 3000 l/t depending on the mining method and the mineralogical constituents of the dust particles. There is a corresponding large range in the rate of possible heat removal by dust suppression water. However, because of the high efficiency of the technique and the utilization of existing water lines, it may be regarded as the first approach to the supply of 'coolth' to a rockbreaking area.

The fraction of total heat load that can be removed by chilled service water is eroded at greater depths owing to the increased effects of autocompression and geothermal heat, while the supply of service water remains constant for any given mining method.

Hydropower

In section 18.3.4, we discussed the utilization of potential energy made available in shaft water pipes. In addition to driving shaft station turbines or hydrolift systems, the water pressure available at spray chambers can also be used to activate fans and pumps at those locations. The concept of **hydropower** takes this idea further and combines it with service water cooling (see previous section).

The use of electrical or diesel power involves the production of heat. Expenditure is then incurred to remove this heat from the mine by ventilation and/or cooling systems. The utilization of machine power, therefore, invokes both an environmental and an associated cost penalty. However, if the machine is activated by chilled water under pressure provided by the shaft column, then the hydraulic machine power and climatic control act in concert rather than in opposition. In a hot and deep mine, chilled hydropower provides, simultaneously, a means of machine power, cooling and dust suppression, all from a single system.

The practical development of hydropower systems is still in its infancy. For relatively low powered units such as drills, hydraulic motors may be employed directly while, for larger machines, turbine–electric combinations may be more practical. Furthermore, differing combinations of water flow, pressure and temperature may be required for the demands of power, cooling and dust suppression. A hydropower system must be designed and balanced to satisfy those varying demands.

Ice systems

The limitations of heat rejection capability underground, together with the use of energy recovery devices, have promoted the trend away from subsurface refrigeration plant towards surface units. However, despite water turbines and hydrolift devices, pumping costs remain the factor which limits the mass flow of water that can be circulated through the shaft pipelines of deep mines. Recalling the relationship $q = m_w C_w \Delta t$, heat removal from a mine can be enhanced by increasing the temperature differential, Δt, between supply and return water. This can be accomplished by combinations of surface and underground refrigeration plant (discussed in the section on location of main plant). Despite such advances in mine cooling technology, heat remains the primary limitation on the depth at which mining can take place.

In order to proceed beyond the physical restraints imposed by the circulation of fluids, other means of transmitting 'coolth' have to be considered. One of these is to employ the latent heat of melting ice. Although the use of ice in mines is not new (section 18.1), the first pilot plant employing an ice pipeline was constructed in 1982 at the East Rand Proprietary Mines in South Africa (Sheer *et al.*, 1984). Further full-scale applications followed soon afterwards (Hemp, 1988).

Ice has a specific heat varying from 2040 J/(kg °C) at 0 °C to 1950 J/(kg °C) at − 20 °C. It has a latent heat of melting of 333.5 kJ/kg. The mass flow of water circulated around a mine for cooling purposes can be reduced by a factor of over 5 if it is supplied in the form of ice.

Example A mine requires 10 MW of cooling. Calculate the mass flow of water involved

1. if the water is supplied at 3 °C and returns at 20 °C, and
2. if the water is supplied as ice at − 5 °C and returns at 20 °C.

Solution

1. *For the water system.*

$$q = m_w C_w \Delta t_w$$

or

$$m_w = \frac{q}{C_w \Delta t_w}$$

$$= \frac{10 \times 10^6}{4187 \times (20 - 3)} = 140.5 \text{ kg/s or l/s}$$

2. *For the ice supply.* Heat absorbed as the temperature of m_i kg of ice increases from -5 to 0 °C,

$$q_{ice} = m_i C_i \Delta t_i$$

Choosing the specific heat of ice as 2030 J/(kg °C) gives

$$q_{ice} = m_i \times 2030 \times 5 = 10\,150\, m_i \quad \text{W}$$

Heat absorbed as m_i kg of ice melts at 0 °C

$$q_L = L_{ice} m_i = 333\,500 m_i \quad \text{W}$$

Heat absorbed as m_i kg of liquid water increases in temperature from 0 to 20 °C:

$$q_w = m_i C_w \Delta t_w$$

$$= m_i \times 4187 \times 20 = 83\,740 m_i \quad \text{W}$$

Then, total heat absorbed

$$q = q_{ice} + q_L + q_w = 10 \times 10^6 \text{ W}$$

i.e.

$$m_i \ (10\,150 + 333\,500 + 83\,740) = 10 \times 10^6 \text{ W}$$

giving

$$m_i = 23.4 \text{ kg/s}$$

Hence, in this example, the volume of cooling water is reduced by a factor of 6 when it is supplied as ice at -5 °C. The calculation also illustrates the dominant amount of heat involved in the change of phase from ice to liquid, q_L.

There are four primary items to consider when investigating an ice system for mine cooling; the large-scale manufacture of ice, the means of transporting it underground, how it is best incorporated into a mine cooling system, and the economics of the system. Each of these factors will be considered in turn.

Manufacture of ice

The ice that is utilized in a mine cooling system may be supplied either as particulate ice at subzero temperature or as a slurry of ice crystals within liquid water or brine.

Particulate ice can be manufactured as cubes, cylinders or tubes of clear ice, or as ice flakes that can be compressed into pellets for transportation. Water is sprayed or caused to flow over the surfaces of plates, tubes, concentric cylinders or drums that are cooled on their opposite sides by a refrigerant fluid. The formation of ice creates an additional thermal resistance to heat transfer. This necessitates an evaporator temperature of some −15 to −30 °C giving rise to coefficients of performance in the range 2.3 to 3.5, lower than might be expected from a well-designed water chiller.

The ice is usually removed from the cold surfaces by mechanical scrapers or by periodic cycling of hot refrigerant through the freezer unit. In the production of flake ice, a thin film of ice is removed continuously from the surface of a rotating drum or metal belt. The particles of ice fall into a hopper that feeds into a screw or belt conveyor for removal from the unit.

As an example of large-scale particulate ice manufacture, the plant at East Rand Proprietary Mines (South Africa) produces 6000 t of ice per day from six 1000 t/day units. Each tubular unit consists of 80 doubled-walled tubes with ice formation on both the inner (99 mm diameter) and outer (508 mm) surfaces of the 4.5 m long tubes. The refrigerant passes through the annulus of each tube. The 15 min cycle consists of 13.5 min of freezing and 1.5 min harvesting (Hemp, 1988). The refrigerant is ammonia and ice harvesting is achieved by passing hot ammonia liquid through the annuli in the icemaker tubes. This arrangement also improves the efficiency of the unit by subcooling the liquid refrigerant before it returns to the evaporators.

Slurry icemakers promote the formation of microcrystals of ice within a stream of water or brine and are at an earlier stage of development for mine cooling systems. Slurry ice has one significant advantage over particulate ice. All industrial and domestic supplies of water contain many impurities including dissolved salts of calcium, magnesium and sodium. The freezing process favours pure water. Hence, if the freezing occurs sufficiently slowly or if the ice–water interface is continuously washed with water, then the ice crystals will have a greater purity than the water in which they grow. The dissolved salts remain in the liquid and, hence, become more concentrated. In contast, the formation of plate, tube or flake (particulate) ice occurs quickly and with a more limited washing action by the water, giving rise to the entrapment of salts within the ice matrix. If the microcrystals, usually less than 1 mm in size, are removed from the ice slurry and washed, then they will be relatively free from impurities. This process is known as **freeze desalinization**. The provision of purified water to a mine cooling system assists greatly in minimizing corrosion and scaling within pipes and heat exchangers.

There are at least three methods that have the potential for large-scale manufacture of slurry ice (Shone and Sheer, 1988). In the **indirect process**, water or brine is passed through tubes that are surrounded by cold refrigerant. Microcrystals of ice are formed within the moving stream of water. Careful control of the salinity, flowrate and temperature is required to prevent the formation of solid ice on the sides of the tube. Scraper devices can also be used to prevent ice from accumulating on the walls.

The **vacuum icemaking process** involves the evacuation of water vapour from a vessel that contains brine. The triple point of water is reached at 0.6 kPa and 0 °C when boiling and freezing occur simultaneously (Fig. 18.1). Ice, liquid water and vapour coexist at this point. The ice slurry that is formed is kept in motion by an agitator until it is pumped from the vessel. About 1 kg of water vapour must be removed for each 7.5 kg of nucleated ice. The vapour is compressed and condensed for recycling. In the vacuum icemaking process, the water acts as its own refrigerant fluid (section 18.2.2).

An even less developed method is the **direct process** in which a mixture of brine and immiscible liquid refrigerant is sprayed through a nozzle into a receiving chamber. The evaporation of the refrigerant cools the mixture and promotes the nucleation of ice crystals within the brine. The ice slurry collects at the bottom of the vessel and is pumped out. The refrigerant vapour is evacuated from the top of the vessel for compression, condensing and recycling.

In all three processes, the salinity of the brine has an important influence in the manufacture of ice slurry. In general, a higher concentration of dissolved salts will give a smaller production of ice crystals but at greater purity. The freezing temperature will also be lower, resulting in a reduced coefficient of performance for the unit. For mine cooling systems, salt concentration between 5% and 15% would appear to be appropriate. However, the optimum salinity for any given installation should be determined from tests on the actual water to be used.

Transportation of ice

The mass of ice required for a mine cooling system is such that it must be supplied continuously rather than in a batch transportation system. While slurry ice can be pumped through pipes, particulate ice must be transported from the icemaking plant to the shaft by conveying systems or by a hydraulic or pneumatic pipeline. In the latter case, the supply air temperature should be not more than 8 °C in order to prevent undue agglomeration of the ice particles.

The particulate ice falls through the shaft pipeline either as **dilute flow** in which the particles are separated and retain their individual identity or as **dense** or **slug** flow. In the latter case, the ice particles agglomerate into discrete slugs separated by air spaces. If the ratio of ice to air is too great, then **extrusion** flow will develop accompanied by a greatly increased risk of pipe blockage. The factors involved in the type of flow through an ice pipeline are the ice/air ratio, the shape, temperature and size of the particles, and the type and size of pipe. Unplasticized (U) PVC has been found to be a satisfactory pipe material for ice lines. However, changes in cross-section or lips at expansion joints should be avoided. At shaft stations, radii of curvature of 3 m or more allow the pipelines to be extended horizontally for several hundred metres. To cite the East Rand Mines example again, four uninsulated UPVC pipes are employed of inside diameter 270 mm and wall thickness 22.5 mm (Hemp, 1988). The longest pipe extends through a vertical distance of approximately 2900 m. Each pipe is capable of carrying 200 t/h (55.6 kg/s) allowing for considerable future expansion of the mine cooling duty.

Incorporation into the mine cooling system

Figure 18.25 gives a simplified representation of one layout that utilizes particulate ice. Return water pumped from the mine passes through a surface pre-cooling tower and is further cooled in a water pre-chiller. This may be supplied by 'coolth' from

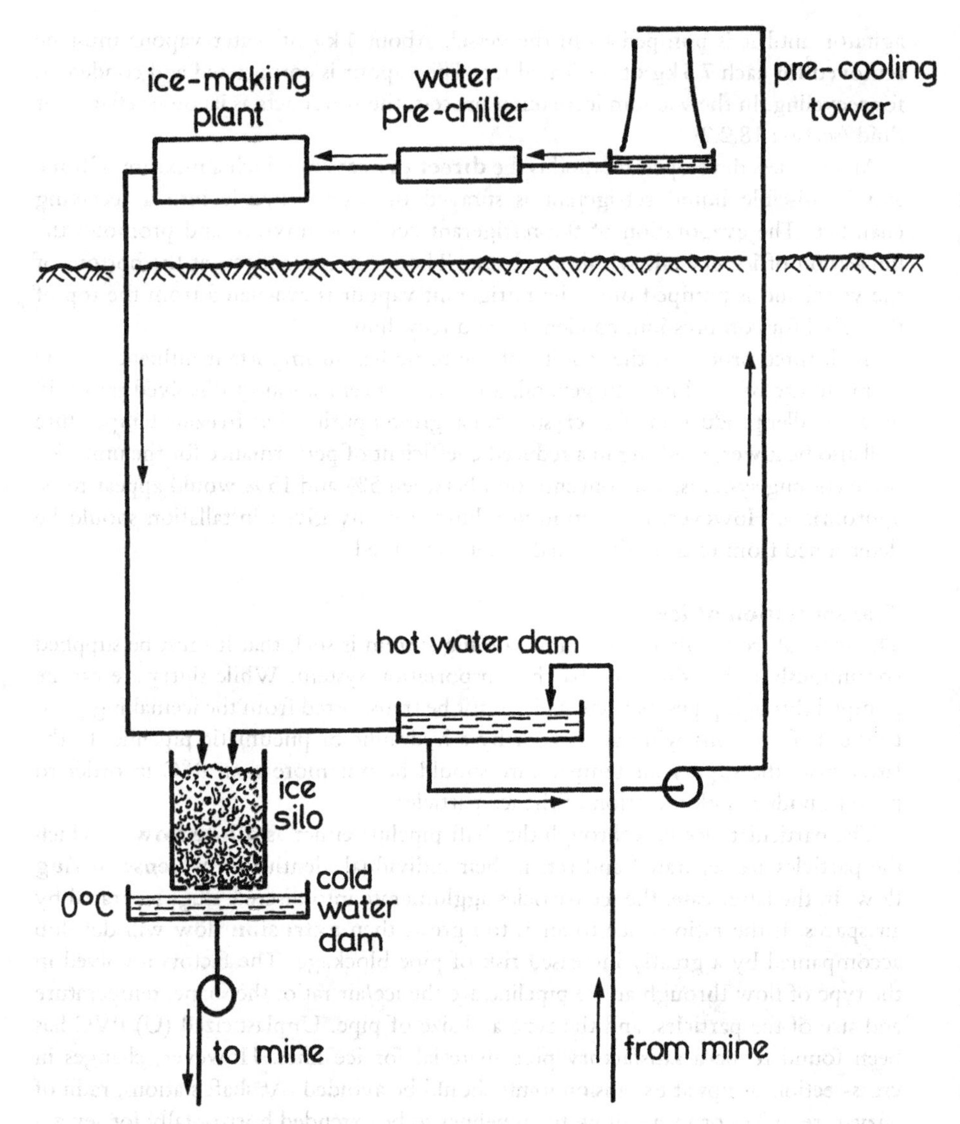

Figure 18.25 Layout of an ice system. Ice silos may be located on several levels.

the icemaking refrigeration plant or, alternatively, be a separate water chiller package. The particulate ice falls through the shaft pipeline and is discharged into an ice–water mixing dam or a silo with a perforated base. Water from the hot water dam is also sprayed into the ice silo. The discharge from the silo enters the cold water dam at 0 °C.

The bed of ice in the silo should be maintained at a thickness of 1 to 2 m. The supply of ice to each silo must be matched to the required cooling duty at that level. Position transducers can be used to monitor the ice level in each silo. The transmitted signals may be employed to control ice feed rates. It is advisable to provide sufficient ice bunkerage to handle short-term fluctuations in demand. Longer-term variations can be handled by activating an appropriate number of individual sets in the icemaking plant or by compressor control on the refrigeration units (section 18.2.3).

Future employment of ice slurries in mine cooling systems may be combined with energy recovery devices. The duties of direct-contact heat exchangers are considerably enhanced when the sprays are supplied with ice slurry of up to 20% ice (Gebler *et al.*, 1988). Droplets in the spray chamber remain at 0 °C until the ice crystals are melted. Further research is in progress into the pumping and pipeline characteristics of ice slurry. Early tests have indicated that, with an ice fraction of only 40%, pipeline pressure drops are about three times those for water (Shone and Sheer, 1988).

Economics of ice systems

The benefits of ice as a medium of heat transfer in mines may be listed as follows:

1. greatly reduced water flow in shaft pipelines; therefore smaller pipes and lower pumping costs
2. water is available at 0 °C underground rather than the 3 to 6 °C common with conventional chilled water systems; therefore less flow and reduced pumping costs in the subsurface (secondary) circuits and improved performance of heat exchangers
3. 'coolth' is stored in ice bunkers and silos to satisfy short-term variations in demand
4. the system is more easily capable of being uprated for future increases in cooling load
5. the subsurface system is simpler having reduced or eliminated the need for turbines
6. the quality of the water is improved by ice making

The major disadvantage of the system is the considerable increase in both capital and operating costs of the icemaking plant. As the depth of a mine increases and the heat load and associated pumping costs both rise, then, at some point, it becomes economically advantageous to convert from water chilling to an ice system. A study in South Africa showed that the cross-over point occurs at a depth of approximately 3000 m (Sheer *et al.*, 1984). However, this will be influenced not only by the cost of hardware but also power charges, surface climate, geothermic gradient, rock properties and heat from equipment or stored materials (in the case of subsurface repositories). It is, therefore, prudent to conduct a site-specific economic study on any proposed ice system.

Summary of design process

Although this chapter has concentrated on the hardware associated with mine air conditioning systems, it is pertinent to remind ourselves that the design of such systems cannot be separated from ventilation planning (sections 9.1 and 16.3.5). In particular, the initial question is how much heat can, or should, be removed by ventilation and how much by cooling systems.

This leads us into the first stages of the design procedure for a mine cooling system.

1. Establish the expected heat load for the entire mine and for each identifiable district, level or zone in the mine. The techniques for doing this are discussed in Chapters 15 and 16.
2. Again, for the whole mine and each individual area, determine the airflows required to deal with dust and polluting gases, taking into account air velocity limits (section 9.3). Ventilation network analyses should be conducted in order to establish airflow distribution and leakage patterns (Chapter 7).
3. Determine the heat removal capacities of the airflows (section 9.3.4). If this is greater than the corresponding heat loads for individual sections of the mine, then the heat can be removed entirely by ventilation and there is no need to consider a cooling system.
4. Conduct exercises to determine the feasibility of removing the excess heat by increased airflows without exceeding velocity limits set by physiological, economic or legislative considerations (section 9.3.6). If the additional airflows are unacceptable, then the need for air conditioning is established and the design of the cooling system can proceed.
5. Determine the heat to be removed by cooling as the difference between the heat load and the heat removal capacity of the air (section 9.3.4).

   Preferably, a climatic simulation program should be used in order to conduct a detailed analysis of the positions and duties of heat exchangers and application of chilled service water (Chapter 16). Interactive studies should also be conducted to optimize between ventilation and air conditioning (Figs. 9.1 and 16.4) (Anderson and Longson, 1986).

   These investigations will establish the required distribution of air cooling devices. If this is limited to a few localized areas, then the feasibility of employing spot or district coolers (section 18.3.1) should be investigated. If, however, the heat problem is more widespread, then the study should be widened to examine centralized refrigeration plant.
6. Investigate the alternative locations of refrigerating plant (discussed in the section on location of main plant) and the feasibility of employing energy recovery devices (section 18.3.4).
7. By summing heat exchanger capacities and allowing for line losses, establish the duty of the refrigeration plant.
8. Estimate evaporator and condenser temperatures on the basis of the desired temperature of the cooled medium and the heat rejection facilities respectively.
9. Determine the required flowrates of chilled water and, hence, the sizes of pipelines (section 18.3.3) and pumping duties.
10. Invite tenders from manufacturers of refrigeration plant and ancillary equipment, including valves, pipelines, pumps, heat exchangers, instrumentation and controls.
11. Establish capital, installation and operating costs of the cooling system.

18.4 AIR HEATING

In the colder countries of the world, the temperatures of intake air entering a mine may fall well below 0 °C for a major part of the year. This can result in severe problems along intake routes, and particularly in surface connecting downcast shafts, slopes or adits. First, a build-up of ice on shaft walls and fittings can cause a significant increase in resistance to airflow. Furthermore, large pieces of ice dislodged by shaft operations or melting can present a hazard to personnel working in or near the shaft. Repeated cycles of freezing and thawing that may occur on a daily or seasonal basis can result in severe damage to concrete linings in shafts or other intake airways, particularly when a seepage of ground water occurs through the lining.

Secondly, subzero temperatures can have adverse effects on the operation and maintenance of equipment. This may result in poor productivity as well as a reduction in safety.

Thirdly, personnel who are inadequately protected against a cold environment will suffer the symptoms described in section 17.6. Here again, this will lead to diminished levels of productivity and a deterioration in safety standards.

Winter temperatures can fall below −40 °C at the surfaces of mines in cold climates (Hall *et al.*, 1988; Moore, 1985). In these conditions, it becomes necessary to heat the intake air. This is the situation that obtains at the majority of Canadian mines (Hall *et al.*, 1989).

The methods of air heating suitable for mine applications may be listed as follows:

1. utilization of waste heat produced from plant or processes on the mine surface;
2. heat recovery from warm return air;
3. direct heating;
4. indirect heating;
5. ice stopes;
6. geothermal or cycled storage heating.

Each of these techniques is examined in the following subsections. To choose the method most applicable for any given facility, it is necessary to consider technical feasibility, reliability, flexibility and economics. Because of the large airflows that are usually involved, the costs of heating may far exceed those of producing the ventilation (Hall, 1985). In very cold climates, it is often the aim simply to increase the intake air temperature to about 0 °C and to allow autocompression, geothermal and other sources of heat to produce more comfortable temperatures in the main working areas.

In order to minimize operating costs, air heating systems may be employed only during main working shifts and when the surface temperature drops below a preset value. Furthermore, two or more of the systems listed above may be combined to minimize energy demand or to provide additional capacity for periods of abnormally low temperature.

There is, however, one situation in which air heating should be avoided or used with great caution. This occurs when the mine workings are in permafrost. Allowing air temperatures to exceed 0 °C may cause partial melting of the permafrost and give rise to problems of ground control in the mine. Indeed, at the Polaris Mine, 1400 km from the North Pole, refrigeration techniques are employed during the short summer in order to maintain the intake temperatures below the melting point of permafrost (van der Walt, 1984).

18.4.1 Utilization of waste heat

The first matter to investigate for any proposed air heating project is whether heat is available from other plant or processes on the surface of the mine.

Secondly, the costs and technical feasibility of using that thermal energy to raise the temperature of intake air should be studied. Common sources of waste heat are compressor stations, generators and mineral processing operations (de Ruiter *et al.*, 1989). If the waste heat is in the form of hot water or steam, then it may be pumped directly through insulated pipes to heat exchanger coils located within all, or part, of the intake airstream. Ice accumulations on the heat exchanger may be avoided by using a single array of coils only.

In water systems, an automatic procedure should be incorporated in order to empty the coils and associated pipework during a shut-down period or if the water temperature approaches freezing point. A further safeguard against frost damage is to employ a glycol–water mixture. A 50% glycol mixture lowers the freezing point to − 40 °C.

If the available heat source provides relatively low temperature thermal energy, then it may be necessary to employ a heat pump (section 18.2). Heat pumps may also allow the utilization of natural heat sources such as a large lake or the ocean, provided that these remain unfrozen through the winter months.

18.4.2 Heat recovery from exhaust air

This method may be considered if the downcast and upcast shafts are close to each other and there is a significant difference between the intake and return air temperatures. Figure 18.26 shows that during cold periods, the warm air returning from the mine is diverted through a direct contact spray chamber, transferring useful heat to the descending water droplets. Water from the sump is pumped through a heat exchanger located within the intake airstream. Again, it is advisable to employ a single array of coils to minimize ice build-up, and to allow for automatic dumping of water from the heat exchanger and pipework when the flow ceases.

The spray chamber may be replaced by a tube coil heat exchanger. This reduces the overall heat transfer efficiency of the system and also introduces the dangers of ice accumulation and corrosion of the tubes, particularly if the return air is humid.

Heat exchange from upcast to downcast airflows is seldom practicable by itself. However, it may be employed in combination with an external source of heat as shown on Fig. 18.26. If that source is waste heat from other plant, then it may be kept in continuous operation. However, if the external source of heat is gas, oil or

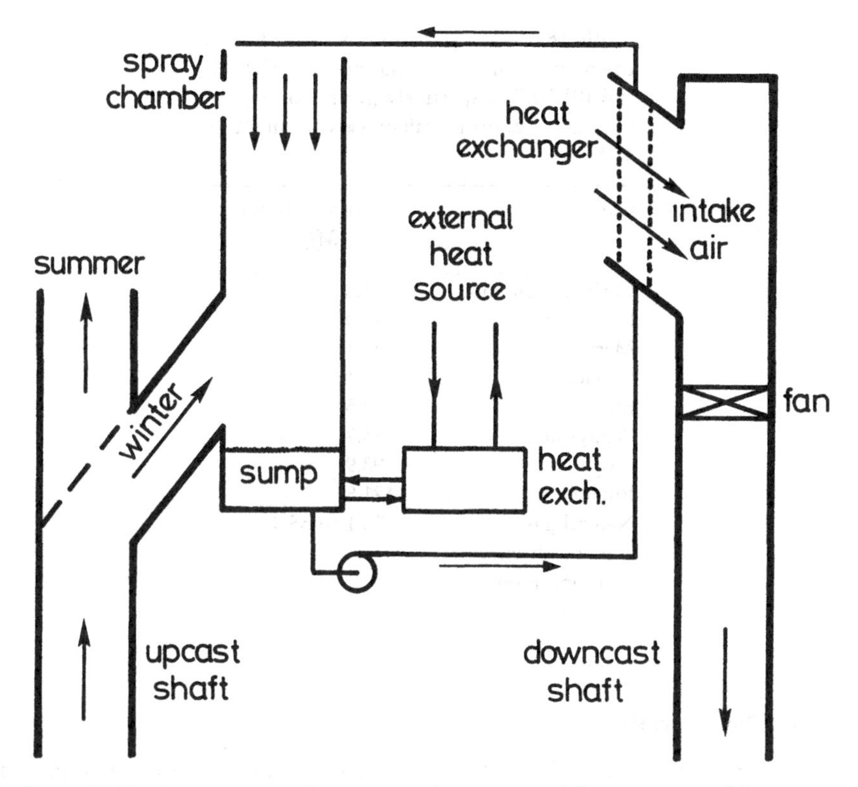

Figure 18.26 Heat recovery from return air, augmented from an external heat source.

electricity, then it may be used only during working shifts and when the surface air temperature falls below a set value.

A more direct means of recovering some of the heat from warm return air is by controlled partial recirculation (Hall *et al.*, 1988, 1989). This technique gives 100% efficiency of heat transfer from return to intake for that fraction of the air which is recirculated. This is limited by the rate at which gaseous and particulate pollutants are added to the mine air (section 4.5). The system should be capable of reverting to normal through-flow ventilation when monitors indicate a trend towards unacceptable concentrations of pollutants or following blasting operations. The volume of air that is recirculated may be controlled automatically by the pollutant concentrations in the return air or set at a fixed fraction of the return airflow. The recirculating cross-cut should be located as close to the surface connection as practicable in order to take advantage of the pollutant dilution arising from air leakage.

If the level of contamination by airborne pollutants is sufficiently low to allow a reduction in the through flow of fresh air, then the costs of heating the intake air will also be reduced. However, any proposed system of controlled partial recirculation must incorporate all of the safeguards and precautions discussed in section 4.5.

Table 18.4 Gross calorific values of gases based on a temperature and pressure of 288 K and 101.3 kPa respectively (for petroleum fuel oils, gross calorific values vary from 42 to 47 MJ/kg)

Gas	*Gross calorific value (MJ/m^3)*
Carbon monoxide	11.8
Hydrogen	11.9
Methane	37.1
Ethylene	58.1
Ethane	64.5
Propylene	85.7
Propane	93.9
Butane	121.8
Natural gas (depends on composition)	26.1 to 55.9

18.4.3 Direct heating

The employment of electrically heated elements to raise the temperature of mine air is limited by high operating costs. It is practical only for very light duties or where electrical power is particularly cheap.

Natural gas, propane or other light hydrocarbon fuels may be injected through nozzles and burned in flame jets within the intake airstream. As no intermediate heat exchangers are involved, this is the most efficient way of using the fuel. The products of combustion, carbon dioxide and water vapour, are added to the airstream in relatively small amounts. However, the gas nozzles should be well maintained to prevent the formation of carbon monoxide through incomplete combustion. A prudent precaution is to locate a gas monitor downstream from the jets with alarms and automatic fuel cut-off at a predetermined concentration of carbon monoxide.

The rate of fuel consumption for any given heating duty may be calculated from the calorific value of the fuel. At the temperatures to which the products of combustion are cooled in mine air, the majority of water vapour will condense. Hence, it is the gross calorific value that should be used. Table 18.4 gives the gross calorific values of several gases. For natural gas and other mixtures, an approximate value may be obtained by summing the calorific values of the constituents weighted according to percentage composition.

Example An airflow of 100 m^3/s and density of 1.4 kg/m^3 is to be heated from $-20\,°C$ to $2\,°C$ by propane jets.

1. Determine the rate of fuel consumption in m^3/h.
2. If the air is supplied at 21% oxygen and 0.03% carbon dioxide, determine the corresponding concentrations downstream from the jets.

Solution

1. Mass flow of air

$$m_a = Q\rho$$
$$= 100 \times 1.4 = 140\ \text{kg/s}$$

Heat required

$$q_a = m_a C_p \Delta t$$
$$= 140 \times 1005 \times [2 - (-20)]$$
$$= 3.095 \times 10^6\ \text{W or } 3.095\ \text{MW}$$

(To be precise, we should take the presence of water vapour into account. However, the effect is very small at these low temperatures and may be neglected.)

The gross calorific value of propane is selected from Table 18.4 as 93.9 MJ/m^3. Hence, heat produced at the jets is

$$q_p = 93.9 \times \text{propane flowrate} \quad \text{MJ/m}^3\ \text{m}^3\text{/s or MW}$$

As $q_p = q_a = 3.095$ MW, the required

$$\text{propane flowrate} = \frac{3.095}{93.9} = 0.0330\ \text{m}^3\text{/s}$$

or $0.0330 \times 3600 = 118.7\ m^3/h$ at standard temperature and pressure.

2. The chemical balance for burning propane is

$$C_3H_8 \quad + \quad 5O_2 \quad \rightarrow \quad 3CO_2 \quad + \quad 4H_2O$$

$$1\ \text{m}^3\ \text{propane} + 5\ \text{m}^3\ \text{oxygen} \rightarrow 3\ \text{m}^3\ \text{carbon dioxide} + 4\ \text{m}^3\ \text{water vapour}$$

Therefore $0.0330\ m^3/s$ of propane will consume $5 \times 0.0330 = 0.165\ m^3/s$ oxygen to produce $3 \times 0.0330 = 0.099\ m^3/s$ carbon dioxide. At entry, the air passes $100 \times 0.21 = 21\ m^3/s$ of oxygen and $100 \times 0.0003 = 0.03\ m^3/s$ of carbon dioxide. Downstream from the jets,

$$\text{oxygen flowrate} = 21 - 0.165 = 20.835\ \text{m}^3\text{/s}$$

and

$$\text{carbon dioxide flowrate} = 0.03 + 0.099 = 0.129\ \text{m}^3\text{/s}$$

These values give close approximations to the concentrations of the gases down-stream from the jets.

18.4.4 Indirect heating

In order to prevent products of combustion from entering the mine ventilation system, fuels may be used in a separate burner to heat a glycol–water mixture. This is then recirculated through an indirect heat exchanger located within the intake air. Again, a single array of coils will minimize the risk of ice accumulation on the tubes. This technique of **indirect** or **off-line** heating inevitably gives a lower thermal efficiency than open flame jets within the intake air. However, the flowrate of the glycol mixture provides an added measure of control and flexibility.

18.4.5 Ice stopes

Figure 18.27 illustrates a metal mining technique that allows air to be heated to a temperature approaching 0 °C at very low cost. Water is sprayed into the top of an abandoned open stope. Cold air coming directly from the downcast shaft or slope

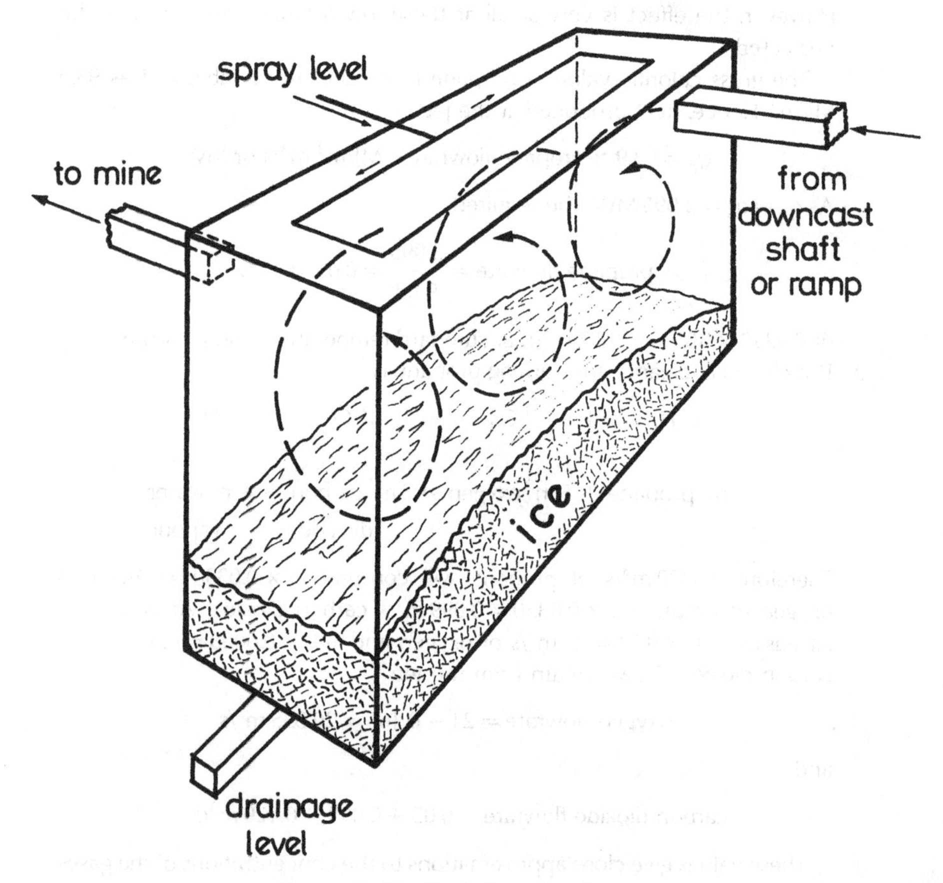

Figure 18.27 Depiction of an ice stope.

enters near the top and circulates through the stope, freezing the water, before exiting at the other end.

The air is heated by three mechanisms:

1. directly from the liquid water droplets,
2. by latent heat of fusion (333.5 kJ/kg) as the water freezes, and
3. by strata heat from the wall rock.

The ice particles fall and form an ice bed that gradually fills the stope through the winter months.

In summer, warm intake air may be cooled by circulating it through the same stopes and melting the ice. A drain level at the base of the stope carries water to a main sump. However, the air cooling may be enhanced by circulating the liquid melt back through the sprays.

The ice stope technique is used very successfully at the Stobie Mine of Inco Ltd. in Canada (Stachulak, 1989). In this mine an airflow of 307 m^3/s is heated in two ice stopes at a cost estimated at about one-ninth that for equivalent gas heating. In that installation, the water is supplied to the nozzles at a pressure of between 550 and 760 kPa. Compressed air is added to the line shortly before the nozzles. This provides a finely divided spray and prevents icing up of the jets. The compressed air is also used to clear water from the pipes when the system is to be shut down.

The disadvantages of the ice stope technique are that it requires the availability of large open excavations and also necessitates an increased pumping capacity for the mine drainage system.

Example Water at 7 °C is sprayed at a rate of 30 l/s into an ice stope. Calculate the maximum mass flow of air through the stope that will allow an increase in air temperature of 20 °C.

Solution Mass flowrate of water,

$$m_w = 30 \text{ kg/s}$$

Heat released by water in cooling from 7 °C to 0 °C

$$q_w = m_w C_w \Delta t_w$$

$$= \frac{30 \times 4187 \times 7}{1000} = 879 \text{ kW}$$

Heat of fusion released by ice formation (latent heat of fusion, $L_i = 333.5$ kJ/kg):

$$q_i = L_i m_w$$

$$= 333.5 \times 30 = 10\,005 \text{ kW}$$

(This illustrates the dominant effect of the latent heat of ice formation.) Total heat transferred to air

$$q_a = q_w + q_i = 879 + 10\,005 = 10\,884 \text{ kW}$$

However, this is also equal to $m_a C_p \Delta t_{air}$, giving

$$m_a = \frac{10\,884 \times 1000}{1005 \times 20} = 541 \text{ kg/s}$$

If the air density were 1.3 kg/m^3, this would give an air volume flowrate of 541/1.3 = 417 m^3/s.

18.4.6 Geothermal and cycled storage heat

Another low cost method of controlling the temperature and humidity of intake air is to course it through one or more sets of old workings or fragmented strata before admitting it to current work areas. This takes advantage of a combination of natural geothermal energy and the 'thermal flywheel' effect of heat storage within the envelope of rock surrounding a mine opening (section 15.2.2).

The flow of geothermal heat into a mine airway is greatest at the moment of excavation and reduces with time until a near equilibrium is attained when temperatures in the surrounding envelope of rock no longer change significantly with respect to time (Fig. 15.3). If, however, the inlet air is cycled between hot and cold, then the surrounding rock will act as a storage heater, absorbing heat during the hot periods and emitting it during the cold periods. This cyclic behaviour is superimposed upon the longer term flux of geothermal heat.

Let us assume a situation in which the virgin rock temperature (VRT) at a given level is higher than the mean annual dry bulb temperature of air entering that level from the downcast shaft or slope. In order to moderate the extremes of winter temperature, the air is passed through a set of old workings that we may refer to as the **control district**. Figure 18.28 illustrates the seasonal variation of air temperature at inlet and exit of a control district, and ignoring the shorter-term diurnal changes. The moderating effect of the control district is shown clearly. The inlet peaks and troughs of summer and winter temperatures are moderated very considerably by the time the air reaches the exit.

However, the figure also shows a longer-term trend over which the average exit temperature falls until a repeated cyclic variation is attained. If the winter temperatures of the air leaving the control district are acceptable in this condition of dynamic equilibrium, then that district can be used indefinitely to moderate the extremes of seasonal variations. If, however, after a few years the winter exit temperatures fall below tolerable levels, then the intake air should be diverted through a younger control district.

Control districts that are no longer effective may be allowed to regenerate by sealing them off. The surrounding rock temperature will gradually tend towards the VRT. After a regeneration period which may be several years, the panel can again be opened up as a control district. If several old districts are available, then some of them may be used as control districts at any one time while others are regenerating (Moore, 1985).

The actual form of Fig. 18.28 for any given mine depends on the extremes of surface

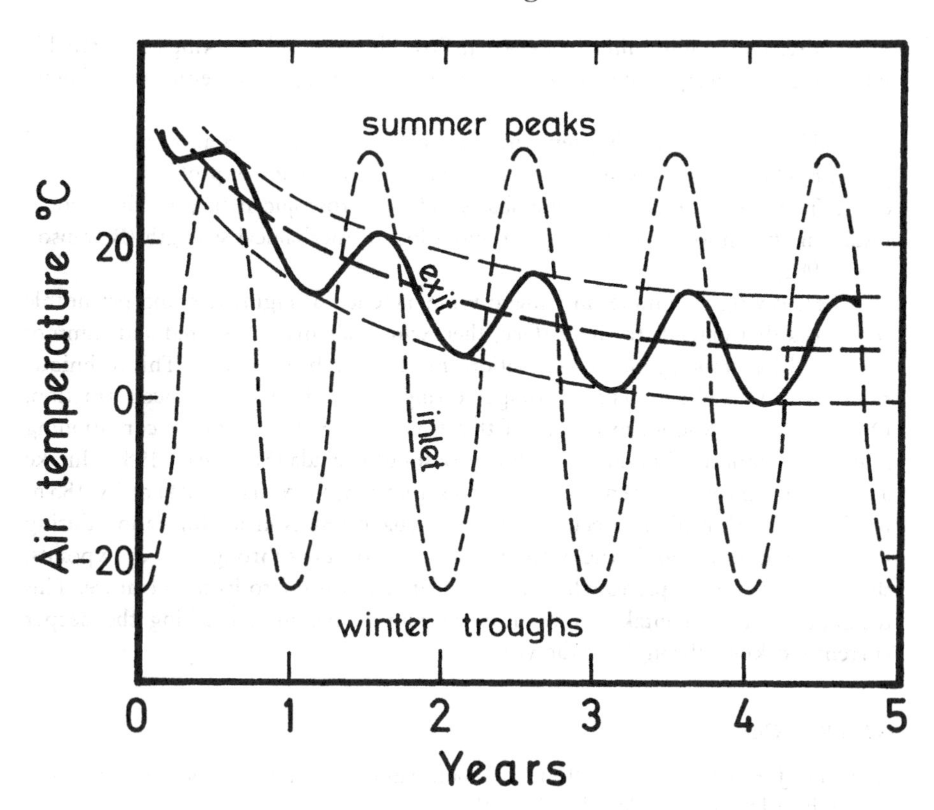

Figure 18.28 Example of the variation in air temperature at the inlet and outlet of a climate control district.

climate, the rate of airflow through the control district and the thermal properties of the surrounding strata. Simulation programs are most helpful in predicting the cyclic behaviour of the air temperatures (Moore, 1985; MVS, 1990). Manual methods of estimating the effects of varying inlet temperatures on strata heat flow have also been developed (Hemp, 1982).

Although the use of a control district can yield a useful degree of air conditioning, it does have some disadvantages and limitations. First, the passage of air through old workings requires additional fan energy. This can be minimized by arranging for the air to move fairly slowly along multiple flow paths, requiring only a small applied pressure differential. Conversely, heat transfer from the strata is enhanced by greater velocities and longer flowpaths connected in series. Simulation tests coupled with site-specific measurements allow the optimum layout to be determined.

Secondly, old workings are liable to changes in resistance because of strata closure and falls of roof. Hence, in order to ensure continuity of ventilation, control districts should be capable of being bypassed at short notice. Also, they should not be relied on as emergency escapeways.

Thirdly, the conditioning of intake airflow through old workings is untenable where there is a danger of spontaneous combustion or significant emissions of strata gases.

In addition to moderating temperatures, control districts can also serve a useful purpose where variations in intake humidity cause roof control problems. This can occur in deposits of salt or some shales. The hygroscopic nature of these rocks results in absorption of water vapour and a loss of mechanical strength (Robinson *et al.*, 1981).

Where caving techniques in a metal mine have left a fragmented and permeable rock mass that extends to the surface, then winter air may be heated and summer air cooled by drawing intake ventilation through the broken rock. This technique can combine the phenomena of storage heating and the latent heat of ice formation. One of the best known examples of this essentially free form of air conditioning is at the Creighton Mine in the Sudbury Basin of Canada (Stachulak, 1989). Intake air is drawn through a bed of broken rock that is approximately 244 m by 183 m in plan and 137 m high. Precipitation and seepage results in ice formation during the winter and melting in the summer. The air progresses through old drawpoints, slusher drifts and ore passes where it is regulated according to its temperature. This technique results in intake air of near-constant temperature reaching the deeper current workings throughout the year.

REFERENCES

Anderson, J. and Longson, I. (1986) The optimization of ventilation and refrigeration in British coal mines. *Min. Eng.* **146**, 115–20.

Arkle, K., *et al.* (1985) Use of thermal insulation materials in mines. *J. Mine Ven. Soc. S. Afr.*, **38** (4), 43–5.

ASHRAE (1988) Cooling towers. *Handbook on Equipment*, Chapter 20, American Society of Heating, Refrigerating and Air Conditioning Engineers.

Baker-Duly, C., *et al.* (1988) Design of a large flexible underground refrigeration installation. *Proc. 4th Int. Mine Ventilation Congr., Brisbane*, 443–9.

Bluhm, S. J. (1980) Predicting performance of spray chambers for cooling air. *Heat., Refrig. Air Condition., S. Afr.*, 27–39.

Bluhm, S. J. (1983) Spot cooling of air in direct contact heat exchangers. *Report of Environmental Engineering Laboratory*, Chamber of Mines Research Organization.

de Ruiter, H., *et al.* (1989) Geothermal heat recovery at Lac Minerals Ltd., Macassa Division. *91st CIMM Annu. General Meet., Quebec, May*.

Gebler, W. F., *et al.* (1988) The design and evaluation of a large direct-contact ice–air cooler for Vaal Reefs Gold Mine. *Proc. 4th Int. Mine Ventilation Congr., Brisbane*, 425–32.

Hall, A. E. (1985) The use of controlled recirculation ventilation to conserver energy. *2nd US Mine Vent. Symp. Reno, NV*, 207–15

Hall, A. E., *et al.* (1988) The use of controlled recirculation to reduce heating costs in Canada. *Proc. 4th Int. Mine Ventilaton Congr., Brisbane*, 301–7.

Hall, A. E., *et al.* (1989) Controlled recirculation investigation at Central Canada Potash Division of Noranda Minerals, Inc. *Proc. 4th US Mine Ventilation Symp., Berkeley, CA*, 226–33.

Hancock, W. (1926) Local air-conditioning underground by means of refrigeration. *Trans. Inst. Min. Eng.* **72**, 342–66.
Heller, K., *et al.* (1982) Development of a high performance, low maintenance, in-line water spray cooler for mines. *1st US Mine Ventilation Symp., Tuscaloosa, AL.*
Hemp, R. (1967) The performance of a spray-filled counterflow cooling tower. *J. Mine Vent. Soc. S. Afr.* **25**, 159–73.
Hemp, R. (1972) Contribution to the design of underground cooling towers by A. Whillier. *J. Mine Vent. Soc. S. Afr.* **25** (6), 85–93.
Hemp, R. (1982) Sources of heat in mines. *Environmental Engineering in South African Mines*, Chapter 22, 569–612.
Hemp, R. (1988) A 29 MW ice system for mine cooling. *Proc. 4th Int. Mine Ventilation Congr., Brisbane*, 415–23.
McAdams, W. H. (1954) *Heat Transmission*, McGraw-Hill, New York, 3rd edn.
Moore, D. T. (1985) Geothermal air heating in a southwestern Wyoming trona mine. *Proc. 2nd US Mine Ventilation Symp., Reno, NV*, 561–70.
Mücke, G. and Uhlig, H. (1984) Performance of finned coil and water spray coolers. *Proc. 3rd Int. Mine Ventilation Congr., Harrogate*, 231–7.
MVS (1990) Investigation of geothermal air heating for General Chemical's Green River soda ash operations. *Report to General Chemical Corp.*, May, 1–33, Mine Ventilation Services, Inc.
Ramsden, R. (1985) Insulation used on chilled water pipes in South African gold mines. *J. Mine Vent. Soc. S. Afr.* **38** (5), 49–54.
Ramsden, R. and Bluhm, S. J. (1984) Air cooling equipment used in South African gold mines. *Proc. 3rd Int. Mine Ventilation Congr., Harrogate*, 243–51.
Ramsden, R. and Bluhm, S. (1985) Energy recovery turbines for use with underground air coolers. *Proc. 2nd US Mine Ventilation Symp., Reno, NV*, 571–80.
Reuther, E. U., *et al.* (1988) Optimization of spray coolers for cooling deep coal mines. *Proc. 4th Int. Mine Ventilation Congr., Brisbane*, 451–8.
Robinson, G., *et al.* (1981) Underground environmental planning at Boulby Mine, Cleveland Potash Ltd. *Trans. Inst. Min. Metall.* (July) (reprinted in *J. S. Afr. Mine Vent. Soc.* **35** (9), 73–88).
Rose, H. (1989) Personality profile. *J. Mine Vent. Soc. S. Afr.*, **42** (6), 110–3.
Sheer, T. J., *et al.* (1984) Research into the use of ice for cooling deep mines. *Proc. 3rd Int. Mine Ventilation Congr., Harrogate*, 277–82.
Shone, R. D. C. and Sheer, T. J. (1988) An Overview of research into the use of ice for cooling deep mines. *Proc. 4th Int. Mine Ventilation Congr., Brisbane*, 407–13.
Stachulak, J. (1989) Ventilation strategy and unique air conditioning at Inco Limited. *Proc. 4th US Mine Ventilation Symp., Berkeley, CA*, 3–9.
Stroh, R. M. (1980) Refrigeration practice on Anglo American gold mines. *Mine Ventilation Society of South Africa Symp. on Refrigeration and Air Conditioning.*
Stroh, R. M. (1982) Refrigeration practice and Chilled water reticulation. *Environmental Engineering in South African Mines*, Chapters 24 and 25.
Thimons, E. D., *et al.* (1980) Water spray vent tube cooler for hot stopes. *US Bureau of Mines Tech. Prog. Rep. TPR 107.*
van der Walt, J. (1984) Cooling the world's coldest mine. *J. S. Afr. Mine Vent. Soc.* **37** (12), 138–41.
van Vuuren, S. P.J. (1975) The optimization of pipe sizes in a refrigeration system. *J. Mine Vent. Soc. of S. Afr.* **28** (6) 86–90.

Weuthen, P. (1975) Air coolers in mines with moist and warm climatic conditions. *Proc. 1st Int. Mine Ventilation Congr., Johannesburg*, 299–303.

Whillier, A. (1967) Pump efficiency determination from temperature measurements. *S. Afr. Mech. Eng.* (October), 153–60.

Whillier, A. (1977) Predicting the performance of forced-draught cooling towers. *J. Mine Vent. Soc. S. Afr.* **30** (1) 2–25.

Whillier, A. (1982) Heat transfer. *Environmental Engineering in South African Mines*, Chapter 19.

FURTHER READING

Bluhm, S. J. (1981) Performance of direct contact heat exchangers. *J. Mine Vent. Soc. S. Afr.* **34** (8 and 9).

Burrows, J. (1982) Refrigeration—theory and operation. *Environmental Engineering in South African Mines*, Chapter 23.

PART FIVE

Dust

19

The hazardous nature of dusts

19.1 INTRODUCTION

The natural atmosphere that we breathe contains not only its gaseous constituents but also large numbers of liquid and solid particles. These are known by the generic name **aerosols**. They arise from a combination of natural and industrial sources including condensation, smokes, volcanic activity, soils and sands, and microflora. Most of the particles are small enough to be invisible to the naked eye. **Dust** is the term we use in reference to the solid particles. The physiologies of air-breathing creatures have evolved to be able to deal efficiently with most of the aerosols that occur naturally. However, within closed industrial environments, concentrations of airborne particulates may reach levels that exceed the ability of the human respiratory system to expel them in a timely manner. In particular, mineral dusts are formed whenever rock is broken by impact, abrasion, crushing, cutting, grinding or explosives. The fragments that are formed are usually irregular in shape. The large total surface area of dust particles may render them more active physically, chemically and biologically than the parent material. This has an important bearing on the ability of certain dusts to produce lung diseases.

Respiratory problems caused by dust are among the oldest of industrial ailments. However, the first legislation for mine dust appears to have been formulated in 1912 when the Union of South Africa introduced laws governing working conditions in the gold mines of the Witwatersrand. Other countries introduced similar legislation in the 1920s and 1930s. However, those laws were concerned primarily with silicosis and required proof of employment in siliceous rock mining (Carver, 1975). The early legislation reflected medical opinion of the time, namely that hardrock dust caused silicosis which led to tuberculosis and eventual death. As that time coal dust was not regarded as particularly harmful. However, the number of recognized cases of coal workers' pneumoconiosis (CWP) increased dramatically through the 1930s. The British Medical Research Council initiated an investigation into respirable disease within the anthracite workers of South Wales. In 1936, the need for protective legislation in the United States was acknowledged by a National Silicosis Conference.

Both in Europe and the United States, the hazards of coal dust were identified first in anthracite mines, but by 1950 it was confirmed that workers in bituminous coal mines were also subject to coal workers' pneumoconiosis (known, also, in America as **black lung**).

It took many years for a quantitative and definitive link to be established between the 'dustiness' of mine atmospheres and respiratory disfunctions. In retrospect, there were three reasons for this. First, it may take several years of exposure before the victim becomes aware of breathing impairment. Secondly, the lung reactions to dust are often similar to those of naturally occurring ailments and, thirdly, the commonly used measure of dust concentration was the number of particles in a unit volume of air. Correlations between dust concentration measured in this way and the incidence of pneumoconiosis were not clear.

A turning point occurred in 1959 at the International Pneumoconiosis Conference held in Johannesburg, South Africa (Orenstein, 1959). Recommendations were made at that conference which resulted in a redirection of pneumoconiosis studies, particularly with regard to the methods and strategies of dust sampling. It was recognized that those particles of equivalent diameter less than 5×10^{-6} m (5 μm) were the ones most likely to be retained within the lungs. These were named **respirable dust**. It was further established that the mass concentration of respirable dust in any given atmosphere was a much better measure of the potential health hazard than the earlier particle count methods. Instruments began to be developed that mimicked the dust retention selectivity of the human lung and, furthermore, could be used continuously to give the integrated effect over an 8 h period.

In this chapter we shall concentrate on the effects of mineralogical dusts on the human respiratory system and the techniques that are now employed for the sampling and measurement of airborne dust.

19.2 CLASSIFICATIONS OF DUST

There are a number of ways to classify aerosol particles depending on the purpose of any given study. Two such classifications are particularly relevant to the subsurface environmental engineer, first with regard to size distributions of the particulates and, secondly, in terms of physiological effects.

19.2.1 Size of aerosol particles

Mine dusts vary widely in shape, dependent largely upon the prevalent mineral constituents. The simplest method of quantifying the size of a non-spherical particle is the **projected area** or **equivalent geometric diameter**. This is the diameter of a sphere that has the same projected area as the actual particle. Other measures of equivalent diameter are defined in Chapter 20.

Typical size ranges of some common aerosols are given in Table 19.1. In general, the size distribution within each range follows a log–normal curve. Particles do not become visible to the naked eye until they are more than 10 μm equivalent diameter.

Table 19.1 Size ranges of common aerosols

Type of aerosol	*Size range (10^{-6} m)*	
	Lower	*Upper*
Respirable dust	—	7
Coal and other rock dusts	0.1	100
Normal atmospheric dusts	0.001	20
Diesel smokes	0.05	1
Viruses	0.003	0.05
Bacteria	0.15	30
Tobacco smoke	0.01	1
Pollens causing allergies	18	60
Fog	5	50
Mist	50	100
Light drizzle	100	400

It follows that the harmful respirable dusts are invisible. Nevertheless, it must be assumed that heavy visible concentrations of dust in a mine atmosphere are accompanied by high levels of respirable dust (section 20.3.1). Another unfortunate aspect of the small sizes of respirable dusts is that they have a very low settling velocity and, indeed, can remain suspended in air indefinitely.

19.2.2 Physiological effects of dusts

A classification of dusts with respect to potential hazard to the health and safety of subsurface workers may be divided into five categories.

1. *Toxic dusts.* These can cause chemical reactions within the respiratory system or allow toxic compounds to be absorbed into the bloodstream through the alveolar walls. They are poisonous to body tissue or to specific organs. Some metalliferous ores fall into this category. The most hazardous include compounds of arsenic, lead, uranium and other radioactive minerals, mercury, cadmium, selenium, manganese, tungsten, silver and nickel (Walli, 1982).
2. *Carcinogenic (cancer-causing) dusts.* The cell mutations that can be caused by alpha, beta and gamma radiation from decay of the uranium series make radon daughters the most hazardous of the carcinogenic particulates (Chapter 13). A combination of abrasion of lung tissue and surface chemical action can result in tumour formation from asbestos fibres and, to a lesser extent, freshly produced quartz particles. Exposure to arsenic dust can also cause cancers. Work is in progress to investigate the potential carcinogenic properties of diesel exhaust particulates.
3. *Fibrogenic dusts.* The scouring action of many dusts causes microscopic scarring of lung tissue. If continued over long periods this can produce a fibrous growth of tissue resulting in loss of lung elasticity and a greatly reduced area for gas

exchange. The silica (quartz, chert) and some silicate (asbestos, mica, talc) dusts are the most hazardous of the fibrogenic dusts and may also produce toxic and carcinogenic reactions. Welding fumes and some metalliferous ores produce fibrogenic dusts. Long and excessive exposure to coal dusts also gives rise to fibrogenic effects.

4. *Explosive dusts.* These are a concern of safety rather than health. Many organic materials, including coals other than anthracite, become explosive when finely divided at high concentrations in air. Sulphide ores and many metallic dusts are also explosive. Hazards associated with the explosive nature of some dusts are discussed more fully in Chapter 21.
5. *Nuisance dusts.* Quite apart from adverse effects on the health of personnel, all dusts can be irritating to the eyes (Gibson and Vincent, 1980), nose and throat, and when in sufficiently high concentration may cause reduced visibility. Some dusts have no well-defined effects on health but remain in the category of a nuisance dust. These include the evaporites (halite, potash, gypsum) and limestones. The soluble salts of halite (NaCl) and potash (KCl) can occasionally cause skin irritations, particularly around hatbands or tightly fitting dust masks.

19.3 DUST IN THE HUMAN BODY

The lungs are the organs where the oxygen necessary for metabolic activity is introduced into the body. Through repetitive inhalations and exhalations air is brought into very close proximity to flows of blood, the two being separated by a very thin membrane about 0.5 μm in thickness. Oxygen diffuses across the membrane from the air to the blood while carbon dioxide diffuses in the other direction. The exchange is maintained by a difference in concentration across the membrane for each of the two gases.

The respiratory system is equipped with defence mechanisms against those gaseous or aerosol pollutants that may exist in the inspired air. However, these mechanisms can be defeated by toxic or carcinogenic agents. Furthermore, after years of exposure to unnaturally high concentrations of dust, the defence system can simply become overloaded, allowing the lungs to become much less efficient as gas exchangers and, also, more susceptible to bronchial infections and pulmonary illnesses.

In this section, we shall outline the structure and normal operation of the human respiratory system, the mechanisms of dust deposition within that system, and the processes that lead towards those ailments known collectively as dust diseases.

19.3.1 The respiratory system

Figure 19.1 is a simplified illustration of the human respiratory system. Air enters through the nostrils where it passes through a mat filter of hairs in order to enter the **nasopharynx**. This filter is the first line of defence and removes the larger dust particles. Those remain trapped until they are blown out or pass back through the nasopharynx to be swallowed. Within the larger volume of the nasopharynx, the air

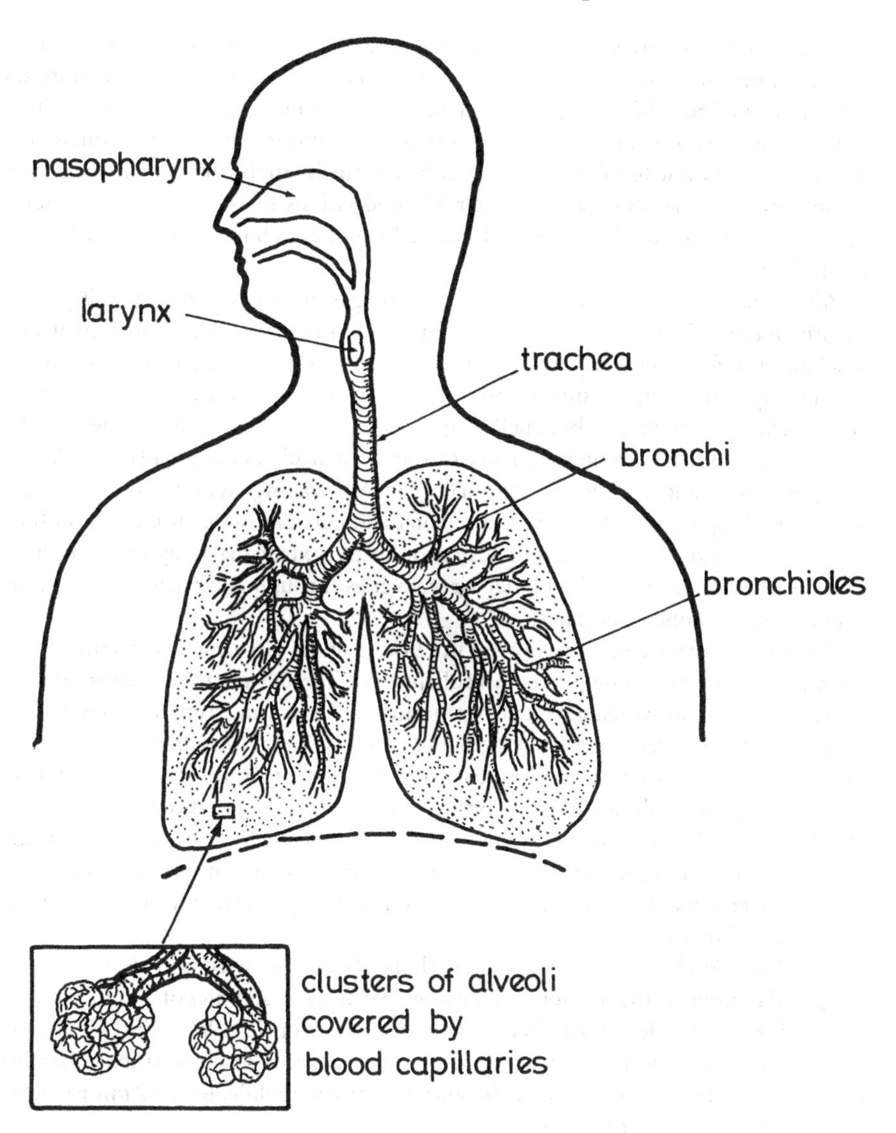

Figure 19.1 Diagrammatic representation of the human respiratory system.

velocity is reduced. In this zone and, indeed, throughout all of the branched air passages leading to the alveoli, the walls are lined with ciliated and mucus-secreting cells. The hairlike **cilia** wave to and fro with a directional bias that promotes movement of the mucous towards the throat where it can be swallowed. Most dust particles greater than 10 μm in size are captured by the hair filter or mucus before inhaled air reached the larynx. Air that is breathed through the mouth bypasses the protection offered by the nostrils and nasopharynx.

The air then progresses through the **trachea**, a tube some 20 mm in diameter, 120 mm long, and composed of a series of tough cartilage rings. This divides into the left and right **bronchi** averaging about 12 mm in diameter and 48 mm long. There are, of course, considerable variations in these dimensions between individuals. The air passages continue to subdivide repeatedly into the **bronchioles** in a manner that is analogous to the roots of a tree. After 16 subdivisions the bronchiole diameter is about 0.6 mm. The total number of branched passages in human lungs may be over 80 million.

Although the diameter of individual air passages decreases with branching, their greatly increased number produces a significant rise in total cross-sectional area available for flow. Hence, the air velocity decreases to the extent that airflow is completely laminar in the smaller bronchioles. All of these passages are coated with mucus which is continuously expelled upwards towards the trachea by the motions of cilia. The mucus secretions and any captured particulates are usually expelled by coughing or swallowing within one day. Mucous coatings within the air passages are normally quite thin. However, bronchial ailments can cause an increase in both the viscosity and thickness of mucous layers to the extent that they can restrict air movement significantly. The higher air velocities through constricted passages can lead to audible noise (wheezing).

The smallest bronchioles terminate in clusters of air sacs or **alveoli** some 0.2 to 0.6 mm in size. The walls of the alveoli provide the 0.5 μm thick membrane, or **epithelium**, across which gas exchange occurs. There are an estimated three to four hundred million alveoli in the average human lungs with a total area of some 75 m^2 available for gas exchange in a healthy adult. Involuntary muscular movements of the diaphragm and ribcage induce rhythmic changes of pressure differential between the alveoli and the external atmosphere giving rise to normal cycles of inhalation and exhalation. Breathing rates vary from 12 to 40 inhalations per minute depending on the level of activity. Table 11.2 indicates corresponding gas exchange rates for oxygen and carbon dioxide.

Few dust particles greater than 3 μm reach the alveoli. Numerous models have been proposed to indicate the relationship between particle size and sites of deposition (e.g. Walli, 1982; Schröder, 1982; National Research Council, 1980). For very small captured particles, clearance from the alveoli may be slow or non-existent. Particles of 0.2 μm diameter have a 25% probability of retention while, for 0.02 μm particles, the retention rate rises to about 55%.

Mucus-producing cells within the alveoli lubricate the surfaces and facilitate their freedom to dilate and contract. Removal of dust particles from the alveoli is undertaken by relatively large (10 to 50 μm) cells known as **macrophages** or **phagocytes**. These nomadic cells are free to move over the alveolar walls and are capable of engulfing particles up to 10 μm in diameter. The commonly accepted view is that the macrophages and any encapsulated dust particles migrate out of an alveolar cluster into the bronchioles where they can be expelled by ciliary motion. Another theory is that macrophages are transported from the alveoli through fluid-filled capillaries that lead to lymphoids at bronchiole junctions (Brundelet, 1965).

The average life expectancy of an macrophage is about one month. However, these scavenger cells may die rapidly if they engulf toxic particles. The number of macrophages increases when the lungs are subjected to increased entry of dust particles. However, there remains considerable doubt about the relationship between respirable dust concentration and macrophage production. It appears to depend on the mineralogical composition of the particles. Furthermore, dust-loaded cells are cleared less rapidly than the normal turnover of dust-free cells (National Research Council, 1980).

Soluble particles dissolve in mucus. The resulting ions either are removed in the mucus or diffuse through the epithelium for elimination through the bloodstream.

19.3.2 Mechanisms of dust deposition in the respiratory system

Deposition of dust particles within the zones of the respiratory system varies with the size and aerodynamic characteristics of the particles, the geometry of the air passages and the patterns of airflow. The three most important mechanisms of deposition are impaction, sedimentation and diffusion, while interception and electrostatic precipitation can become significant for particular types of dust. These five modes of deposition are discussed in the following subsections.

Impaction

The density and, therefore, the momentum of dust particles are greater than those of a comparable volume of air. At each bend of the tortuous paths followed by air as it pulses in and out of the lungs, dust particles will tend to follow a straight line and impact into the mucus-coated walls of the air passages.

The effectiveness of deposition by impaction increases with the acuteness of the bend and the velocity of the air. There appears to be considerable anatomical variation between individuals in the tortuosity of the branching bronchioles. Velocities are greater in the upper air passages. However, the latter also have a larger diameter and particles must travel further across the streamlines before impact occurs. During exertion, breathing rates and, hence, air velocities increase throughout the system. The subject takes air deeper into the lungs and, furthermore, the resistance of the nostrils and hair filter will promote breathing through the mouth. For all of these reasons, heavy physical labour will increase deposition by impaction in the respiratory system.

Constriction of air passages by thickening of the mucus layer, bronchial infections or lung damage will also result in higher air velocities and, hence, increased deposition by impaction.

Sedimentation

This term refers to the gravitational settlement of particles and is effective at low air velocities for dust particles greater than 0.5 μm. Smaller particles become subject to Brownian motion and diffusion effects. Sedimentation assists in the deposition of larger particles in the nasopharynx during the reversal points of the breathing cycle. More importantly, however, sedimentation is an effective mechanism of deposition in the low velocity laminar flows within the finer bronchioles and the alveoli.

Another factor is that the full capacity of the lungs is seldom used. During normal quiescent breathing the **tidal volume** may utilize some 65% to 75% of lung capacity. Sedimentation of dust particles will occur in the near stagnant air of the unused dead-space. A phase of heavy breathing followed by a quiescent period will first draw dust particles into the deeper recesses of the lung and then encourage deposition by sedimentation in the dead-space as breathing becomes more shallow.

Diffusion

Submicron particles are subject to random displacement by bombardment from gas molecules (Brownian motion). The effect increases as the size of the particles decreases and becomes significant for particle diameters of less than 0.5 μm. Although Brownian motion occurs throughout the respiratory system, it becomes an effective mode of dust deposition only when the mean displacement becomes comparable with the size of the air passage. Hence, it is particularly important in the alveoli and finer bronchioles.

Interception and electrostatic precipitation

Interception becomes significant for fibrous particles. A dust fibre is often defined as a particle where the length-to-diameter ratio exceeds 3. Such particles tend to align themselves with the direction of airflow and fibres 200 μm long can penetrate deeply into the lung. Nevertheless, the ends of fibres are likely to contact the walls of air passages, particularly at bends and bifurcations, and accumulations of fibres can occur at these locations. This is the mechanism of interception. The effect is accentuated by curly fibres such as those of chrysotile asbestos while the straight fibres of amphibole asbestos have a greater probability of penetrating to the alveoli.

Within the working areas of a mine, newly produced particles of mineral dust may carry a substantial electrostatic charge. The moving electromagnetic fields that surround such particles can induce charges of opposite sign on the walls of air passages in the respiratory system. This results in the electrostatic precipitation of particles onto the walls and capture by the film of mucus.

19.3.3 Dust diseases

Pneumonoconiosis, now more usually shortened to pneumoconiosis, is a generic term for damage to cardiorespiratory organs caused by the inhalation of dust. More specific names have been given to illnesses caused by particular types of dust. Some of those are discussed in this section and are often known as the dust diseases.

While the toxic and carcinogenic properties of some dusts cause specific physiological reactions (section 19.2.2), most dusts that occur in mines are not in themselves fatal. However, through progressive incapacitance of the lungs, they render the victim more susceptible to mild respiratory ailments such as colds or influenza as well as tuberculosis, chronic bronchitis and emphysema which may, indeed, result in hastened death. Chronic bronchitis is an inflammation of the linings of the respiratory system and is accompanied by periodic or constant coughing. Emphysema involves the rupture of interalveolar walls caused by excessive pressure within the alveoli. This is often a result of constrictions in the bronchioles. The rupture frequently occurs during bouts of coughing. Adjacent alveoli break through into each other. This leads to progressive abnormal breathlessness, particularly on even light exertion. Respiratory difficulties may also cause excessive strain on the heart with resultant cardiac complications.

Fibrogenic effects

The fibrogenic dusts introduced in section 19.2.2 promote the abnormal development of fibrous tissue within the alveolar clusters. They may commence at discrete foci with the fibrous tissue and radiate outwards to form fibroblasts. These, in turn, can merge into nodules and conglomerates. The permanent scarring and change in the alveolar structure can have severe secondary effects. Gas exchange across the alveolar walls is inhibited and the loss of natural elasticity can cause a significant reduction in tidal volume. Furthermore, the production of macrophages is reduced, allowing uncontrolled accumulations of dust particles to occur in the alveoli. Large fibrous masses can also distort or damage blood vessels, causing functional impairment of the heart (National Academy of Sciences, 1976).

Coal workers' pneumoconiosis

Many dusts, including coal, produce a low biological response. However, over sufficiently long periods of exposure a build-up of retained dust occurs in the form of soft plaques within the lung tissue. These can be observed as small black spots on chest X-rays. Similar early diagnoses can be made for other mineral dusts including ores of iron (siderosis), tin (stannosis) and aluminium (aluminosis). In the case of coal, such indication of coal workers' pneumoconiosis may not be revealed for some 10 to 15 years after initial employment in coal mines. Furthermore, the subjects may not be aware of any incapacitance at that time.

In more advanced cases, the opacities grow in size and number until they coalesce. This is likely to be accompanied by fibrosis and all of the consequences described in the previous subsection.

Silicosis

This is one of the most dangerous of the dust diseases and is caused by particles of free crystalline silica (quarts, sandstones, flint) but not by the silicates in clays or fireclay. The hazard is greatest from freshly produced dust in operations involving mining or comminution of silica-bearing rocks, or from sandblasting. It is suspected that the more severe cases of coal workers' pneumoconiosis may be associated with quartz particles mixed with coal dust.

The early stages of the disease produce, again, local foci of dust accumulations that may be observed on X-ray films. However, there appears to be a level of silica accumulation above which progressive massive fibrosis occurs. A number of theories have been advanced to explain this behaviour (Schröder, 1982). It is probable that the initial accumulations cause simple microscopic abrasion from the hard cutting edges of the particles. The initiation of progressive fibrosis may be the result of a toxic reaction to silicic acid or electrochemical surface energy on newly cut and charged particles. A further suggestion is that the breakdown of poisoned macrophages may invoke an auto-immune reaction that produces fibrous antibody structures.

Asbestosis

Asbestos is an inorganic mineral fibre composed mainly of silicate chains. The two common forms of asbestos are chrysotile containing tough curly fibres, and amphibole with long, straight and brittle fibres. Asbestos fibres are captured in the respiratory system primarily by interception (section 19.3.2) and accumulations are most likely to occur at bends and bifurcations. However, the aerodynamic characteristics of fibres are determined by their diameter rather than length and long fibres may reach the alveoli.

Asbestosis is associated with fibrosis, but of a different type from that given by advanced cases of silicosis or coal workers' pneumoconiosis. The initial plaques are more brittle and contain sharp raised ridges that may become calcified. During further progression of the disease, fibrous bands radiate throughout the lungs and cause a significant loss of elasticity. This reduces the tidal volume and, together with the fibrotic loss of gas exchange area, leads to abnormal breathlessness. The reduction in oxygen transfer causes blood pressure in the pulmonary artery to rise putting the right ventricle of the heart under strain. Cardiac failure may follow.

Cancers of the bronchial system, lung tissue and abdominal organs have been linked with excessive exposure to asbestos fibres. However, it is considered that it is not the silicate chains themselves that cause these cancers but rather carcinogens that are adsorbed on the fibre surfaces prior to inhalation.

Precautions against dust diseases

The problems of dust diseases have probably attracted more research funding throughout the twentieth century than any other environmental hazard in underground mining. Three distinct areas of research have been undertaken in countries with major mining industries, namely

1. medical studies into the development, treatment and diagnosis of dust diseases,
2. the techniques of sampling and measurement of airborne dust, and
3. dust suppression and control in mines.

The primary precautions against dust diseases mirror those same three areas. Experience has shown that, although highly skilled interpretation of X-ray films is required, this remains the most important tool in discovering the onset of a dust disease and in monitoring its progress. Personnel who are required to work in atmospheres that contain any of the toxic, carcinogenic or fibrogenic dusts should be given free access to chest X-ray examinations. An examination should be given before the commencement of employment in order to identify any existing condition. Further chest X-rays are recommended at intervals of not more than two years. Workers should be assured that early identification of coal workers' pneumoconiosis or silicosis does not imply physical impairment nor loss of employment but should be regarded as an indication that reassignment to work in less dusty conditions would be prudent. Legislation may guard against significant reductions in financial remuneration.

The second group of precautions concerns organized and routine procedures of dust sampling, preferably by means of instruments that measure mass concentration of respirable dust. This area is discussed further in section 19.4 and is intended to test compliance with set threshold limit values. In most countries the latter are mandatory values enforced by law.

The most effective means of protecting personnel against dust diseases is the control of dust in mines. This involves minimizing the production of dust, suppressing it at the source, removing airborne dust and separating workers from dusty areas. Great improvements have been made in these areas since the 1950s. The subject is handled in greater detail in Chapter 20.

19.3.4 Threshold limit values

National experience of mining systems, geological conditions, legislation, litigation, labour organizations, research and social consciousness in various countries has resulted in wider variations in recommended or mandated threshold limit values for dusts than for other airborne contaminants in mines. The threshold limit values given in Table 19.2 are based on recommendations of the American Conference of Governmental Industrial Hygienists (ACGIH) and are simply a guide to airborne concentrations that are currently thought to produce no adverse effects from a daily working time exposure. The comments made in section 11.2.1 on threshold limit values for gases apply equally well here. In particular, subsurface ventilation engineers must become familiar with limiting concentrations mandated by their own national or state laws.

There is a further reservation that applies to the use of threshold limit values for dust concentrations. Despite the move from particle count to gravimetric methods of measurement, there remain considerable variations in the results given by differing instruments (e.g. Phillips, 1984). These arise from differences in the design of instruments, the particle size distribution curves they are intended to follow and the efficiency of dust capture. Even a single instrument can indicate a different result if the rate of sample airflow through the device is altered. For these reasons, individual

Table 19.2 Guideline threshold limit values for selected aerosols given in milligrams per cubic metre of air (based on recommendations of the American Conference of Governmental Industrial Hygienists; considerable variations occur in the laws of other countries)

Aerosol	*TLV (TWA)*[a] *(mg/m³)* *(unless otherwise stated)*	*Comments*
Aluminium oxide	10	
Arsenic	0.2	Carcinogen
Asbestos	2 fibres per millilitre (STEL = 10 fibres/ml)	Carcinogen. Fibres longer than 5 μm
Borates		
decahydrate	5	
others	1	
Calcium		
carbonate	10	Marble, limestones
sulphate	10	Gypsum
Carbon black	3.5	Carcinogen
Coal	2	Respirable (< 5% quartz)
Fibrous glass	10	
Fluorides	2.5	
Graphite (natural)	2.5	Respirable fraction
Iron oxide fumes	5	
Kaolin	10	
Magnesite	10	
Mica	3	Respirable fraction
'Nuisance' dusts	10	Non-hazardous material
Oil mist		
mineral	5	Excluding vapour
vegetable	10	
Perlite	10	
Portland cement	10	
Radon daughters	4 WLM/year	See section 13.3.3
Quartz	0.1	Respirable fraction
Silicon	10	
Soapstone	3	Respirable fraction
Talc	2	Respirable fraction
Welding fumes	5	

[a] See section 11.2.1 for definitions of threshold limit value and time-weighted average.

countries or mining industries have tended to 'adopt' a particular instrument as a standard and to employ empirical conversion factors in order to compare data obtained by other devices (Rogers, 1991). It follows that, while such variations in instrument performance exist, the application of dust threshold limit values for purposes of legislation or enforcement should be referred to a particular instrument used according to a specified procedure.

Another factor should be borne in mind when comparing threshold limit values imposed by differing legislative authorities. Laws governing dust concentrations usually specify not only the limiting concentrations of respirable dust but also the sampling locations. Hence, a measurement required to be taken in a return airway should not be compared with one that is obtained at the position of a machine operator.

For coal mines, legislation may require reduced threshold limit values in certain areas. For example, the 2 mg/m^3 TLV for respirable dust in American coal mines is reduced to 1 mg/m^3 for intake airways within 61 m (200 ft) of a working face. Further restrictions are applied when quartz particles are present in the coal dust. One method is to reduce the TLV to 10 divided by the quartz percentage when the quartz content in the dust exceeds 5%. This is based on a quartz TLV of 0.1 mg/m^3.

Non-hazardous material that forms 'nuisance' dusts (section 19.2.2) is normally allocated a threshold limit value of 10 mg/m^3.

19.4 THE ASSESSMENT OF AIRBORNE DUST CONCENTRATIONS

19.4.1 Background

The accurate assessment of dust concentrations in relation to the health of personnel in mines is beset with difficulties. First, the fact that physiological consequences develop very slowly is compounded by dust concentrations varying across wide limits with respect to both time and place in a mine. Readings of dust concentrations measured over a short time interval have very little relevance to the long-term health of the workforce. It was this difficulty that masked correlations between most particle count data and incidence of pneumoconiosis until the 1959 Johannesburg Conference. Secondly, as indicated in the previous section, it is not only the concentration of particles that matters but also their size distribution and mineralogical composition. It is not surprising, therefore, that even modern instruments may vary quite significantly in the quality and type of data they produce.

In choosing dust instrumentation it is important to define the primary purpose of the intended measurements. These may be part of long-term investigations with the aim of establishing environmental standards. Other research surveys may be aimed at the spatial and temporal variations of dust concentration to which specific groups of personnel are exposed, or to investigate the effects on airborne dust of particular mining equipment or dust suppression techniques. Routine measurements in a mine are made to protect the health of the workforce and to check or ensure compliance with mandatory standards.

The earliest attempts to measure airborne dust quantitatively are reported to have occurred in the eighteenth century using observations of dust deposition on a polished surface (Walli, 1982). However, the major developments in this field have taken place since the introduction of the konimeter, a hand-held jet-impact device invented by Kotzé, a South African government mining engineer, in 1916. Modern versions of the konimeter remain in use in that country although its remaining life must be very limited.

Since the Johannesburg Conference of 1959, the vast majority of new instruments have indicated the mass concentration of respirable dust ($< 5\,\mu m$). Most of these instruments pass the dust in a continuous air sample through combinations of elutriators (settling chambers), cyclones, or jet impactors and filters. Furthermore, in order to meet the need for longer-term readings, most gravimetric instruments have been designed to operate over several hours. Such data are necessary from the viewpoint of health protection but mitigate against rapid readout, continuous monitoring and short-term control. Alternative methods of dust assessment combined with modern electronic integrating circuitry allow short-term and long-term data collection to proceed simultaneously. The most popular of these utilize the principle of light beam scattering by dust particles (the Tyndall effect).

A recognition of the differing objectives of health protection and more immediate concerns such as dust explosibility, visibility and nuisance effects have led to two distinct applications of dust monitors. First, lightweight personal samplers are increasingly being worn by underground workers. These give a record of the respirable dust concentrations to which the person has been exposed, on a shift-by-shift basis. Secondly, heavier and more sophisticated instruments may be set up to measure dust concentrations either for several hours or, on a more permanent basis, to allow continuous monitoring for local indication and recording or transmission to a central control station.

Modern trends in dust instrumentation are toward

1. an increase in the use of personal samplers,
2. cascade devices or other means of selection to indicate mass concentrations in each of a number of particle size ranges,
3. an immediate indication of mineral content (particularly quartz) in addition to mass concentration, and
4. an increase in light scattering methods for continuous monitoring.

In the following subsections, we shall introduce the main principles employed in instruments for the assessment of airborne dust. However, because of the rapid evolution and number of new devices, we shall not attempt detailed descriptions of individual monitors. For these, the reader is referred to trade journals and manufacturers' literature.

19.4.2 Particle count methods

Although particle count methods were used for many years and models were produced in a number of countries, they are now nearing obsolescence. There were essentially two families of particle count instruments. One of these relied on the impact principle in which a short-lived but high velocity pulse of air was induced through a jet and directed at a receiving surface of treated glass, film or liquid. The most enduring of this type of instrument is the konimeter (section 19.4.1).

The second type of particle count instrument was the thermal precipitator. Molecular bombardment from a heated wire diverted dust particles from a moving sample stream on to an adjacent glass slide (section 20.2.4). The slides, films or liquids produced by particle count instruments were subjected to microscope analysis for counting the number of particles in each size range. For many years this was carried out manually and was somewhat subject to individual bias. Latterly, it has been conducted by 'light assessors' that really measure surface area (Martinson, 1982). Computer-controlled particle analysers are now available that can perform such a task much more rapidly and efficiently.

19.4.3 Gravimetric methods

Figure 19.2 illustrates the operation of one of the earlier and very successful gravimetric samplers, the MRE, developed at the Mining Research Establishment of the National Coal Board in the United Kingdom (Dunmore *et al.*, 1964). This instrument continues to be a standard in several countries. Air passes through the instrument at a rate of 2.5 l/min under the action of a diaphragm pump. The air velocity between the parallel plates of the elutriator drops to the extent that particles larger than 7 μm settle out. The remaining finer particles pass on to be collected on a 5 μm membrane filter. The filter is weighed before and after a sampling period which may, typically, be 8 h.

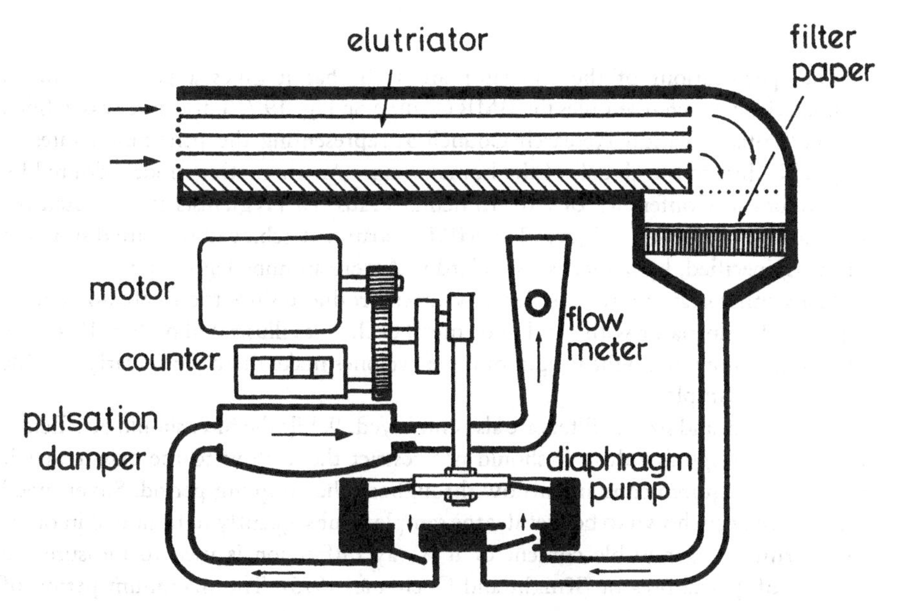

Figure 19.2 Schematic of an MRE gravimetric dust sampler.

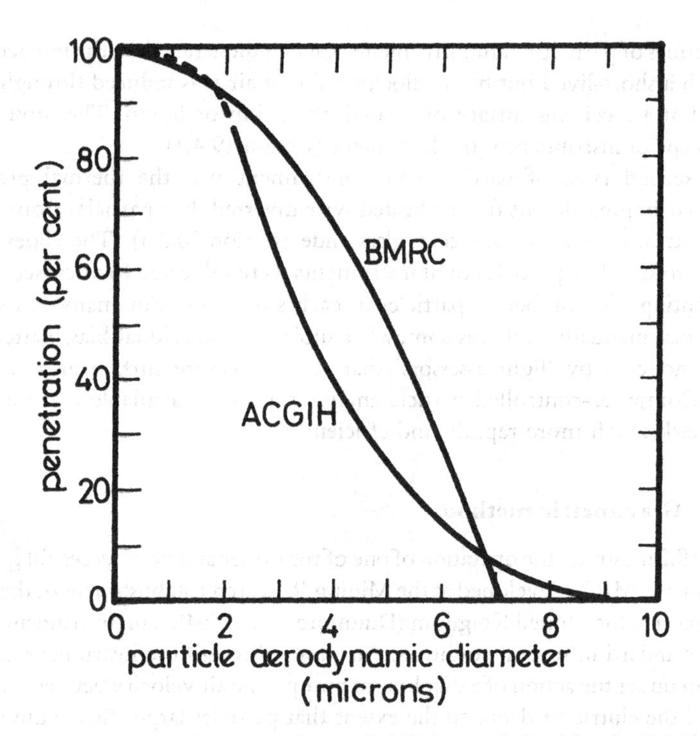

Figure 19.3 Lung penetration curves for respirable dust as defined by the British Medical Research Council (BMRC) and the American Conference of Governmental Industrial Hygienists (ACGIH).

The specifications of the elutriator are such that it gives a particle removal characteristic which resembles the BMRC curve on Fig. 19.3. This curve was defined by the British Medical Research Council as representing the penetration rates of respirable dust to the alveoli of the human lungs. An alternative model adopted by the American Conference of Governmental Industrial Hygienists is illustrated for comparison on the same figure. The MRE instrument achieved the rare distinction of being specified, by name, as a standard in American mine legislation.

Many other gravimetric samplers have been produced since the 1960s. Some have replaced the elutriator with a nylon or metal cyclone as illustrated on Fig. 19.4. The small dimensions and light weight of these cyclones make them particularly suitable for personal samplers.

Other types and sizes of filters are also employed. While the filter should be efficient in trapping respirable dust, it should not restrict the airflow to the extent that it inhibits the required constant airflow throughout the sampling period. Silver metal filters have been shown to be useful if the sample is subsequently to be heated in order to determine combustible content or if X-ray diffraction is used to measure the amount of quartz present (Knight and Cochrane, 1975). The maximum period of

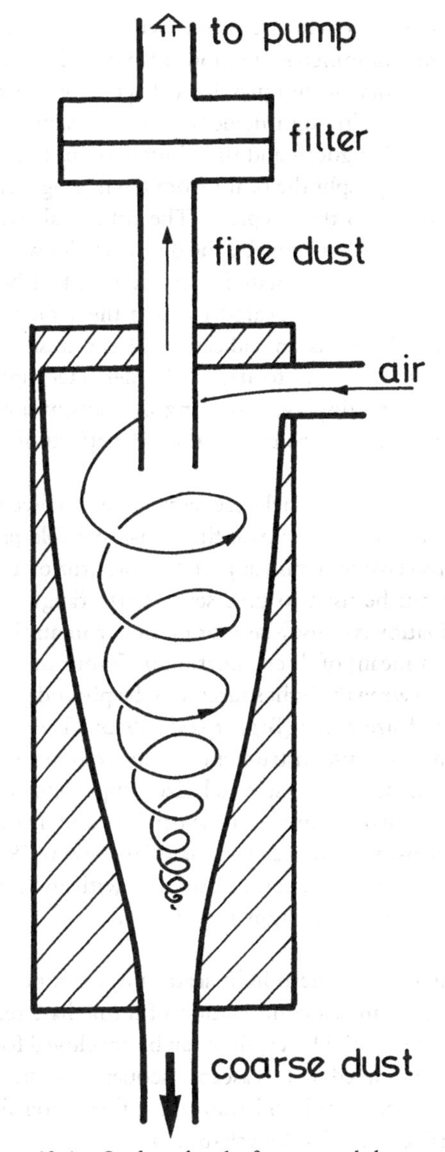

Figure 19.4 Cyclone head of a personal dust sampler.

sampling through any given filter is reduced because of premature blockage in the presence of diesel fumes or mists of water or oil (Gardiner, 1988).

19.4.4 Photometric (light-scattering) methods

It was the English physicist John Tyndall (1820–1893) who first explained the scattering of sunlight by dust particles in the atmosphere. The blue component of white light is scattered most which explains why the sky appears blue. The same effect is utilized in the photometric methods of measuring dust concentration. The corresponding instrumental techniques have been developed particularly by the German mining industry although devices using the same principle are now also produced in the United Kingdom and the United States of America.

The early Tyndalloscope split the beam from a white light source. Half proceeded through polarizing filters to the eyepiece. The other half was diverted through a sample chamber where light reflected by the dust particles was collected and directed to the eyepiece. The filters were rotated until the two half-beams that were visible simultaneously in the eyepiece appeared to have the same intensity. The angle of filter rotation was employed as an indication of the dust concentration. Modern Tyndallometers use photosensors to detect the deflected light and to produce an electrical output for visual display, recording or transmission. Preferred angles of forward light scatter can be chosen to minimize the effects of particle shape or index of refraction.

There are two techniques of particle size discrimination used in these devices. Some photometric dust instruments eliminate the non-respirable particles by passing the sample through a pre-classifier consisting of an elutriator or a cyclone. A number of cyclones in cascade can be used to give several size ranges. The other method of particle size discrimination is to use a laser or monochromatic light source. The choice of wavelength gives a means of discriminating in favour of the desired size range of particles. The angle between the light source and the photodetector is another method of selecting a preferred size range (Breuer *etal.*, 1973; Thaer, 1975).

Although photometric dust instruments really respond to light reflected from particle surfaces, using them for selected size ranges enables the readings to be interpreted in terms of total volume of those particles and, hence, mass concentration (assuming constant density of the particle material). Figure 19.5 illustrates an instrument that combines a forward scattering laser unit with an 8 h filtration system. This permits direct calibration of the light unit with respect to mass concentration of respirable dust.

The major advantage of modern light scattering instruments is that the addition of electronic circuitry permits a combination of immediate readout and integration over any chosen time interval. Hence, they can be employed for both short-term and long-term sampling. Coupled with cascade (sequential) interrogation of a sample stream they can indicate, record and transmit information for each of a series of particle size ranges (Breuer, 1975; Oberhalzer, 1987).

19.4.5 Personal samplers

Personal samplers are lightweight gravimetric devices that attach to the clothing of individual workers and are powered by the same battery used for the caplamp. Current personal samplers employ cyclone and filter units. The pump is switched on before the employee enters the mine and remains running until the caplamp is replaced on

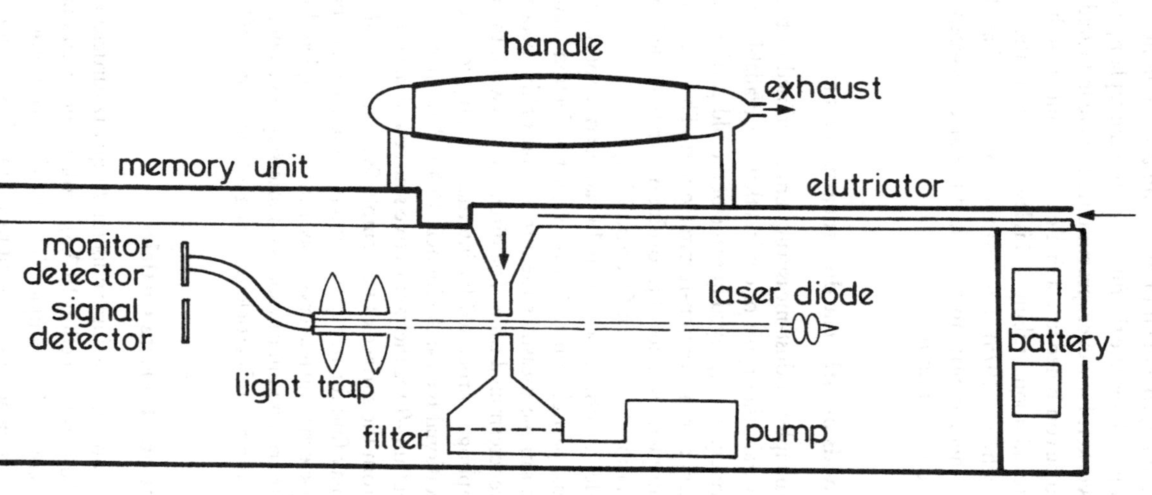

Figure 19.5 Schematic of a SIMSLIN II dust sampler.

the charging rack at the end of the shift. The filter is replaced at that time. The wider employment of personal samplers not only provides a history of individual exposure to respirable dust but, coupled with the type of employment of the worker, also gives a wealth of data for the improvement of dust conditions in mines.

Some personal samplers integrate the filter unit into the caplamp in order to sample near the face of the worker. However, practical trials have indicated a preference for the filter and pump to be combined into a single unit mounted on the belt or caplamp battery. This eliminates additional tubing and is less intrusive to the worker (Gardiner, 1988; Knight and Cochrane, 1975). Furthermore, provided that the device is not covered by clothing, there appears to be little difference between results given by belt- and hat-mounted filteration units.

19.4.6 Other methods of airborne dust measurement

In addition to the principles of dust-measuring units outlined in the preceding subsections, there are several other types of devices that are available or under development. The impact principles employed in some of the old particle count devices still have a role as classifiers to remove non-respirable dust. This may further reduce the weight of personal samplers. A series of jet orifice sizes can be used to simulate a lung penetration curve (National Research Council, 1980).

Another type of device utilizes a carbon-14 radioactive source to pass beta rays through a mass of dust particles on a filter or impact collection plate. A detector on the other side of the sample measures the attenuation of the beta rays. In use, the device is run for a predetermined time. The difference between the flux of beta rays before and after sampling is processed by internal circuitry to indicate the mass of particles collected (National Research Council, 1980). A similar device developed in Poland employs strontium-90 as the radioactive source and measures reflected backscatter of beta rays from the surface of the sample (Krzystolik *et al.*, 1985).

A technique that may find further application is to measure the change in resonant frequency of a piezo-electric quartz crystal as a mass of sampled dust collects on it. Here again, such units arranged in cascade with interstage cyclones or impactors can be used to indicate a number of particle size ranges (Sem *et al.*, 1977).

19.4.7 Discrimination of quartz particles

Mine dust is seldom composed of a single mineral. As quartz particles occur commonly and are particularly dangerous (section 19.3.3), efforts have been directed at quantifying the quartz content of dust samples. This is of concern not only in hardrock mines but also because of the increased use of mechanized extraction and roof bolting in coal mines.

The employment of X-ray diffraction gives well-defined peaks for quartz on the output spectrum. This method has been used for a number of years as a means of mineralogical analysis of mine dust (Knight and Cochrane, 1975; Bradley, 1975). A further development is to incorporate the principle into dust-monitoring equipment in order to display the quartz content concurrently with respirable dust concentration.

19.4.8 Sampling strategy

Throughout the history of organized dust measurements in mines, investigators have been faced with the problems of large variations in the observed results. Furthermore, the significance of these results has often been subject to debate and interpretation. The very real variations in the concentration and mineralogical content of dusts that exist with respect to time and location in any mine are compounded by the differing efficiencies with which alternative instruments simulate the dust retention characteristics of the human lung. These difficulties have led to coining of the phrase **sampling strategy** which really means the 'why, where, when and how' dust samples are taken.

The 'why', i.e. the objectives of dust sampling, was outlined in section 19.4.1. It is, indeed, important to have a clearly defined purpose for any given set of dust measurements, particularly where the setting or checking of mandatory standards is involved. The objectives of a dust survey will then usually influence the 'where, when and how'.

When checking compliance with threshold limit values, the locations of area measuring points are normally dictated by law—for example at specific points in intake and return airways or with respect to the positions of machine operators. Research into dust minimization techniques in various countries has been greatly influenced by the wording of the relevant national law. For example, spray fan or other air diversion techniques can produce significant reductions in dust concentrations at the operator position of a continuous miner but may have little or no effect on respirable dust counts in the return airway. If, on the other hand, the purpose of the measurements is to check the dust production of a particular piece of equipment or operation, then the instruments should be located at specified distances upwind and downwind from that equipment or operation.

The times and durations of sampling will, again, usually be specified by law, often at intervals of two or three months at each mandatory sampling point. The samples may be required to be submitted for analysis to specified laboratories operated or authorized by government agencies. Furthermore, check samples may be taken by inspectors appointed by government.

For short-term sampling, it is easy to introduce conscious or inadvertent bias into dust measurements. The maximum dust make from a machine will be obtained when the equipment is running at full load. However, the measured concentration of non-toxic dust should not necessarily be interpreted as an epidemiological (health) hazard. It becomes so only if personnel are working in or downstream from the dust make and are exposed to concentrations that exceed the set standards on the basis of an 8 h time-weighted average.

The importance of adhering to agreed types of instruments and sampling procedures was emphasized in section 19.3.4, particularly when checking compliance with statutory requirements. Indeed, the instruments and measurement procedures may again be specified within the regulations. A factor that is often overlooked is the effect of air velocity on the sampling ports of the instrument. At the higher air velocities, dust particles can be diverted around the sampling ports (Baskhar and Ramani, 1987). Isokinetic devices can be fitted to the instrumentation in order to match the inlet port sampling velocity with that of the approaching airstream.

The variability of measured dust concentrations coupled with the intrinsic uncertainties of instrumentation and sampling procedures should dictate that those measurements be subjected to statistical analysis. In practice this is often relegated to a rejection of 'unrepresentative or suspect' samples and the straight averaging of the remainder (National Academy of Sciences, 1976; Martinson, 1982). Such a procedure is clearly open to bias and is not to be recommended. Where any given sampling practice or location consistently yields a significant number of suspect samples, then an investigation should be carried out in order to determine whether the unexpected variations are, indeed, real and, if not, the weaknesses of the sampling technique. Statistical examinations can be carried out to test whether the number of samples is sufficiently large to be representative of the particular type of work, time or location. The data produced by wider use of personal samplers, together with computer-based records and powers of statistical computation, facilitate such tasks.

This returns us to the question of personal samplers. The question of sampling strategy became of particular importance with the advent of 8 h gravimetric sampling units and the relationship between the results given by these instruments and the earlier, short-term, particle count units. Personal samplers have revolutionized the philosophy of dust sampling strategy. Provided that a correlation has been established between the results given by specified types of personal sampler and corresponding levels of epidemiological hazard, then there appears to be little reason why those types of personal samplers should not be used for both individual exposure monitoring and compliance with appropriately worded regulations. This separates sampling for health reasons from those measurements that may be taken for purposes of control or planning.

The question arises on how many underground personnel should be asked to wear personal samplers. This would appear to depend on the degree of perceived exposure to dust and the type of mining. In heavily mechanized workings, there will be relatively few persons within the most dusty areas. Furthermore, the mobility and differing occupations of those workers may subject them to wide variations of dust exposure. In such circumstances, it would be in the interests of those workers to wear personal samplers. The initiation of the practice should be accompanied by suitable explanations of the benefits of the devices, how they should be treated, and the manner in which the data will be used.

In labour-intensive mines with much larger numbers of personnel, many of them engaged in similar occupations, it may be unnecessary to ask all persons to wear personal samplers. A minimum number that is deemed to be representative of each work group should be decided, including the more mobile supervisory staff (Quilliam, 1975). The actual persons wearing the samplers may be changed on a rota basis.

However, after each shift the exposure record of each individual should be updated according to the relevant representative value for that group and location. Provided that the instruments are compatible, the data obtained can be used to compare dust indices for differing occupations, locations and mines (van Sittert, 1988).

The modern approach is to combine full-shift personal samplers with both long- and short-term dust monitoring that is incorporated into mine environmental electronic surveillance systems. The data obtained from such means will help to safeguard the health of workers and be valuable in the planning and design of future mining operations.

REFERENCES

Baskhar, R. and Ramani, R. V. (1987) A comparison of the performance of impactors and gravimetric dust samplers in mine airflow conditions. *Proc. 3rd US Mine Ventilation Symp., Penn State*, 502–8.

Bradley, A. A. (1975) The determination of the quartz content of gravimetric mass samples of airborne dust by an X-ray technique. *Proc. 1st Int. Mine Ventilation Congr., Johannesburg*, 455–8.

Breuer, H. (1975) TBF 50 and Tyndallometer TM digital —two instruments supplementing each other for the occupational hygienic and technical assessment of dust conditions. *Proc. 1st Int. Mine Ventilation Congr., Johannesburg*, 445–52.

Breuer, H., Gebhart, J. and Robock, K. (1973) Photoelectric measuring apparatus for determination of the fine dust concentration. *Staub Reinhalt. Luft* **33**, (in English).

Brundelet, P. J. (1965) Experimental study of the dust-clearance mechanism of the lung. *Acta Pathol. Microbiol. Scand. Suppl.* **175**, 1–141.

Carver, J. (1975) Respirable dust regulations in the United Kingdom. *Proc. Int. Mine Ventilation Congr., Johannesburg*, 399–405.

Dunmore, J. H., Hamilton, R. J. and Smith, D. S. G. (1964) An instrument for the sampling of respirable dust for subsequent gravimetric assessment. *J. Sci. Instrum.* **41**, 669–72.

Gardiner, L. R. (1988) Personal gravimetric dust sampling for the South African gold mining industry. *Proc. 4th Int. Mine Ventilation Congr., Brisbane*, 507–15.

Gibson, H. and Vincent, J. H. (1980) An investigation of 'fly-dust' nuisance in mines. *Proc. 2nd Int. Mine Ventilation Congr., Reno, NV*, 620–2.

Knight, G. and Cochrane, T. S. (1975) Gravimetric dust sampling with quartz analysis and its use in metal and mineral mines. *Proc. 1st Int. Mine Ventilation Congr., Johannesburg*, 407–14.

Krzystolik, P. A., *et al.* (1985) Portable coal dust/stone dust analyzer. *Proc. 2nd US Mine Ventilation Symp., Reno, NV*, 171–9.

Martinson, M. J. (1982) Sampling pathogenic airborne particulates. *Environmental Engineering in South African Mines,* Chapter 14, 357–78, Mine Ventilation Society of South Africa.

National Academy of Sciences (1976) *Mineral Resources and the Environment: Coal Workers' Pneumoconiosis*—Medical Considerations, *Some Social Implications*, Washington, DC.

National Research Council (1980) *Measurement and Control of Respirable Dust in Mines*, US National Academy of Sciences, Washington, DC, NMAB-363.

Oberhalzer, J. W. (1987) Assessment of colliery dust levels using a computerized dust measuring system. *Proc. 3rd US Mine Ventilation Symp., Penn State*, 617–24.

Orenstein, A. J. (ed.) (1959) *Proc. Int. Pneumoconiosis Conf., Johannesburg*, Churchill, London, 632 pp.

Phillips, H. R. (1984) New methods and standards for respirable dust monitoring in the New South Wales coal industry. *Proc. 3rd Int. Mine Ventilation Congr., Harrogate*, 203–8.

Quilliam, J. H. (1975) A review of dust sampling techniques in South African gold mines. *Proc. 1st Int. Mine Ventilation Congr., Johannesburg*, 419–21.

Rogers, W. G. (1991). Silica; crystalline quartz exposure standards··· what is the most appropriate standard to use? *J. Mine Vent. Soc. S. Afr.* (March), 38–43.

Schröder, H. H. E. (1982) The properties and effects of dust. *Environmental Engineering in South African Mines*, Chapter 12, 313–36, Mine Ventilation Society of South Africa.

Sem, G. J., Tsurubayashi, K. and Homma, K. (1977) Performance of the piezo-electric microbalance respirable aerosol monitor. *Am. Ind. Hyg. Assoc. J.* **38**, 580.

Thaer, A. (1975) The Leitz Tyndallometer TM digital as applied to the determination of fine dust concentration under the auspices of occupational and environmental medicine. *Kungl Arbetarskyddsstyrelsen Semin., Stockholm*, April 23.

van Sittert, J. M. O. (1988) An overview of the proposed guidelines to standardise the respirable dust sampling strategy for risk determination in collieries. *Proc. 4th Int. Mine Ventilation Congr., Brisbane*, 527–31.

Walli, R. A. (1982) Mine dusts. In *Mine Ventilation and Air Conditioning* (ed. Hartman), Chapter 5, 84–130.

20

The aerodynamics, sources and control of airborne dust

20.1 INTRODUCTION

The physical characteristics of aerosols have been subjected to intensive study for the free surface atmosphere. This is an important area in meteorology and investigations of the behaviour of contaminant plumes in the atmosphere. Somewhat less attention has been paid to the aerodynamic characteristics of dust when the carrying airstream is confined within the boundaries of ducts or tunnels.

The first main section in this chapter outlines the several phenomena that govern the manner in which airborne dust is transported through the branches of a ventilation network and the deposition of dust particles on the roof, floor and sides of mine airways.

A prerequisite to the successful control of airborne dust in a mine is an understanding of the potential sources of the dust. These are discussed in the second main part of the chapter. While some sources are obvious such as a power loader or tunnelling machine, others are less so including the crushing of immediate roof strata by modern powered supports. The final section outlines the methods of dust control in mining operations. These include prevention of the formation of dust, suppression and removal of dust particles from the air, isolating personnel from concentrations of dust and the diluting effects of airflow. The latter was introduced in section 9.3.3.

20.2 THE AERODYNAMIC BEHAVIOUR OF DUST PARTICLES

The very large size range of dust particles that exist in the ventilation system of an active mine results in a variety of differing phenomena influencing the behaviour of the particles. The smallest particles act almost as a gas and react to molecular forces while the larger particles are influenced primarily by inertial and gravitational effects. In this section we shall consider the influence of gravitational settlement, molecular diffusion, turbulent or eddy diffusion, coagulation, impingement, re-entrainment and computer simulations.

20.2.1 Gravitational settlement

The rate at which a particle falls through air under the action of gravity depends not only on the size and density of the particle but also on its shape. In section 19.2.1, the concept of an equivalent geometric diameter based on projected area was introduced. The majority of analyses in this subject assume that each particle is a homogeneous sphere. In the study of particle aerodynamics this has given rise to further alternative definitions of equivalent diameter including

1. Stokes' diameter, the diameter of a sphere that has same density as the actual particle and falls through air at the same rate, and
2. aerodynamic diameter, the diameter of a sphere of density 1 g/cm^3 that falls through air at the same rate as the actual particle.

Despite these additional definitions, the geometric diameter remains the one that is most commonly used in practice.

Stokes' law and terminal velocities

When any body is suspended in a fluid, at least two forces act upon it (Fig. 20.1). One is the weight of the body within the prevailing gravitational field. The volume

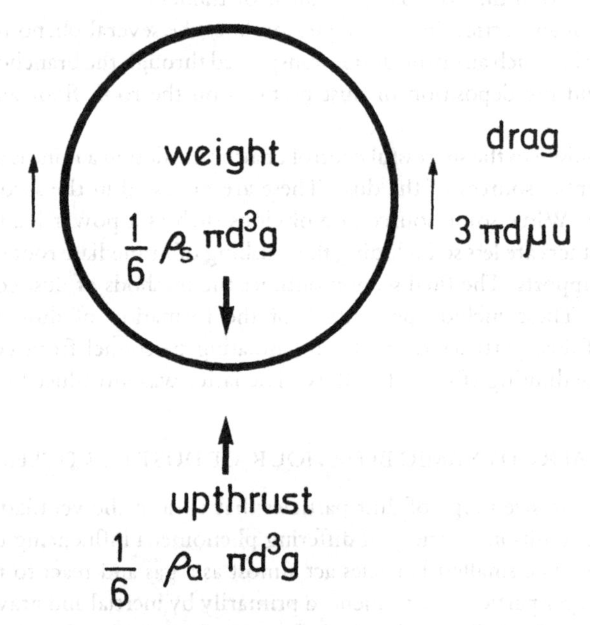

Figure 20.1 Forces on a falling particle in air. At equilibrium: $(1/6)\pi d^3 g(\rho_s - \rho_a) = 3\pi d\mu u$.

of a sphere of diameter d is

$$\tfrac{1}{6}\pi d^3 \quad \mathrm{m}^3$$

If this has a density of $\rho_s\,(\mathrm{kg/m^3})$ then its weight becomes

$$\tfrac{1}{6}\rho_s \pi d^3 g \ \mathrm{N} \tag{20.1}$$

where g = gravitational acceleration (m/s^2).

However, the sphere displaces its own volume of fluid and will experience an upthrust equal to the weight of fluid displaced, i.e.

$$\tfrac{1}{6}\rho_a \pi d^3 g \quad \mathrm{N} \tag{20.2}$$

where ρ_a = density of the fluid (kg/m^3).

The net force causing downward movement is the combination of the two:

$$\tfrac{1}{6}\pi d^3 g(\rho_s - \rho_a) \quad \mathrm{N} \tag{20.3}$$

If the particle is moving relative to the fluid then it will experience a further resistance or **drag** because of viscous shear and conversion of some of its kinetic energy into turbulent eddies within the fluid. A general expression for drag was given in section 5.4.6 as

$$\mathrm{drag} = C_D A_b \rho_a \frac{u^2}{2} \quad \mathrm{N} \tag{20.4}$$

where C_D = coefficient of drag (dimensionless), u = relative velocity between the particle and the fluid, m/s, and A_b = projected area ($= \pi d^2/4$), m^2.

Many investigators have investigated relationships between C_D and Reynolds' number, Re, for fully submerged bodies (e.g. Prandtl, 1923). In the case of spheres, the diameter is used as the characteristic length in the calculation of Reynolds' number. For the particular case of laminar flow around a particle, Sir George G. Stokes (1819–1903), the Cambridge physicist, showed that

$$C_D = \frac{24}{\mathrm{Re}} \tag{20.5}$$

Now, as

$$\mathrm{Re} = \frac{\rho_a u d}{\mu_a}$$

where μ_a = dynamic viscosity of the fluid (N s/m^2),

$$C_D = \frac{24\,\mu_a}{\rho_a u d}$$

Substituting for A_b and C_D, equation (20.4) gives

$$\text{drag} = \frac{24\,\mu_a}{\rho_a u d}\frac{\pi d^2}{4}\rho_a\frac{u^2}{2}$$

$$= 3\pi\mu_a u d \quad \text{N} \tag{20.6}$$

As the particle accelerates downwards, its velocity, u, increases until the drag equals the downward force quantified in equation (20.3):

$$\tfrac{1}{6}\pi d^3 g(\rho_s - \rho_a) = 3\pi\mu_a u d \quad \text{N} \tag{20.7}$$

At that point of dynamic equilibrium, the velocity of fall becomes constant and is renamed the terminal velocity, u_t. Equation (20.7) may now be rearranged as

$$u_t = \frac{d^2 g(\rho_s - \rho_a)}{18\mu_a} \quad \frac{\text{m}}{\text{s}} \tag{20.8}$$

Equations (20.5) to (20.8) have all been referred to as Stokes' law.

Stokes' law applies with good accuracy to particles that are above the respirable range (5 μm). Smaller particles become sensitive to slippage and molecular forces. Stokes' law is based on the assumption of laminar flow. If the terminal velocity is sufficiently high to cause the onset of a turbulent wake then the transfer of kinetic energy from the particle to the fluid (inertial effects) can no longer be ignored. The upper limit of Stokes' law occurs at a Reynolds' number, Re, of about 0.1 which, for many mineral particles falling at their terminal velocity through air, is equivalent to geometric diameters of approximately 20 μm.

For larger particles at their terminal velocity, u_t, we may balance equations (20.3) and (20.4):

$$\frac{1}{6}\pi d^3 g(\rho_s - \rho_a) = C_D\frac{\pi d^2}{4}\rho_a\frac{u_t^2}{2}$$

giving

$$u_t = \sqrt{\frac{4}{3}\frac{dg(\rho_s - \rho_a)}{C_D\rho_a}} \quad \frac{\text{m}}{\text{s}} \tag{20.9}$$

For dust particles in air, $\rho_s \gg \rho_a$ and the term $\rho_s - \rho_a$ is usually truncated to ρ_s. Flagan and Seinfeld (1988) suggest the approximations for coefficients of drag, C_D given in Table 20.1

Slip flow

Stokes' law applies to dust particles that are large in comparison with the mean free path of the gas molecules. Hence, those particles see the gas as a continuum. As the particle size approaches the mean free path of the gas molecules this no

Table 20.1 Approximations for coefficients of drag for spherical particles (after Flagan and Seinfeld, 1988)

Reynolds' number, Re	C_D
<0.1	$\frac{24}{\text{Re}}$ (Stokes' law)
$0.1<\text{Re}<2$	$\frac{24}{\text{Re}}\left[1+\frac{3}{16}\text{Re}+\frac{9}{160}\text{Re}^2\ln(2\text{Re})\right]$
$2<\text{Re}<500$	$\frac{24}{\text{Re}}\left[1+0.15\,\text{Re}^{0.687}\right]$
$500<\text{Re}<2\times10^5$	0.44

longer holds. Two effects are then observable; first the jerky dislocations caused by molecular bombardment, known as Brownian motion and discussed in section 20.2.2, and secondly, the drag force reduces as the small particle becomes more able to move or 'slip' through the intermolecular voids.

In order to quantify the very small distances now being considered, let us recall that the **mean free path** of a gas molecule is defined as the average distance it moves between collisions with other gas molecules. Although air is a mixture of gases, it is convenient to treat it as a single gas of equivalent molecular weight 28.966 and gas constant 287.04 J/(kg K).

From the kinetic theory of gases it can be shown that the mean free path, λ, is given by

$$\lambda=\frac{\mu}{0.499\,P\,(8/\pi R T)^{1/2}}\quad \text{m} \tag{20.10}$$

where μ = dynamic viscosity (N s/m^2), P = pressure (N/m^2), R = gas constant (J/(kg K)) and T = absolute temperature (K). For air at $P=100$ kPa, $T=293$ K (20 °C), $R=287.04$ J/(kg K) and $\mu_a=17.9\times10^{-6}$ N s/m^2 (section 2.3.3),

$$\lambda=\frac{17.9\times10^{-6}}{0.499\times10^5[8/(\pi\,287.04\times293)]^{1/2}}$$
$$=6.52\times10^{-8}\,\text{m or } 0.0652\,\mu\text{m}$$

When particle diameters fall below 5 μm, the effect of slippage becomes significant. In order to extend the applicability of Stokes' law, a correction factor, C_c, can be introduced to reduce the calculated value of drag. Thus, for small particles, equation (20.6) is corrected to

$$\text{drag}=\frac{3\pi\mu_a ud}{C_c}\quad \text{N} \tag{20.11}$$

A number of relationships between C_c and d have been suggested (e.g. Allen and Raabe, 1982), based mainly on a series of classical experiments on liquid aerosols carried out by Millikan between 1909 and 1923. Values of the slip correction factor for air at 25 °C and 101 kPa are given in Table 20.2.

Table 20.2 Slip correction factor for dust particles in air (after Flagan and Seinfeld, 1988)

d (μm)	0.01	0.05	0.1	0.5	1.0	5.0	10.0
C_c	22.7	5.06	2.91	1.337	1.168	1.034	1.017

The equation

$$C_c = \frac{9.56 \times 10^{-8}}{d^{1.045}} + 0.99$$

with d expressed in metres, gives C_c within an accuracy of 2%.

Incorporating the slip correction factor into Stokes' law for terminal velocity, equation (20.8) gives a relationship that can now be extended down to a particle size of 0.01 μm:

$$u_t = \frac{d^2 g(\rho_s - \rho_a) C_c}{18 \mu_a} \quad \frac{\text{m}}{\text{s}} \tag{20.12}$$

Figure 20.2 gives a graphical representation of this equation for particles of varying diameter and density falling through air of temperature 20 °C. The curvature of the lines on this log–log plot is due to the effects of slippage.

Example Determine the terminal velocities and time taken for particles of geometric equivalent diameter 0.1, 1, 10 and 100 microns to fall a distance of 2 m, through air of density $\rho_a = 1.1$ kg/m^3 and dynamic viscosity, $\mu_a = 18 \times 10^{-6}$ N s/m^2. The density of the dust material is 2000 kg/m^3.

Solution The slip-corrected Stokes' equation (20.12) gives

$$u_t = \frac{d^2 g(\rho_s - \rho_a) C_c}{18 \mu_a}$$

$$= \frac{9.81\,(2000 - 1.1)}{18 \times 18 \times 10^{-6}} d^2 C_c$$

$$= 6.052 \times 10^7 d^2 C_c \quad \text{m/s}$$

Applying this relationship to each of the given particle diameters and reading corresponding values of C_c from Table 20.2 (remembering to multiply microns by 10^{-6} to convert diameters to metres) gives

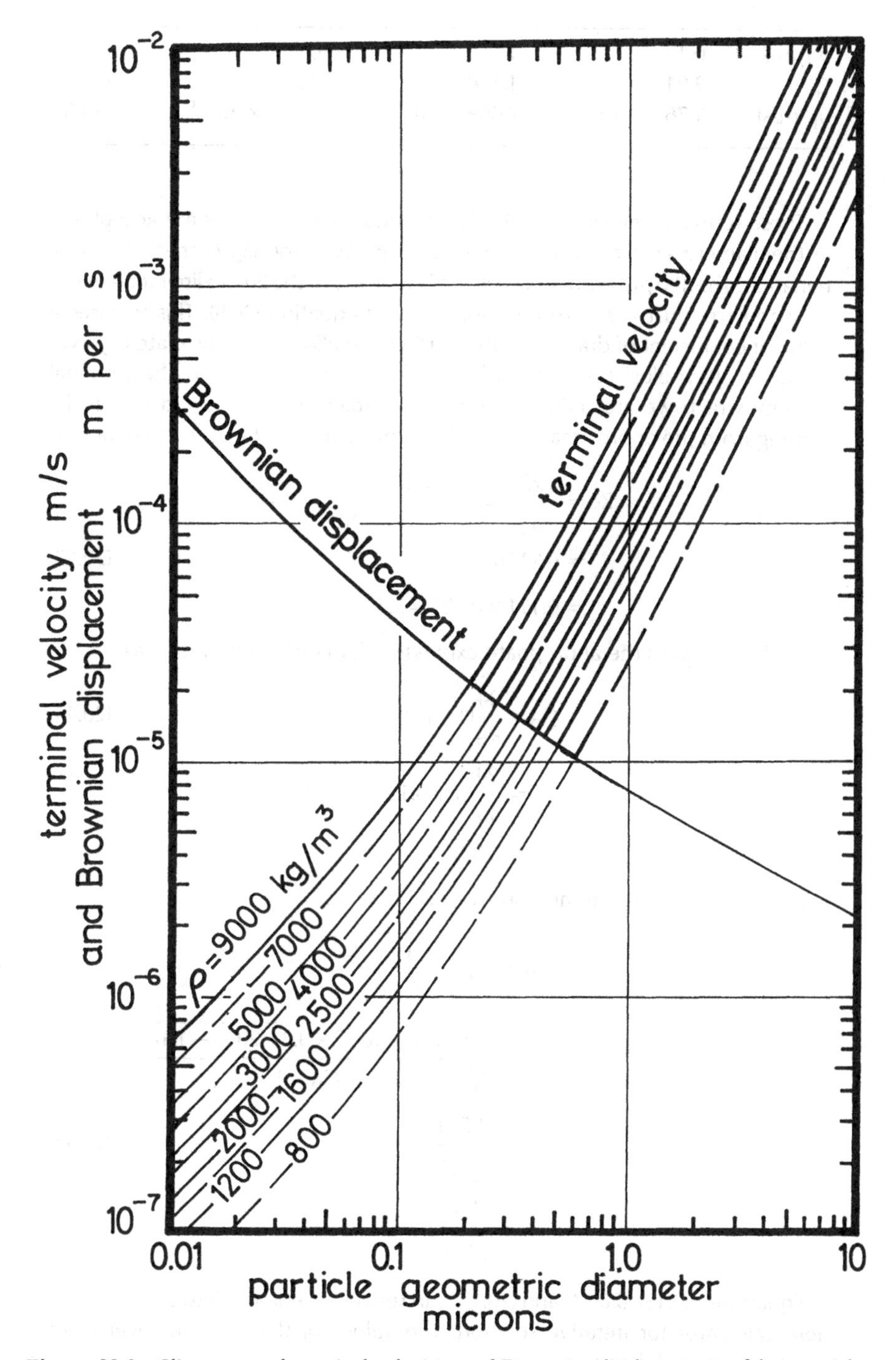

Figure 20.2 Slip-corrected terminal velocities and Brownian displacements of dust particles falling through still air of viscosity 17.9×10^{-6} Ns/m^2 (20° C). Based on $g = 9.81$ m/s^2.

d (μm)	0.1	1	10	100
C_c	2.91	1.168	1.017	1
u_t (m/s)	1.761×10^{-6}	7.069×10^{-5}	6.155×10^{-3}	0.605

These terminal velocities for the 0.1, 1.0 and 10 μm particles are acceptable as the diameters fall into the range of applicability of the slip-corrected Stokes' equation. The 100 μm particle, however, is well above the 20 μm limit for laminar flow and we must revert to the more general equation (20.9). This requires a value of coefficient of drag, C_D. Table 20.1 would allow us to calculate C_D if we knew the Reynolds' number. Unfortunately, that depends on the terminal velocity which we are trying to find. The problem can be solved iteratively, starting from the approximation $u_t = 0.605$ m/s given by the Stokes' equation:

$$\mathrm{Re} = \frac{\rho_a d\, u_t}{\mu_a} = \frac{1.1 \times 100 \times 10^{-6} \times u_t}{18 \times 10^{-6}}$$

$$= 6.111\, u_t \tag{20.13}$$

$$= 6.111 \times 0.605 = 3.7$$

Table 20.1 gives the appropriate expression for coefficient of drag as

$$C_D = \frac{24}{\mathrm{Re}}(1 + 0.15\,\mathrm{Re}^{0.687}) \tag{20.14}$$

$$= \frac{24}{3.7}(1 + 0.15 \times 3.7^{0.687})$$

$$= 8.882$$

Equation (20.9) now gives an improved value of u_t:

$$u_t = \sqrt{\frac{4}{3}\frac{dg(\rho_s - \rho_a)}{C_D \rho_a}}$$

$$= \sqrt{\frac{4}{3}\frac{100 \times 10^{-6} \times 9.81\,(2000 - 1.1)}{C_D \times 1.1}}$$

$$= \frac{1.5417}{\sqrt{C_D}} \tag{20.15}$$

$$= \frac{1.5417}{\sqrt{8.882}} = 0.517 \text{ m/s}$$

Equations (20.13), (20.14) and (20.15) can readily be entered into a programmable calculator for iterative solution. The values of the variables over eight iterations are as follows:

u (m/s)	Re	C_D
0.605	3.70	8.88
0.517	3.16	10.10
0.485	2.96	10.66
0.472	2.89	10.90
0.467	2.85	11.00
0.465	2.84	11.05
0.464	2.84	11.06
0.463		

The procedure converges to $u_t = 0.463$ m/s.

The time taken for each of the particles to fall through 2 m can now be determined as $t = 2/u_t$:

Diameter (μm)	u_t(m/s)	t
0.1	1.761×10^{-6}	315 h
1.0	7.069×10^{-5}	7.9 h
10	6.155×10^{-3}	5.41 min
100	0.463	4.32 s

It is clear from this example that little gravitational settlement of respirable dust (<5 μm) can be expected within the retention times of ventilated areas underground. Indeed, coupled with the effects of Brownian motion, submicron particles can be considered to remain in permanent suspension.

20.2.2 Brownian motion

For very small particles, the bombardment by fluid molecules is no longer balanced on all sides. The result is that the particles undergo random and jerky displacements. This is known as Brownian motion and can be seen under an optical microscope.

Brownian displacements

As Brownian movements are random, it is necessary to analyse their effect statistically on a complete population of particles. If we consider a vertical plane in still air of uniform dust concentration and with only Brownian motion causing horizontal movement of the particles, then the average displacement of particles moving through the plane in one direction ($+\bar{x}$) will be equal to the average displacement of particles in the opposite direction ($-\bar{x}$). Hence, the net displacement is zero—not a very useful result. However, if we square the displacements (positive or negative) then the sum is always a positive number. We can then quantify Brownian motion in terms of **mean-square displacement**, $\bar{x}^2$.

A relationship for mean-square displacement was first derived by Einstein in 1905 (see, also, Seinfeld, 1986) and has been verified by numerous observers:

$$\bar{x}^2 = 2\frac{MR}{A}T\frac{C_c}{3\pi\mu d}t \quad \text{m}^2 \tag{20.16}$$

where M = molecular weight of gas, R = gas constant (J/(kg K)) (note that $MR = R_u$ = universal gas constant, 8314 (J/(K mol), section 3.3.1), A = Avagadro's constant (6022×10^{23} molecules in each mole) and t = time over which the displacement takes place (s). (The ratio

$$\frac{R_u}{A} = \frac{8314}{6022 \times 10^{23}} = 1.381 \times 10^{-23} \text{ J/K}$$

is known as Boltzmann's constant.)

For air at 20 °C (T = 293 K), the viscosity is 17.9×10^{-6} N s/m². Equation (20.16) can then be simplified to

$$\bar{x}^2 = \frac{2 \times 1.381 \times 10^{-23} \times 293}{3\pi \times 17.9 \times 10^{-6}} \frac{C_c t}{d}$$

or

$$\bar{x} = 6.925 \times 10^{-9} \sqrt{\frac{C_c t}{d}} \quad \text{m} \tag{20.17}$$

Using the values of C_c given in Table 20.2 and $C_c = 1$ for $d > 10\ \mu$m, equation (20.17) has been superimposed on Fig. 20.2 with t set at 1 s. This allows the Brownian displacement to be compared with the terminal velocity curves. Inspection of the figure indicates that at some point, as particle diameter decreases, Brownian displacement becomes predominant. This occurs within the range 0.2 to 0.6 μm, dependent upon the density of the material. At all lower diameters, gravitational settlement is nullified.

Brownian diffusivity

A consequence of random Brownian displacements is that migration of particles will occur from regions of higher to lower dust concentrations. We can describe the process as a form of diffusion and obeying Fick's law:

$$N_b = D_b\frac{dc}{dx} \tag{20.18}$$

where N_b is the **flux** of particles through an area of 1 m² in 1 s (particles/(m² s)) by Brownian diffusion, c is the concentration (particles/m³), x is the distance (m) in the direction considered and D_b is a coefficient known as the Brownian diffusivity (m²/s).

Let us now attempt to find a relationship that will allow us to quantify the Brownian diffusivity. Consider the 1 m cube shown on Fig. 20.3. It contains c particles, i.e. a concentration of c particles/m³. In time Δt, a net number of those

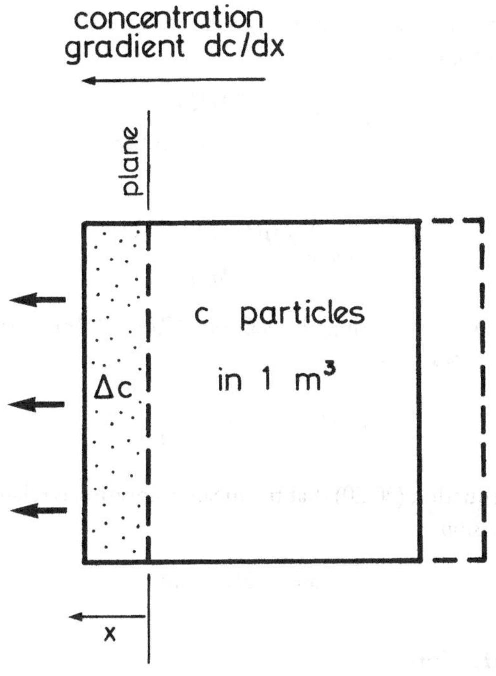

Figure 20.3 A number of particles, Δc, diffuse from a concentration of c particles in $1\,m^3$, through an orthogonal area of $1\,m^2$, and over a Brownian dislocation distance, x.

particles, Δc, will diffuse across a $1\,m^2$ plane by Brownian dislocations and as a consequence of the concentration gradient dc/dx.

Let us take the average Brownian dislocation, $\bar{x}$, to be the distance, x, through which the particles move in time Δt. Hence, their average velocity in the x direction will be $x/\Delta t$. Furthermore, the flux across the $1\,m^2$ plane will be the number of particles involved multiplied by their average velocity, i.e.

$$N_b = \Delta c \frac{x}{\Delta t} \quad \frac{\text{particles}}{\text{m}^2\text{s}} \tag{20.19}$$

Combining equations (20.18) and (20.19)

$$\Delta c \frac{x}{\Delta t} = D_b \frac{dc}{dx}$$

Now, over the very small distance of a Brownian dislocation, $x(=\bar{x})$, we can state that $\Delta c = dc$ and $\Delta t = dt$, giving

$$x\,dx = D_b\,dt$$

Integrating both sides between corresponding boundary limits,

$$\frac{x^2}{2} = D_b t \tag{20.20}$$

As we chose the distance x to be the average Brownian dislocation, $\bar{x}$, we can combine with equation (20.16) to give

$$\bar{x}^2 = 2D_b t = \frac{2MRT}{A}\frac{C_c}{3\pi\mu d}t \qquad (20.21)$$

from which

$$D_b = \frac{MRT}{A}\frac{C_c}{3\pi\mu d} \quad \frac{m^2}{s} \qquad (20.22)$$

Again, for air at 20 °C, inserting the values $MR/A = 1.381 \times 10^{-23}$ J/K, $T = 293$ K and $\mu = 17.9 \times 10^{-6}$ N s/m^2 gives

$$D_b = 2.398 \times 10^{-17}\frac{C_c}{d} \quad \frac{m^2}{s} \qquad (20.23)$$

Note, also, from equation (20.20) that the mean dislocation is related to the Brownian coefficient of diffusion

$$\bar{x} = \sqrt{2D_b t} \quad m \qquad (20.24)$$

20.2.3 Eddy diffusion

The previous two sections have considered the effects of gravity and molecular bombardment on dust particles. In ventilated areas, a larger influence is exerted on dust particles by the turbulent nature of the airflow. The transport of dust particles by eddies can also be described by a diffusion equation

$$N_e = \varepsilon\frac{dc}{dx} \quad \frac{\text{particles}}{m^2 s} \qquad (20.25)$$

where ε = eddy diffusivity (m^2/s) and N_e = flux of particles through an area of 1 m^2 in 1 s (particles/(m^2 s)) by eddy diffusion. The total rate of diffusion by both Brownian action and eddies is given by combining equations (20.18) and (20.25). Then

$$N = N_b + N_e = (D_b + \varepsilon)\frac{dc}{dx} \quad \frac{\text{particles}}{m^2 s} \qquad (20.26)$$

(A glance back at equations (15A.2) and (15A.5) reveals the analogy with diffusion for both heat and momentum.)

The flux of particles passing from the turbulent core of an airflow through the buffer boundary layer to the laminar sublayer is of particular interest as these are particles that have a high probability of being deposited on the solid surfaces. Gravitational settlement will, of course, add to such deposition on floors or other upward-facing surfaces. Eddy action can impart sufficient inertia to a dust particle to carry it into the laminar sublayer. Within the sublayer there are no such eddies. Hence only Brownian bombardment can superimpose further transverse forces. Ignoring any effects of re-entrainment, Brownian dislocations at the surface and away from the

Table 20.3 Expressions for eddy diffusivity, ε, as a function of dimensionless distance from a surface, $y*$ (after Owen, 1969)

Dimensionless distance from surface	*Eddy diffusivity* $\varepsilon(m^2/s)$
$0 < y* < 5$ (laminar sublayer)	$0.001 \frac{\rho}{\mu} y*^3$
$5 < y* < 20$ (buffer layer)	$0.012 \frac{\rho}{\mu} (y* - 1.6)^2$
$y* > 20$ (turbulent core)	$0.4 \frac{\rho}{\mu} (y* - 10)$

surface will be zero. Hence there will be a Brownian concentration gradient towards the surface. Coupled with the initial transverse inertia, this will tend to produce deposition of particles that enter the laminar sublayer. As may be expected, this phenomenon is influenced by the same factors that affect the boundary layers of fluid flow through rough ducts, i.e. fluid density, viscosity and velocity (Reynolds' number) as well as the roughness of the surface.

The average size of eddies grows from zero at the edge of the laminar sublayer to a maximum within the turbulent core and, hence, varies with distance, y, from the surface. In order to take the other variables into account, a **dimensionless distance**, $y*$ is defined as

$$y* = y \frac{u \rho}{2f \mu} \tag{20.27}$$

where y = actual distance from the surface (m), u = average velocity of fluid (m/s), f = friction factor for the surface (dimensionless), ρ = fluid density (kg/m^3) and μ = dynamic viscosity (N s/m^2). Note that $yu\rho/\mu$ has the form of a Reynolds' number. (The group $u/2f$ is sometimes referred to as the **friction velocity**.) Values of eddy diffusivity are suggested in Table 20.3.

In order to track the combined Brownian and eddy transverse transportation of dust particles across an airway, it is necessary to carry out integrations of equation (20.26) across each of the zones specified in Table 20.3 (Bhaskar and Ramani, 1988). This is accomplished in a manner similar to that used for convective heat transfer in section 15A.3.

20.2.4 Other forms of dust transportation

The processes of sedimentation, Brownian and eddy diffusion, coupled with coagulation, are the predominant mechanisms leading to the deposition of dust particles. There are, however, other phenomena that play a secondary role in governing the behaviour of airborne dust.

Many particles gain an electrical charge during formation. The effects of frictional flow as air moves through a duct or airway can also induce electrical charges on dust particles. Even particles that are initially uncharged may gain dipole characteristics due to van de Waal's forces. The primary effect of electrostatic forces is to increase rates of coagulation (section 20.2.4).

Suppose a dust particle of charge, q, moves through an electrical field of strength E, then it will experience an electrostatic force, qE. This may occur particularly around electrical equipment. At equilibrium velocity, this force is balanced by fluid drag (equation (20.11) for laminar flow around the particle), giving

$$\frac{3\pi\mu_a d}{C_c} u_e = qE \quad \text{N} \tag{20.28}$$

where u_e = the electrical migration velocity relative to the air (m/s).

The induction of electrical charge on dust particles to assist in deposition is utilized in electrostatic precipitators (section 20.4.2) and in the control of paint or powder sprays. However, the high voltages that are required impose a limit on the use of such devices in underground openings.

Phoretic effects refer to phenomena that impart a preferential direction to Brownian motion. **Thermophoresis** is the migration of particles from a hotter to a cooler region of gas and is caused by the enhancement of Brownian displacement at higher temperatures (equation (20.16)). The dust particles are subjected to greater molecular bombardment from the side of higher temperature. The temperature gradient must be considerable to produce a significant effect and the phenomenon has little influence on dust deposition in mine airways. However, it is utilized in instruments such as the thermal precipitator (section 19.4.2).

Photophoresis occurs when an intense light beam or laser is employed in a dusty atmosphere. The absorption of light by the particle causes an uneven temperature field to exist around that particle. The resulting excitation of nearby gas molecules causes thermophoresis to occur in a direction that depends on the induced temperature field around the surface of the particle.

An effect that encourages dust deposition on wet surfaces in **diffusiophoresis**. The migration of water vapour molecules away from an evaporating surface will result in a replacing flux of the more massive air molecules towards the surface. The result will be a net Brownian force on dust particles also towards the surface.

20.2.5 Coagulation

In any concentration of dust particles, collisions between the particles will occur as a result of Brownian motion, eddy action or differential sedimentation. Dependent upon the surface properties of any two such particles, they may adhere together to form a larger single particle. As the process continues, some particles will grow to the extent that their terminal velocity becomes significant and they will flocculate out of suspension. This phenomenon of **coagulation** is influenced by the number and size distribution of the particles (large particles are more likely to be struck by other particles), temperature and pressure of the air (governing Brownian displacements) and electrical charge distributions. The shape of the particles and the presence of adsorbed vapours on their surfaces will also affect the probability of their adhering upon collision.

Analysis of coagulation is, again, an exercise in statistics. Consider, first, a concentration of n particles in 1 m^3. The average frequency of collisions (dn/dt particles involved in collisions per cubic metre per second) clearly depends on the number of particles in that space. We can write

$$\frac{dn}{dt} = -an \quad \frac{\text{particles}}{\text{s m}^3} \tag{20.29}$$

where a = the probability of any two particles colliding (negative as the number of discrete particles is decreasing with time).

However, the probability of collision is itself proportional to the number of particles:

$$a = Kn \quad (1/\text{s}) \tag{20.30}$$

giving

$$\frac{dn}{dt} = -Kn^2 \quad \frac{\text{particles}}{\text{s m}^3} \tag{20.31}$$

K is known as the **coagulation coefficient** or **collision frequency function** (m^3/(particles s)). Equation (20.31) can be integrated readily:

$$\int \frac{dn}{n^2} = \int -K\,dt$$

$$\frac{1}{n} = Kt + \text{constant}$$

At $t = 0$, $n = n_0$ = original concentration of particles, giving

$$\text{constant} = \frac{1}{n_0}$$

so that

$$\frac{1}{n} = Kt + \frac{1}{n_0} \quad \left[\frac{\text{particles}}{\text{m}^3}\right]^{-1} \tag{20.31a}$$

Values of the coagulation constant can be found for any given dust cloud by plotting the variation of particle concentration with respect to time. For Brownian coagulation of equal sized particles in a continuum, K is given by (Flagan and Seinfeld, 1988)

$$K = \frac{8}{3}\frac{MR}{A}\frac{T}{\mu_a} \quad \frac{\text{m}^3}{\text{particles s}} \tag{20.32}$$

Hence for air at 20 °C, $MR=8314\,\mathrm{J/(K\,mol)}$, $A=6022\times10^{23}$, $T=293\,\mathrm{K}$ and $\mu_a=17.9\times10^{-6}\,\mathrm{N\,s/m^2}$, giving

$$K=0.6\times10^{-15}\,\mathrm{m^3/(particle\,s)} \tag{20.33}$$

Ranges of size distribution and the other matters that influence coagulation result in considerable variations being found in observed values of the coagulation coefficient.

There is a further problem that limits the applicability of this analysis; it has taken no account of the differing sizes of particles or the fact that K changes as the agglomerates grow larger. A somewhat more sophisticated approach concentrates on one size range at a time and considers the appearance of particles of that size by agglomeration of smaller particles. Additionally, their progression out of the size range as they continue to grow should be taken into account. Let us assume, for the sake of explanation, that diameters are additive. (Actually, we should use particle volume rather than diameter.) Then, for example, particles of size 10 μm can appear by coagulation of smaller particles. If we employ subscripts to denote the size of particles, then

$$6_1 \text{ and } 6_9 \rightarrow 6_{10}$$

i.e. six (1 μm particles) agglomerating with six (9 μm particles) yields six (10 μm particles). Similar examples are

$$3_2 \text{ and } 3_8 \rightarrow 3_{10}$$
$$5_3 \text{ and } 5_7 \rightarrow 5_{10}$$
$$3_4 \text{ and } 3_6 \rightarrow 3_{10}$$
$$2_5 \text{ and } 2_5 \rightarrow 2_{10}$$

Totals: 38 particles collide to yield 19 particles of size 10 μm.

In each of these groups, the collisions result in the number of particles being halved. Using the concept of coagulation coefficient and the form of equation (20.31), we can write that the rate of formation of particle size k (10 μm in our example) is

$$\frac{dn_k}{dt}(\text{formation}) = \frac{1}{2}\sum_{i=1}^{k-1} K_{ij}n_i n_j \quad \frac{\text{particles formed}}{\mathrm{s\,m^3}} \tag{20.34}$$

where K_{ij} is the particular coagulation coefficient for colliding particles of size i and j and n_k is the number of particles of size k that are formed from the collisions of n_i particles (size i) and an equal number of n_j particles (size j).

However, while all of this is going on, particles of size k are disappearing because further coagulation causes them to grow out of that size range. This can occur by each particle size k agglomerating with another particle of any size. In this case, we count the number of k size particles that are disappearing rather than being formed. Hence, we no longer require the factor of $\frac{1}{2}$ and can write:

$$\frac{dn_k}{dt}(\text{disappearance}) = -\sum_{m=1}^{\max} K_{km}n_m n_k \quad \frac{\text{particles lost}}{\mathrm{m^3\,s}} \tag{20.35}$$

where max = largest size of particle to be considered relevant to the processes of coagulation. (As n_k has a single value at any given time, t, it can be brought outside the summation sign.)

Combining equations (20.34) and (20.35) gives the overall rate of change of concentration of particle size k:

$$\frac{dn_k}{dt} = \frac{1}{2}\sum_{i=1}^{k-1} K_{ij} n_i n_j - n_k \sum_{m=1}^{\max} K_{km} n_m \quad \frac{\text{particles}}{\text{m}^3\,\text{s}} \tag{20.36}$$

This result was reported by Chung (1981) but attributed to Smoluchowski. Even more complex analyses have been conducted for liquid aerosols involving not only particle size changes by coagulation but also by evaporation. These are of relevance in meteorology and surface atmospheric pollution.

20.2.6 Impingement and re-entrainment

The phenomena of impingement and re-entrainment become significant only in situations of high velocity or excessive turbulence such as may occur in and around ventilation shafts or fan drifts. In such cases, the momentum gained by some dust particles may cause them to be ejected from the curved streamlines of eddies and impinge on the walls or other solid objects. Deposition by impaction of the particles on the walls can then occur. This is the principle employed in impact dust samplers such as the konimeter (section 19.4.2).

Impact deposition in mine airways is counteracted to some degree by re-entrainment in those same conditions of high velocity and turbulence. A particle on any surface and submerged within the laminar sublayer can be made to roll over the surface by viscous drag of the air when a sufficiently high velocity gradient exists across the sublayer. An accelerated rolling action may cause the particle to bounce until it momentarily escapes beyond the sublayer where capture by eddies can re-entrain it into the main airstream. Chaotic turbulence can have the same effect by transient thinning of the sublayer. The phenomena associated with these boundary layer effects are, again, influenced by Reynolds' number and surface roughness. Re-entrainment can be analysed by considering the drag and frictional forces on particles on or very close to solid surfaces (Ramani and Bhaskar, 1984).

20.2.7 Computer models of dust transport

The earlier mathematical models developed to describe dust transport in mine airways were empirical in nature (e.g. Hamilton and Walton, 1961). The growing availability of digital computers since the 1960s combined with a better understanding of aerosol behaviour have led to the development of mathematical models to simulate the behaviour of dust particles in mine ventilation systems (Bhaskar and Ramani, 1988). Such a model may be based on a form of the convective diffusion equation

$$\frac{dc}{dt} = E_x \frac{\partial^2 c}{\partial x^2} - u\frac{\partial c}{\partial x} + \text{sources} - \text{sinks} \tag{20.37}$$

where c = concentration (particles/m^3), t = time (s), x = distance along the airway (m), u = air velocity (m/s) and E_x = turbulent dispersion coefficient in the x direction (m^2/s).

This can be solved numerically between given boundary limits of time and distance (Bandopadhyay, 1982) to track the temporal variations of dust concentration along a mine airway. The 'sinks' term is determined from the relationships given in the preceding subsections and, in particular, the effects of gravitational settlement, Brownian motion, eddy diffusion and coagulation. The 'sources' must be defined as a dust production–time curve or histogram that characterizes the make of dust from all significant sources along the length of airway considered.

20.3 THE PRODUCTION OF DUST IN UNDERGROUND OPENINGS

The majority of dust particles in mines are composed of mineral fragments. Oil aerosols may become significant when drilling operations are in progress. Diesel exhaust particulates can also form a measurable fraction of airborne dust in those mines that utilize internal combustion engines. However, in this section we shall concentrate on the manner and processes through which mineral dusts are formed. Although the primary means of controlling mine dusts are discussed in detail in section 20.4, we shall introduce some of these, for particular operations, in this section.

20.3.1 The comminution process

Mineral dusts are formed whenever any rock is broken by impact, abrasion, crushing, cutting, grinding or explosives. For any given material, the energy input required to break the rock is proportional to the new surface area produced. As dust particles have a large surface area relative to their mass, it follows that any fragmentation process which produces an excessive amount of dust involves an inefficient use of energy. Before discussing specific operations that produce dust, a valuable insight into particles size distribution can be gained from a brief analysis of the comminution process.

Suppose a given brittle material is broken into fragments and the particles classified into a series of size ranges. Commencing with the mass of finest particles and progressively adding on the mass of each next coarser range, a table of cumulative 'mass finer than' can be assembled. If this is plotted against particle diameter on a log–log basis (Fig. 20.4) then a straight line is obtained for the smaller particles and curving over at larger sizes. The curve of Fig. 20.4 follows an equation of the form

$$M = \left[1 - \left(1 - \frac{x}{x_0}\right)^r\right]^m \quad \text{kg} \tag{20.38}$$

where x = particle diameter, (m) (we use x here, temporarily, in order not to confuse diameter with the differential operator, d), x_0 = diameter of the initial fragment (m), M = cumulative mass finer than size x (kg), r is a constant that depends on the

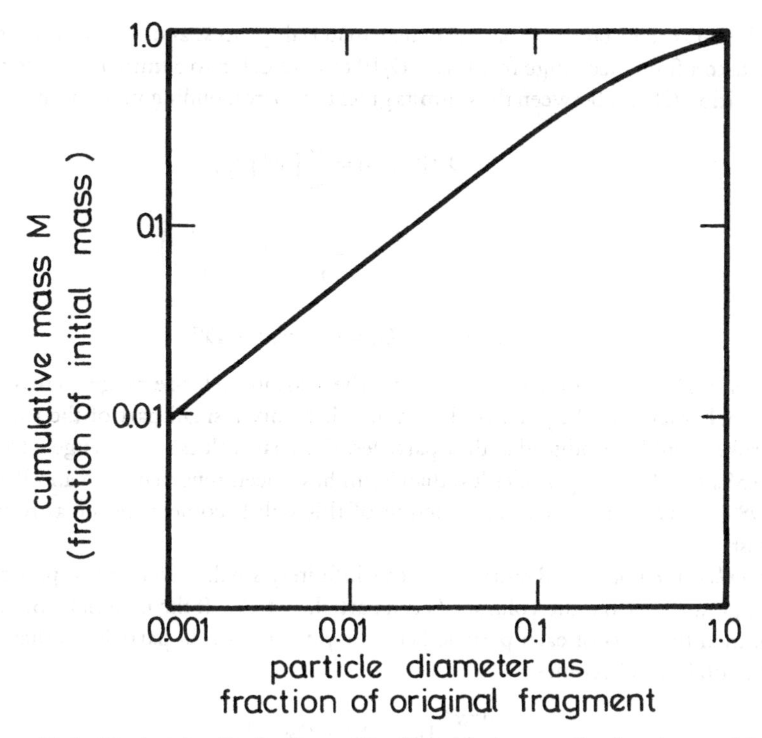

Figure 20.4 Typical size distribution graph of 'cumulative mass finer' against particle size.

particular comminution process and m is a characteristic of the material having values in the range 0.5 to 1 and varying only slightly with the method of comminution. (This is known as the Gaudin–Meloy–Schuhmann equation (Marshall, 1974; Gaudin and Meloy, 1962.)) If the term $(1 - x/x_0)^r$ is expanded by the binomial theorem, then for $x \ll x_0$

$$M = r\left(\frac{x}{x_0}\right)^m \quad \text{kg} \tag{20.39}$$

For dust particles, x is certainly very much smaller than x_0. Equation (20.39) quantifies the straight line portion of Fig. 20.4 and has been shown to hold for particle sizes down to 0.01 μm (National Research Council, 1980).

Let us now try to find a means of determining the mass and the number of particles in each size range.

1. *Mass.* Consider the mass, dM, of particles contained within the incremental range x to $x + dx$. Differentiating equation (20.39) gives

$$\begin{aligned} dM &= \frac{rm}{x_0^m} x^{m-1}\, dx \\ &= C x^{m-1}\, dx \quad \text{kg} \end{aligned} \tag{20.40}$$

where C=constant for that particular material, process and initial size. Now let us take a finite size range from, say, $D/10$ to D (e.g. 0.5 to 5 μm). Then integrating equation (20.40) between those limits gives the corresponding mass for that range:

$$M(D/10 \text{ to } D) = \frac{C}{m}[x^m]_{D/10}^{D}$$

$$= \frac{C}{m}D^m\left(1 - \frac{1}{10^m}\right)$$

or

$$M(D/10 \text{ to } D) = \text{constant} \times D^m \tag{20.41}$$

As m is always positive this shows that the mass in each size range increases with particle diameter. In practice this means that only a small part of the total rock broken will be produced as dust particles. For coal, values in the range 5 to 9 kg/t (0.5% to 0.9%) of particles less than 7 μm have been reported (Qin and Ramani, 1989). However, only a tiny fraction of this will become airborne as respirable dust.

2. *Number of particles.* Returning to our infinitely small increment of particle size range, x to $x + dx$, the volume of each particle is $\pi x^3/6$. If the material is of density, ρ, then the mass of each particle becomes $\rho\pi x^3/6$. For dn particles in that range, the total mass becomes

$$\frac{\rho\pi x^3}{6}dn = dM = Cx^{m-1}\,dx$$

(from equation (20.40)) giving

$$dn = C'x^{m-4}\,dx \quad \text{particles} \tag{20.42}$$

where C' = constant for that material, process and initial size. Integrating over the finite size range $D/10$ to D gives

$$n(D/10 \text{ to } D) = \frac{C'}{m-3}[x^{m-3}]_{D/10}^{D}$$

$$= \frac{C'}{m-3}D^{m-3}\left(1 - \frac{1}{10^{m-3}}\right)$$

i.e.

$$n(D/10 \text{ to } D) = \frac{\text{constant}}{D^{3-m}} \quad \text{particles} \tag{20.43}$$

As m lies in the range 0.5 to 1.0, this shows that the number of particles rises logarithmically as the particle diameter decreases.

Equations (20.41) and (20.43) indicate that in any rock breaking process, the bulk of mass will appear as larger fragments. However, the number of fine dust particles produced may be enormous. Fortunately, most of those particles remain attached to the surfaces of larger fragments. The degree to which dust particles are dispersed into the air would seem to depend on the nature of the rock as well as the comminution process. For brittle materials, the fragmentation becomes more 'explosive' in nature; the resulting surface vibration causes an enhanced dispersion of dust particles into the air. Hence, although comminution of softer materials may generate more dust particles, a greater proportion of those will remain adherent to the surfaces of larger particles and will not become airborne. The production of airborne respirable dust has been reported in the range 0.2 to 3.0 g/t (Qin and Ramani, 1989; Knight, 1985).

20.3.2 Mechanized mining

Machines that break rock from the solid have the potential to be prolific sources of dust. These include longwall power loaders, continuous miners, roadheaders, tunnelling machines, raise borers and drills. Figure 20.5(a) illustrates a pick point acting against a rock face. Compressive forces induce a zone of pulverized material immediately ahead of the pick point. As the pick moves forward into that zone, the resultant wedging action produces tensile failure along a curved plane—a chip is broken away. The process is repeated continuously as the pick advances. The majority of the pulverized

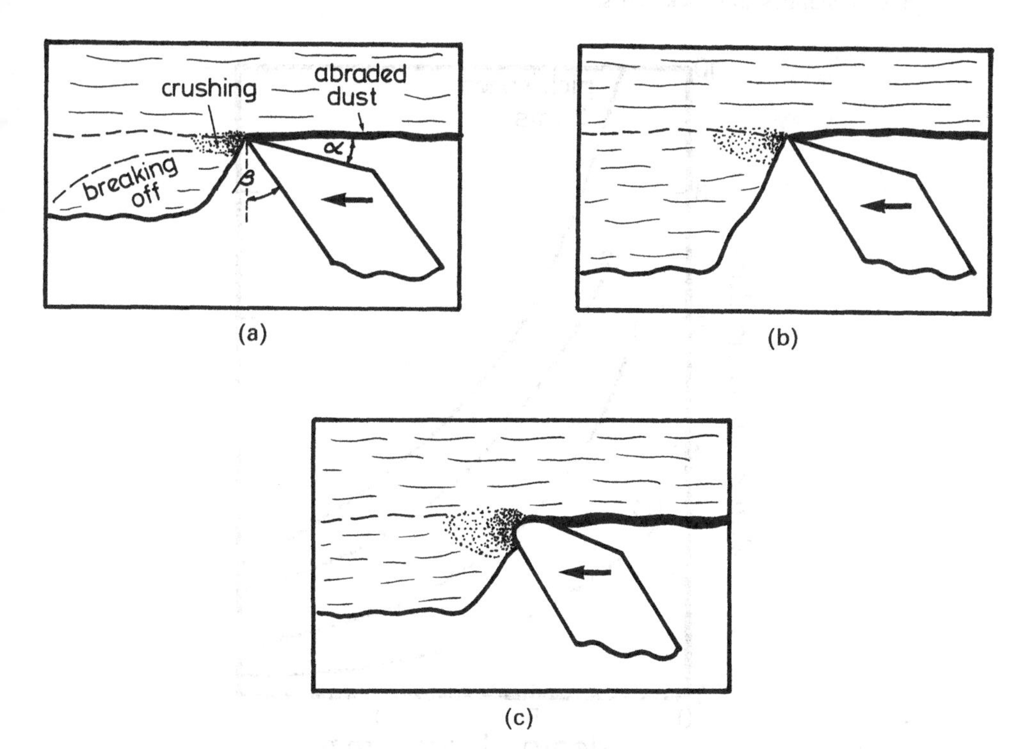

Figure 20.5 A zone of pulverized rock forms ahead of a cutter pick: (a) shallow cut (α = rake angle; β = clearance angle); (b) deep cut; (c) blunt pick.

material is abraded onto the surfaces of the rock face and the chip. The amount of dust produced at the tensile failure plane itself may be quite small in homogeneous brittle material and is influenced by the presence of preformed dust in natural cleavage planes. However, the explosive nature of that tensile failure is a major factor in determining the amount of dust that is projected into the air.

A machine that takes a greater depth of cut will require higher torque and may be subject to greater vibration and bit breakage. However, a comparison of Fig. 20.5(a) and 20.5(b) indicates that more of the broken material will be in the form of chips and, hence, the amount of dust produced in terms of grams per tonne will be reduced. The specific energy (per tonne mined) will also fall. Figure 20.5(c) shows that the greater area of contact given by a blunt pick will create additional dust in the pulverized zone. If such wear causes a significant reduction in the rake angle (Figure 20.5(a)) then the back of the bit will rub against the newly formed face, absorbing additional energy and producing further pulverized rock. Furthermore, as the clearance angle reduces, the chip may not be ejected efficiently but remain in place to be crushed against the unbroken rock. The design of a rock-cutting bit is a compromise between the efficiency of cutting (energy absorbed per tonne), wear characteristics and dust production. Considerable diversity of opinion exists on preferred bit geometries for given machines and rock types.

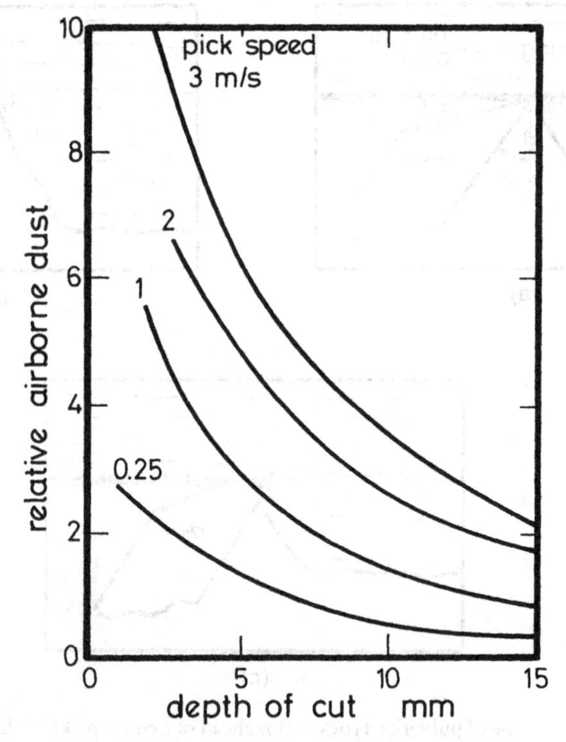

Figure 20.6 The effect of pick speed and depth of cut on dust production.

Another factor that influences the proportion of dust which becomes airborne is the speed at which the pick moves. For any given depth of cut, an increased speed results in greater rate of comminution and, hence, dust production. Additionally, movement of the cutter drum causes a higher relative velocity to be induced between the local airstream and the material on the face (or fragments broken from the face). This assists in entrainment of dust particles into the air (section 20.2.5). The effect of pick speed on airborne dust is illustrated on Fig. 20.6.

20.3.3 Supports

Crushing of roof and floor strata by roof supports may liberate significant amounts of dust when the support is moved. This can be a particular problem on mechanized longwall faces that are equipped with powered hydraulic supports. As setting and yield loads of the supports increase so, also, does the amount of dust produced. The repeated lowering and raising of these supports can give a near-continuous source of dust on longwall faces. Unless roof coal is left, this may be high in quartz content. The effect can be minimized by using wide-web roof beams or cushioning materials. Sheets of flexible material linking adjacent canopies have also been used to mitigate against roof dust.

20.3.4 Blasting

Drill and blast remains the predominant method of mining in metal (hardrock) mines. The peak concentrations of dust and gases (section 11.3.4) that are produced by the larger blasts are usually too high to be diluted effectively by the normal ventilating airflow. This necessitates the mine, or part of the mine, being evacuated of personnel for a **re-entry period** during and after the blast. The length of the re-entry period can vary from half an hour to several hours for stoping areas, dependent upon the layout of the ventilation network and the velocities of the air. This is a classical example of isolating personnel from the dust.

The amount of dust produced depends upon a number of factors including

1. the mining method,
2. the type of rock,
3. the choice of explosive,
4. the charge density and drilling pattern, and
5. the type of stemming.

Blasts that eject the fragmented material into an air space (e.g. open stoping) will tend to produce sharper but shorter-lived peaks of dust than caving techniques. However, the latter may produce more pulverized material capable of being entrained into the airstream during subsequent loading and transportation operations. Water ampoules have been employed as stemming in an attempt to reduce dust emissions from blasting operations.

Another technique is to place very fine but high capacity water sprays (fog machines) upwind of the blast before and during the re-entry period. The combination of increased humidity and fine water droplets assists in the agglomeration and sedimentation of dust particles. Spraying the muckpiles produced by blasting is advisable before loading commences.

Secondary blasting also produces short peaks of dust concentration. This is yet one further reason for employing methods of mining that minimize the need for secondary blasting.

20.3.5 Loading operations

This is another part of some mining cycles that can produce a great deal of dust whether the loading operations are carried out by slushers, load–haul–dump (LHD) vehicles or loading machines in headings. The dust arises from a combination of particles produced previously from the mining process and held within the muckpile, and those that are generated by further comminution during loading.

In addition to adequate (but not excessive) airflows, the primary means of combating dust from loading operations are water sprays and ensuring as little disturbance as possible to the loaded material. The air velocity should not be less than 0.5 m/s at loading points. Abrasion of the floor by heavy slusher buckets should be minimized. It is preferable to employ lighter buckets in tandem operating at a speed of some 0.6 m/s (Sandys and Quilliam, 1982). Spray bars should be located at intervals along slusher paths and, particularly, at points of transfer between buckets.

Muckpiles in headings should be sprayed with water continuously or frequently during mucking operations except where hygroscopic minerals inhibit the copious use of water. In hot mines, pre-chilling of this water produces cooling as well as dust suppression (section 18.3.5). Steam injection into muckpiles and the addition of wetting agents into the water have also been found to be beneficial in some cases (Knight, 1985). Exhaust auxiliary ventilation is preferred for dusty operations in headings, employing a force overlap, if necessary, to deal with gas emissions at the face (section 4.4.2).

The skill of the driver of an LHD can have considerable influence on dust production. Choosing the best point to insert the bucket into the muckpile will result in filling the bucket with a minimum number of thrusts and with least disturbance to the material. Similarly, at the dump point, the muck should be tipped gently and not dropped from a height. This should also be borne in mind during the design of tipping operations from rail-mounted dump cars. Cones and shutes at dump points should be designed to minimize impact forces on tipped material.

20.3.6 Transportation and crushing

Dust is produced throughout most mineral transportation arrangements, including conveyors, transfer points, bunkers, skips, airlocks and vehicular traffic. Dust on the surfaces of conveyors may be re-entrained into the air because of vibration of the belt as it passes over rollers. Spillage returning on the bottom belt, if not cleared, will generate dust as the material is crushed against rollers. Similarly, an excessive use of water can result in dust adhering to the belt surface. This may subsequently be deposited under the conveyor during the return journey of the bottom belt. Belt scraper devices or brushes at the drive heads should be properly maintained and all accumulations of debris or dust should regularly be cleaned from under the conveyor and at return rollers. Conveyor structure should be inspected routinely and attention paid to damaged idlers and centring devices.

Vehicle arrestors on rail transportation systems should incorporate deceleration devices in order to avoid impact loads on either the vehicles or the transported material. Tracks should be adequately maintained and not allowed to develop sudden changes in direction or gradient.

The mineral transportation routes and mine ventilation system should be planned in liaison in order to avoid, wherever possible, minerals being transported through an airlock. The high velocities that can occur over belt conveyors at airlock leakage points can cause excessive production of dust. This can be minimized by employing side plates and attaching a length of flexible material (such as old belting) on the conveyor discharge side of the airlock so that it drags over the surface of the conveyed material.

Unless the mineral is hygroscopic, it should be kept damp throughout its transportation through the mine. Bunkers and, wherever possible, conveyor transfer points and stage loaders should be shrouded and fitted with internal sprays. It is also useful to duct the air from such shrouds directly into return airways. Sprays or dribbler bars onto conveyors some 5 to 10 m before a transfer point are often more effective than sprays actually at the transfer point itself.

Ore passes in metal mines should avoid lengthy segments of free fall. Air leakage at dump and draw points should be into the ore pass and, hence, pull dust laden air away from personnel. This can be arranged by an opening into the ore pass and connected either directly or via ducting to a return airway. If this is not practicable then dusty air drawn by a fan from an intermediate point in an ore pass can be filtered and returned to the intake system.

Crushers in any mine are prolific sources of dust. Here again, sprays may be used on the material before, during and after the crushing process. This is another situation where it is particularly valuable to draw air from the crusher enclosure and filter it.

20.3.7 Workshops

Aerosols produced in underground workshops are likely to occur as oil mists, diesel particulate matter and welding fumes. The latter may be handled by exhaust hoods extracting air from welding bays and directing it into a return airway. Indeed, all of the airflows through workshops should, preferably, pass into return airways. The general arrangements for diluting and removing airborne contaminants from workshops are discussed in section 9.3.5.

20.3.8 Quartz dust in coal mines

The availability of instrumentation that can discern the quartz content of mine dusts within each of a range of particle sizes (section 19.4.7) has led to the observation that airborne dust in coal mines often has a quartz content that is significantly higher than that of the coal seam being worked. Furthermore, the percentage of quartz becomes particularly high in the finer sizes including the respirable range (Ramani *et al.*, 1988; Padmanabhan and Mutmansky, 1989). Coupled with the special danger to health of quartz dust, this has led to research aimed at discovering the causes of such anomalous appearances of quartz in airborne dusts of coal mines.

There would appear to be at least two explanations. First, roof and floor strata usually have a much higher quartz content than the coal seam. Hence any fragmentation of those strata will cause emissions of quartz dust. This can occur by rock-winning machines cutting into the roof or floor, cross-measures drilling for roof-bolting or other purposes, development drivages out of the seam or exceeding the height of the seam, hydraulic roof supports and fracturing of roof or floor strata.

A second, less obvious, cause of the apparently anomalous percentages of quartz in the dust of coal mines is hypothesized to be the different comminution characteristics of coal and quartz (section 20.3.1). Fragmentation of the stronger and more brittle quartz minerals may result in a greater proportion of that dust being ejected into the air than is the case for coal. The greater degree of entrainment would favour the finer particles.

20.4 CONTROL OF DUST IN MINES

The initial decisions that affect the severity of dust problems are made during the stages of design and planning for the mining of any geological deposit. The methods of working, rate of mineral production and equipment chosen will all influence the amount of dust that is generated and becomes airborne. The layout of the mine, sizes and numbers of airways, and the efficiency of the ventilation system dictate the rate at which airborne contaminants, including dust, are diluted and removed from the mine.

For an existing mine, there are four main methods of controlling the production, concentration and hazards of airborne dust:

1. suppression
   —the prevention of dust becoming airborne,
2. filtration and scrubbing
   —the removal of dust from the air,
3. dilution by airflow, and
4. isolation
   —separation of personnel from the higher concentrations of dust.

In general, good management and housekeeping at a mine assist greatly in maintaining control of the dust problem. These measures includes planned maintenance schemes for equipment, quantitative ventilation planning, cleaning up spillage, rock debris and local accumulations of dust, and adequate supervision of work practices.

20.4.1 Dust suppression

It is difficult and often expensive to remove respirable dust from the air. Hence, every attempt should be made to prevent it from becoming airborne in the first place. Methods of achieving this are known collectively as **dust suppression** and are discussed in this section.

Pick face flushing and jet-assisted cutting

Figure 20.5 gives a visual impression of how a rock face is pulverized in advance of a moving cutter pick. Pick face flushing involves directing a jet of water at a pick point during the cutting process. This has been found to give a markedly improved dust suppression when compared with conventional water sprays on the drums of shearers, continuous miners or tunnelling machines. The water feeding each jet can be channelled through conduits drilled in the bit holder and via a phasing valve that activates the jet only while the bit is cutting rock. Water filters are required to prevent blockage of the nozzles. A further advantage of pick face flushing is that the streak of incendiary sparks that often appears behind the pick in dry cuttings is quenched. Hence, the incidence of frictional ignitions of methane is reduced greatly. Interlock switches may be employed to ensure that the machine cannot operate without the dust suppression water being activated.

A number of researchers have investigated the extension of pick face flushing to much higher water pressures, not only to improve dust suppression further but also in an attempt to produce a higher efficiency of cutting. The use of high pressure water jets alone, with or without the addition of abrasive particles, has had only limited success as a practical means of mining. However, combining the mechanism of cutter picks with high pressure water jets directed at the pick point has led to significant improvements in machine performance and the extension of mechanized mining to much harder material that, previously, could be mined only by drill and blast techniques. This technique is known as **jet-assisted cutting**.

In addition to environmental enhancements, jet-assisted cutting permits the same rate of comminution with reduced loading on the cutter pick. This results in a significant reduction in wear rates and, hence, less production time lost because of bits having to be changed. Furthermore, the total specific power (per tonne mined) required by the combination of a high pressure water pump and the cutting machine can be less than that of a conventional machine.

The benefits of jet-assisted cutting are attainable by increasing the water pressure but reducing the nozzle size in order to keep the flow rate no greater than that employed in conventional pick face flushing. This can be important in hot mines or where floor strata react adversely to water. However, it has been reported that there is little apparent improvement in levels of airborne dust until the water pressure attains some critical value (Taylor *et al.*, 1988). This would appear to be in the range 10 to 15 MPa for cutting coal. After the critical water pressure is attained, a dramatic reduction in airborne dust can be expected. However, this levels out again at water pressures in excess of 20 MPa. Indeed, if the velocities of the jet and resulting spray are too high then re-entrainment can exacerbate dust concentrations. Work continues on the preferred location of the jet. Distances as small as 2 mm between the nozzle and the pick point have been suggested (Hood *et al.*, 1991).

The environmental and operational benefits of jet-assisted rock cutting arise from at least seven mechanisms (Hood, 1991).

1. The pulverized rock immediately ahead of the pick point is wetted before it has an opportunity to become airborne.
2. The cooling action of the jet reduces wear: the bits remain sharp for significantly longer periods of time and bit breakage is less frequent.
3. Impact of the high velocity jet will produce an aerosol of very fine water droplets around the cutting head, thus enhancing the agglomeration and capture of airborne dust particles.
4. The washing action of the high energy jet removes the cushion of pulverized material quite efficiently. This allows the pick point to act on a much cleaner surface. The effect of a cushion of pulverized rock is to distribute the force exerted by the pick over a broader front, i.e. similar to that of a blunt pick (Fig. 20.5(c)). It is to be expected that the total amount of finely crushed rock would be reduced although this remains to be demonstrated.
5. Penetration of the water into natural cleavage planes in the material and ahead of the mechanical effect of the bit assists in pre-wetting dust particles that already exist within those planes.
6. Frictional ignitions of methane are virtually eliminated.
7. The total specific energy required for the rock cutting process may be reduced.

Water infusion

A technique of dust suppression that has been employed by some coal mining industries since the 1950s is pre-infusion of the seam by water, steam or foam. One or more boreholes are drilled into the seam in advance of the workings through which the fluid is injected. The migration of water through the natural fracture network of the coal results in pre-wetting of included dust particles. The success of the method is dependent upon the permeability of the seam and the type of coal-winning equipment employed. Good results have been reported where coal ploughs are used—these relying more on coal breakage along natural cleavage than the cutting and grinding action of shearers or continuous miners (Heising and Becker, 1980).

In practice, some *in situ* experimentation is usually necessary to determine the optimum injection pressure and flowrate, and the time period of injection. Water pressures in the range 2 to 34 MPa have been reported with water volumes of 7 to

20 l/t in South African coal mines (Sandys and Quilliam, 1982). Best results are obtained at fairly modest pressures but applied over as long a period as possible. British experience in coals of limited permeability indicated water pressures of 1.5 to 2.5 MPa and flowrates of 0.2 to 2 l/min. If too high a pressure is used then the water flows preferentially along major planes of weakness. Hydrofracturing may occur, resulting in weakened roof conditions during mining and, possibly, backflow along bed separation routes to give water inflows at the current working faces. Water infusion is not recommended in areas of weak roof/floor strata or in the proximity of faults or other geological anomalies. Steam and wetting agents have been employed in attempts to improve pre-saturation of the zone. Water infusion must also be expected to influence the migration of strata gas (section 12.3.2). Holes drilled initially for in-seam methane drainage may subsequently be used for water infusion (Stricklin, 1987).

Wetting agents, foams and roadway consolidation

Worldwide experience of surfactants used as wetting agents in dust suppression water has been highly variable. The technique has been employed since at least 1940 (Hartman and Grenwald, 1940). In addition to the use of wetting agents to enhance the effects of water infusion, they may be employed to improve the performance of sprays and also, at sufficiently high concentration, to produce a foam around a rock fragmentation process.

Rocks vary considerably in their wettability characteristics. If surfactants added to muckpile sprays are to be effective then they must be at a high enough concentration to cause penetration of the fragmented material within an acceptable time period (Knight, 1985). The potential effects of such concentrations on mineral processing should be considered carefully. Wetting agents added to sprays intended to remove airborne dust are considered to have three beneficial effects. First, the reduced surface tension allows greater atomization of the water—the droplets are smaller and greater in number, hence improving the probability of capturing dust particles (section 20.4.2). Secondly, the existence of a liquid coating on dust particles will improve the chances of coagulation when two particles collide. Third, the molecular structure of surfactants tends to counteract electrostatic forces that may keep particles apart (Wang *et al.*, 1991).

If a wetting agent is in sufficient concentration within a spray directed at a rock cutting device then a foam can be formed that enshrouds the comminution process. This assists in coating the fragments with a wetting fluid and in inhibiting entrainment of the dust into the air. Again, this approach has met with mixed success (Bhaskar *et al.*, 1991). It also interferes with ventilation of the cutting head and should be used with caution in gassy conditions.

Accumulations of dust on roadway floors used for travelling in both underground and surface mines can become airborne when disturbed by traffic. **Roadway consolidation** involves the use of water, hygroscopicopic salts and binders to encapsulate the dust and to maintain the floor in a firm but moist state. Flakes of calcium chloride or magnesium chloride may be employed with lignin sulphonate as a binder. The process involves raking and levelling the surface dust, and spraying it lightly with water until it is wetted to a depth of some 2 to 3 cm. The addition of a wetting agent may be necessary. The total amount of water required can be of the order of 40 l/m^2. Free-standing pools of water should be avoided. The hygroscopic salt should be spread evenly at a rate that depends on the mean humidity of the air. For flake calcium chloride this will vary from about 3.8 kg/m^2 at a relative humidity of 40% down to 0.1 kg/m^2 for a relative humidity of 90%. It is advisable to apply three-quarters of the salt during the initial application and the remainder about one week later. The treatment will normally last for about six months although re-spraying with water may be required after three months. Sodium chloride (common salt) will be effective while the relative humidity remains above 75%. In all cases, care should be taken against corrosion of equipment and, in particular, within the vicinity of electrical apparatus.

20.4.2 Removal of dust from air

The larger dust particles will settle out by gravitational sedimentation in the air velocities typical of most branches in a mine ventilation system. Unfortunately, the more dangerous respirable particles will effectively remain in suspension. Removing these from the air for large flowrates can be expensive. The choice of a dust removal system is dictated by the size distribution and concentration of particles to be removed, the air flowrate and the allowable dust concentration at outlet. The size of any unit is governed primarily by the air volume flow to be filtered. Operational costs can be determined from the product of the pressure drop and air flowrate through the unit, pQ (section 5.5), and the means of supplying and filtering water in the case of wet scrubbers. Where high efficiency is required for large flowrates over a wide range of particle sizes such as the emergency filters needed on nuclear waste repositories (section 4.6), two or more types of filters may be arranged in series, each taking out progressively smaller particles. This prevents the finer filters from becoming clogged quickly and, hence, prolongs the life of the system before cleaning or renewal of filters becomes necessary.

The efficiency of any dust removal facility, η, may be expressed either in terms of number of particles per m^3 of air,

$$\eta_p = \frac{\text{number of particles in/m}^3 - \text{number of particles out/m}^3}{\text{number of particles in/m}^3} \tag{20.44}$$

or in terms of mass of particles,

$$\eta_m = \frac{\text{mass of particles in/m}^3 - \text{mass of particles out/m}^3}{\text{mass of particles in/m}^3} \tag{20.45}$$

In both cases, it is usual to restrict further the count of particles or mass to a specified size range. Hence, for protection against pneumoconiosis, it is preferable to employ equation (20.45) for respirable particles only, i.e. less than 5 μm equivalent diameter.

Devices to remove dust from air may be fitted to other pieces of equipment such as rock-cutting machinery, along transportation routes, within ventilation ducting or as free-standing units to filter dust from the general airstream. In this section, we shall discuss principles of the devices that are most commonly employed to reduce concentrations of airborne dust from mine atmospheres, namely, water sprays, wet scrubbers and dry filters or separators.

Water sprays

Water is by far the most widely used medium for conditioning mine air, whether it be for cooling (section 18.3), dust suppression or dust filtration. Open sprays can also be employed to direct, control or induce airflows in order to protect machine operators from unacceptable concentrations of dust (section 20.4.4).

The important parameters governing the efficiency of a spray can be highlighted through an analysis of the capture of dust particles by water droplets. Consider Fig. 20.7—in which air passes over a water droplet with a velocity relative to the droplet of u_r. The streamlines of air bend around the droplet. However, the inertia of dust particles causes them to cross those streamlines. Particles that lie closer to the centreline of motion will impact into the droplet and be captured by it. We can conceive a flow tube of diameter y within which all particles will be captured while particles that are further from the centreline will be diverted around the droplet. The efficiency of capture by a single droplet, E, can be defined as the ratio of the cross-sectional area of the capture tube to the facing area of the droplet:

$$E = \frac{y^2}{D_w^2} \tag{20.46}$$

where D_w = droplet diameter (m).

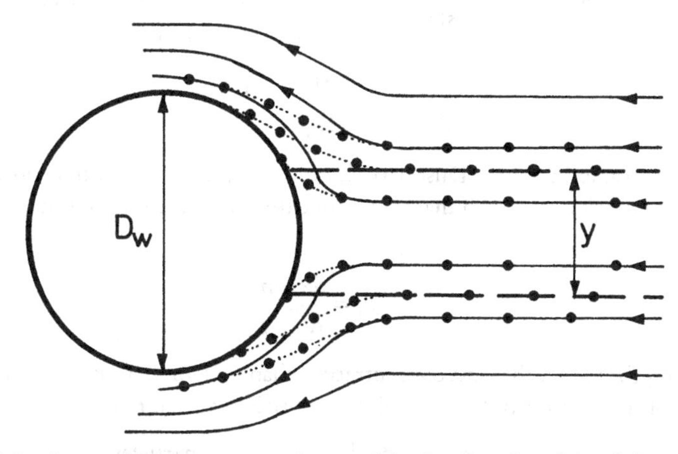

Figure 20.7 A water droplet captures dust particles from a tube of air of diameter y. The relative velocity between the droplet and air is u_r(m/s).

If there is a uniform dust concentration of n particles/m^3 then the rate of capture of particles by one droplet of water is

$$n \times u_r \times \text{area of capture tube} \quad \frac{\text{particles}}{\text{m}^3}\frac{\text{m}}{\text{s}}\text{m}^2 \text{ or particles/s}$$

that is,

$$\text{particles collected per droplet per second} = Enu_r\pi\frac{D_w^2}{4} \tag{20.47}$$

In order to maintain consistency with the definition of dust concentration that we are using here (particles/m^3), it is preferable to restate this latter expression in terms of particles collected per cubic metre of air rather than particles captured per second. We can do this by dividing by the air flowrate Q (m^3/s). Then rate of capture by one droplet (dn/dt = rate of change of dust concentration) becomes

$$-\frac{dn}{dt}(\text{one droplet}) = Enu_r\frac{\pi D_w^2}{4}\frac{1}{Q} \quad \frac{\text{particles}}{\text{droplet s}}\frac{\text{s}}{\text{m}^3} \text{ or } \frac{\text{particles}}{\text{droplet m}^3} \tag{20.48}$$

where t = time (s). This expression is negative (as concentration is falling).

Now if water is dispersed in the spray at a volume flowrate of W, m^3/s, and the volume of each droplet is $\pi D_w^3/6$, m^3, then the rate at which droplets are formed and pass through the spray is

$$\frac{W}{\pi D_w^3/6} = \frac{6W}{\pi D_w^3} \quad \frac{\text{m}^3}{\text{s}}\frac{\text{droplet}}{\text{m}^3} = \frac{\text{droplets}}{\text{s}} \tag{20.49}$$

Multiplying by the particle capture for one particle, equation (20.48), gives the total rate at which particles are captured per cubic metre of air:

$$-\frac{dn}{dt}(\text{all droplets}) = Enu_r\frac{\pi D_w^2}{4}\frac{1}{Q}\frac{6W}{\pi D_w^3} \quad \frac{\text{particles}}{\text{droplet m}^3}\frac{\text{droplets}}{\text{s}}$$

$$= \frac{3}{2}Enu_r\frac{W}{D_w}\frac{1}{Q} \quad \frac{\text{particles}}{\text{m}^3\text{s}} \tag{20.50}$$

Now consider Fig. 20.8. Dust particles and air pass each other in counterflow with a relative velocity of u_r such that they move through a separation distance dx in time dt, i.e.

$$u_r = \frac{dx}{dt} \quad \frac{\text{m}}{\text{s}}$$

During that time, the dust concentration changes from n to $n - dn$, i.e. the rate of change of dust concentration is $-dn/dt$ (particles/(m^3 s)). However,

$$-\frac{dn}{dt} = -\frac{dn}{dx}\frac{dx}{dt} = -\frac{dn}{dx}u_r \quad \frac{\text{particles}}{\text{m}^3\text{s}}$$

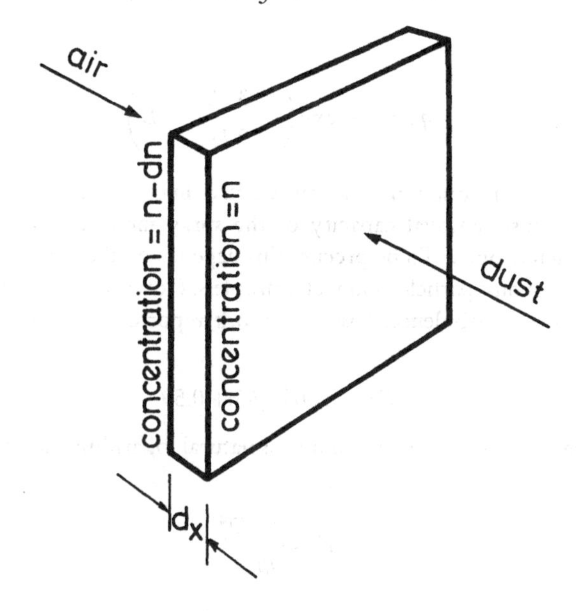

Figure 20.8 Over a separation distance of dx between water droplets and dust particles, the dust concentration falls by dn (particles/m^3).

Combining with equation (20.50) gives

$$-\frac{dn}{dt} = -\frac{dn}{dx}u_r = \frac{3}{2}E\frac{n}{D_w}u_r\frac{W}{Q}$$

or

$$dn = -\frac{3}{2}E\frac{n}{D_w}\frac{W}{Q}dx \quad \frac{\text{particles}}{\text{m}^3}$$

Integrating over the complete effective length of the spray, L (algebraic difference between the distances moved by the particles and droplets in the x direction), gives the total number of particles removed between the inlet concentration, N_{in}, and outlet concentration, N_{out} (particles/m^3):

$$\int_{N_{in}}^{N_{out}} \frac{dn}{n} = -\frac{3}{2}\frac{E}{D_w}\frac{W}{Q}\int_0^L dx$$

$$\ln(N_{out}/N_{in}) = -\frac{3}{2}\frac{E}{D_w}\frac{W}{Q}L$$

$$\frac{N_{out}}{N_{in}} = \exp\left(-\frac{3}{2}\frac{E}{D_w}\frac{W}{Q}L\right) \qquad (20.51)$$

Reference to equation (20.44) shows that the particle removal efficiency of the spray is given by

$$\eta_p = 1 - \exp\left(-\frac{3}{2}\frac{E}{D_w}\frac{W}{Q}L\right) \tag{20.52}$$

Examination of this equation is most instructive in understanding the performance of sprays. The dust removal capacity of the spray increases with E, the capture efficiency of each droplet. To be precise, this depends on the nature of the flow and the relative sizes of dust particles and water droplets. However, a coarse approximation for fully developed turbulence, based on work reported by Jones (1978b), can be assessed as

$$E = 0.266\ln(K) + 0.59 \tag{20.53}$$

over the range $0.2 < K < 4$, where ln means natural logarithm and the dimensionless parameter K is

$$K = \frac{u_r \rho D_p^2}{9\mu D_w} \tag{20.54}$$

where ρ = particle density (kg/m^3), D_p = particle diameter (m) and μ = kinematic viscosity of the air (N s/m^2).

In particular, the capture efficiency increases with the relative velocity between the dust particles and droplets (u_r), the diameter (D_p) and density (ρ) of the particles (these three governing particle inertia) and increases further as the water droplets become smaller (D_w).

Returning to equation (20.52) reinforces the fact that the overall efficiency of the spray improves with smaller water droplets. A coarse spray of large water droplets will have very little effect on airborne respirable air.

A parameter of basic importance in equation (20.52) is the water-to-air ratio (W/Q). Values in the range from 0.1 to over 2 (l water)/(m^3 air) have been reported. The lower values produce poor efficiency of dust capture. However, if too high a value is attempted then the concentration of droplets may become so great that coalescence occurs. The larger droplets then lead to decreased efficiency. A practical range of W/Q for sprays and wet scrubbers in mines is 0.3 to 0.6 l/m^3. Tests on compressed air-powered atomizing nozzles have indicated an optimum W/Q value of 0.45 l/m^3 (Booth-Jones *et al.*, 1984).

The last point to be gleaned from equation (20.52) is confirmation of the intuitive expectation that the spray efficiency is improved as the length (L) and, hence, time of contact between the air and the water droplets are increased.

In order to produce the finely divided sprays necessary to affect respirable dust, a number of methods are employed. The simplest technique is to supply high pressure water to the nozzles. Pressures of some 3000 to 4000 kPa applied across suitable nozzles give smaller droplets at spray velocities high enough to cause air induction—surrounding dust-laden air is drawn into the spray and thus improves the dust removal capacity of the unit. A variety of nozzle designs are available commercially. These control the shape as well as influencing the atomization of the spray. Full cone and hollow cone sprays have good air induction characteristics while fan-shaped sprays are excellent at confining the dust clouds produced by shearers and continuous miners. Atomization is further improved in some nozzles by impinging the high velocity jet against an impact surface located facing and close to the orifice. Another arrangement causes the water to rotate rapidly around an orifice before ejection. In all cases, it is particularly important in mining that nozzle designs should mitigate against blockage from particles either in the water supply or (in the case of machine-mounted sprays), thrown forcibly against the jet from an external source.

Compressed air-assisted sprays can produce fine atomization with droplets in the respirable range. The water feed is connected into the compressed air supply close to the nozzles. The water enters the compressed airstream either by its own applied pressure or by venturi action. It is advisable to insert non-return valves into the water line. The combination of very high turbulence at the nozzles and expansion of the compressed air into the ambient atmosphere produces fine droplets.

Compressed air-assisted sprays can be further enhanced by the addition of a sonic device to the nozzle (Schröder *et al.*, 1984). Air expands through the nozzle into a facing resonator cup where it is reflected back to complement and amplify the initial shock wave at the mouth of the orifice. An intense field of sonic energy is focused in the gap between the nozzle and the resonator cup. Water droplets issuing from the nozzle and passing through the sonic field are further broken down to respirable sizes and, indeed, to submicron diameters. Similar effects can be achieved by high frequency oscillation of pairs of piezoelectric crystals.

A high degree of atomization can be achieved without high pipeline pressures through impingement devices. A free-standing 'fog machine' of this type may consist of a stainless steel disk spinning at about 3000 rpm. A low pressure water supply is fed to the centre of the disk. Centrifugal action causes the water to flow outwards over the surface of the disk to impact at high velocity on a ring of stationary and closely spaced vanes around the perimeter. A fan impellor located behind the disk projects an airstream and the fog forward. The same principle is employed in wetted fan scrubbers.

Wet scrubbers

As the name suggests, these are devices that also employ water to achieve dust removal. However, in this case the water streams (or sprays) and the airflow are controlled within an enclosure designed to maximize those parameters that improve the efficiency of dust capture (equation (20.52)). Wet scrubbers bring dust particles into intimate contact with wet surfaces and within a highly turbulent mixture of air, water droplets and dust. They have become popular for mining applications as they require less maintenance than most other dust filters and can achieve respirable dust capture efficiencies exceeding 90%.

Here again, we shall restrict our discussion to the operating principles employed in the most common wet scrubbers. Many competing devices are marketed and

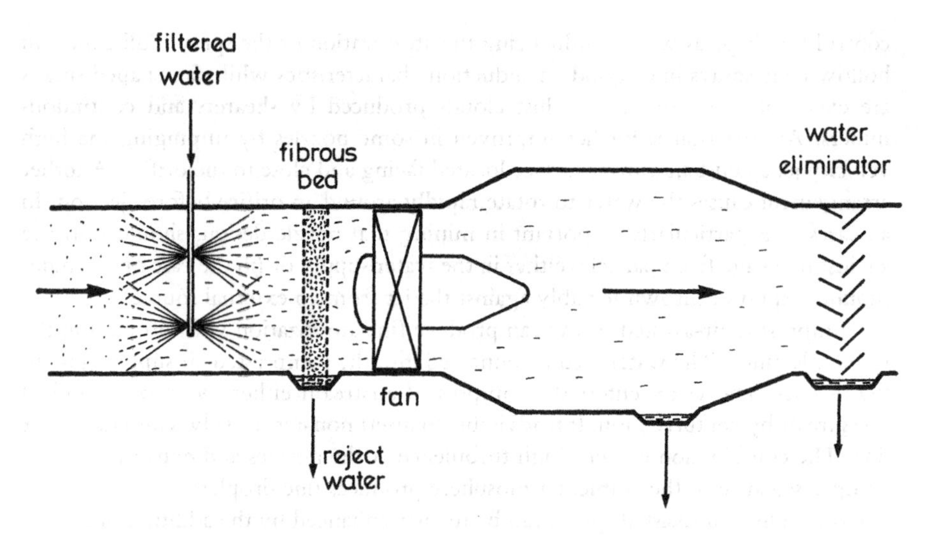

Figure 20.9 Fibrous (flooded) bed scrubber.

manufacturers' literature should be consulted to match performance with required duties, and to compare capital and operating costs.

The **fibrous** (or **flooded**) **bed** scrubber illustrated in Fig. 20.9 is one of the most widely used devices employed in mine dust collectors. Stainless steel or other non-corrosive material is used as the fibre material. Either water is admitted along the top of the fibrous bed and allowed to trickle downwards through it or, preferably, the water is sprayed directly into the air upstream from the fibrous bed. The air follows a tortuous path through the bed while the inertia of the dust particles causes them to strike and adhere to the wet fibres. The efficiency of dust removal increases with the fineness of the fibres, the thickness of the bed and the velocity of the air. This must be balanced by the resistance of the unit to airflow and, hence, the operating cost. Efficiencies exceeding 90% for respirable dust can be attained.

The dust-laden water collects at the bottom of the fibrous bed from where it is drained, filtered and recycled. Arrangements must be made to remove the effluent sludge and to supply make-up water. In all cases, wet scrubbers can be supplied with chilled water to achieve simultaneous cooling and dust collection. Again, filtration within the chilled water cycle is necessary.

A water eliminator is required by most designs of wet scrubber in order to remove residual droplets of water. Several different systems of water elimination are available in practice including a second fibrous mat, a series of wavy or inclined plates, turning vanes to induce swirl into the air and, hence, throwing droplets outwards towards the duct walls, or an egg-tray arrangement. Here again, droplet removal is achieved by impingement.

Figure 20.10 illustrates the principle of the **wetted fan scrubber**. Sprays upstream and/or at the facing boss of a fan produce droplets that are mixed intimately and at high velocity with air across and around the fan impeller blades. The polluted water

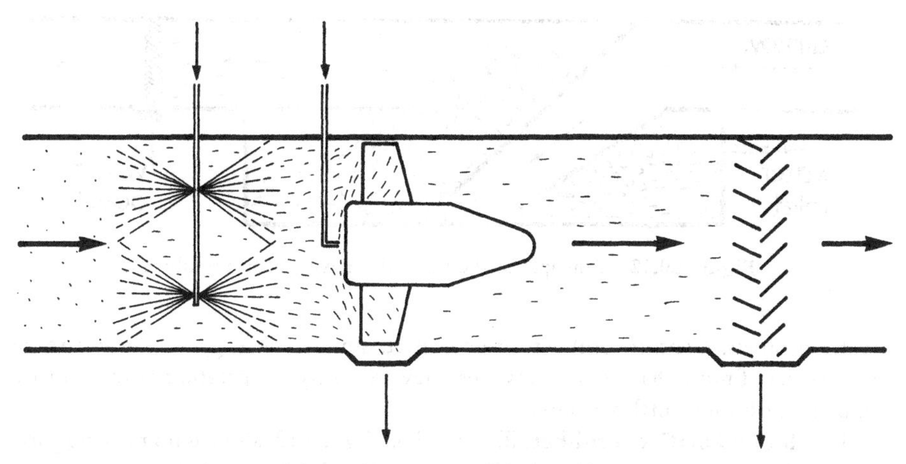

Figure 20.10 Wetted fan scrubber.

collects around the internal surface of the fan casing for removal and recycling. The addition of a fibrous bed downstream from the fan gives a powerful combination of dust collection devices. A disadvantage of wetted fan scrubbers is the pitting that may occur on the impeller blades and requiring additional fan maintenance. Designs employing centrifugal as well as axial fans have been developed. Wetted fan scrubbers are well suited to lower airflows and have an application as in-line dust collectors in auxiliary ventilation ducting.

The **venturi scrubber**, depicted on Fig. 20.11, has no moving parts. Sprays are located upstream and/or at the throttled section of a venturi arrangement. Air velocities through the throat are typically in the range 60 to 120 m/s with a high degree of turbulent mixing. This encourages the impaction of dust particles into

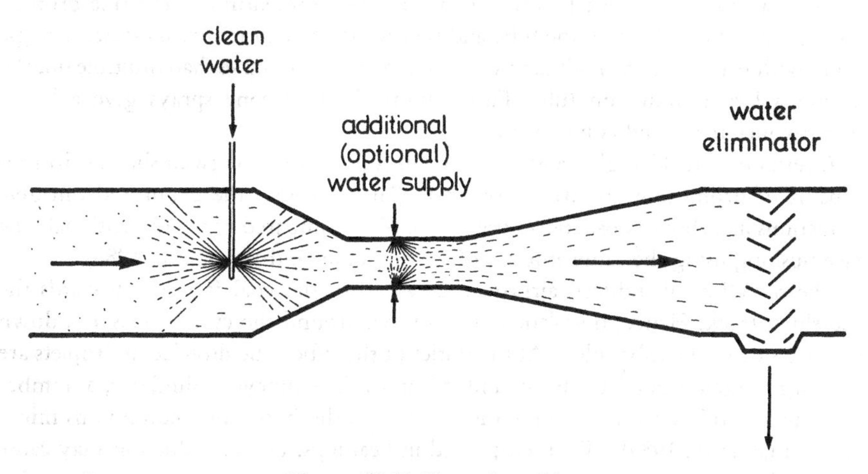

Figure 20.11 Venturi scrubber.

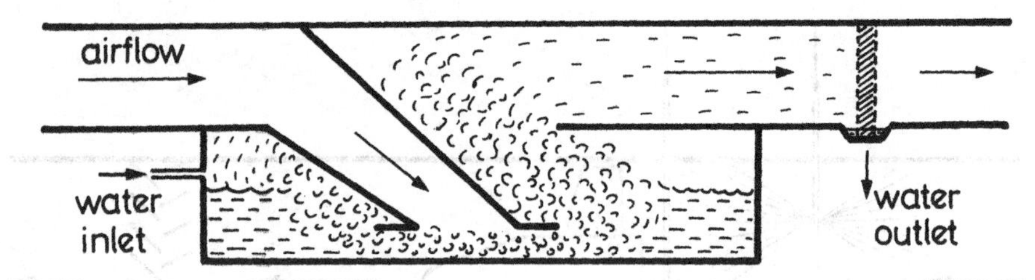

Figure 20.12 Principle of the flooded (or wet) orifice scrubber.

water droplets. Venturi scrubbers are compact, simple and rugged, and can reach efficiencies of more than 90%. However, they are costly in operating power and are suitable for limited airflows only.

The **flooded orifice scrubber**, illustrated on Fig. 20.12, also has no moving parts and has the additional advantage that there are no nozzles that might become clogged. Air from the inlet duct flows outwards beneath a lip that is submerged in water. Movement of the air causes extreme agitation of the water and entrainment of droplets. Collection efficiencies of more than 80% can be achieved with this system.

The preferred location for a dust collection device is as close as practicable to the source of the dust. The types of wet scrubbers outlined in the previous paragraphs are suitable as free-standing units or within ventilation ducts. However, attempts to attach them to coal or rock winning machines have shown them to be somewhat bulky for that application and insufficiently robust to withstand the rigours of a working face. A device that met increasing favour for shearers and continuous miners through the 1980s was the simple **high pressure spray fan** or **induction tube** (Jones, 1978a; Sartaine, 1985; James and Browning, 1988; Jayaraman *et al.*, 1989). This is illustrated by Fig. 20.13 and consists of a water jet spraying into a tube of some 100 mm diameter. The water is supplied at pressures in the range of 6 to 12 MPa through a nozzle of about 1.5 mm diameter. The momentum of the fine droplets induces an airflow through the tube and is very effective in removing dust. A single spray within a relatively small tube appears to be more effective than multiple nozzles within a larger induction tube. Furthermore, hollow cone sprays give a better performance than solid cone sprays.

A series of 9 to 12 high pressure spray fans built into a longwall shearer drum is capable of promoting an airflow of up to 2 m^3/s around the drum and can give reductions in airborne dust concentrations of 80% compared with conventional pick face flushing using the same amount of water (James and Browning, 1988).

The direction of induced airflow is away from the coal face and towards the travelling track. Hence, dust-laden air is drawn around the cutter picks and down the face side to the tube inlets. At the outlet of the tubes, the dust-laden droplets are discharged against deflector plates and fall on to the conveyor. Similarly, a number of induction tubes can be mounted in parallel on the boom of a continuous miner (Jayaraman *et al.*, 1989). When employed in headings, the air induction may cause local recirculation and greatly improved ventilation of the cutter heads. Provided

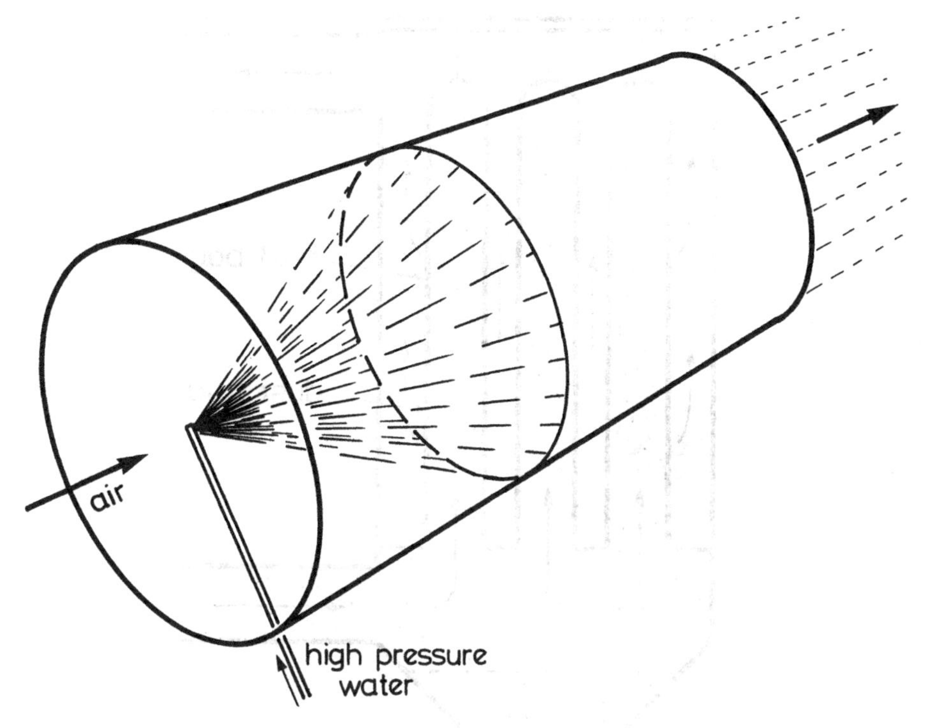

Figure 20.13 The high pressure spray fan or induction tube.

that adequate airflow is supplied to the face end of the heading, this will enhance the overall safety of the environment. However, legislative enforcement agencies should be consulted in industries where recirculation is prohibited.

The advantages of the high pressure spray induction tubes are that

1. they are simple, robust and have no moving parts,
2. they can be built into the machine structure,
3. they promote ventilation of the cutter heads as well as removing dust,
4. they give a good efficiency of dust capture, and
5. provided that the water pressure is maintained, there is little chance of blockage.

Dry filters and separators

There are many situations in subsurface ventilation systems where increasing the humidity of the air by the use of wet scrubbers is inadvisable. These include mines where heat and humidity is already a problem although cycling chilled water through wet scrubbers will reduce temperature, humidity and dust concentration simultaneously. Other difficulties that can arise from increases in humidity include clogging of hygroscopic minerals during transportation, roof control where the overlying strata are subject to rapid weathering, and where the mineral is subject to spontaneous

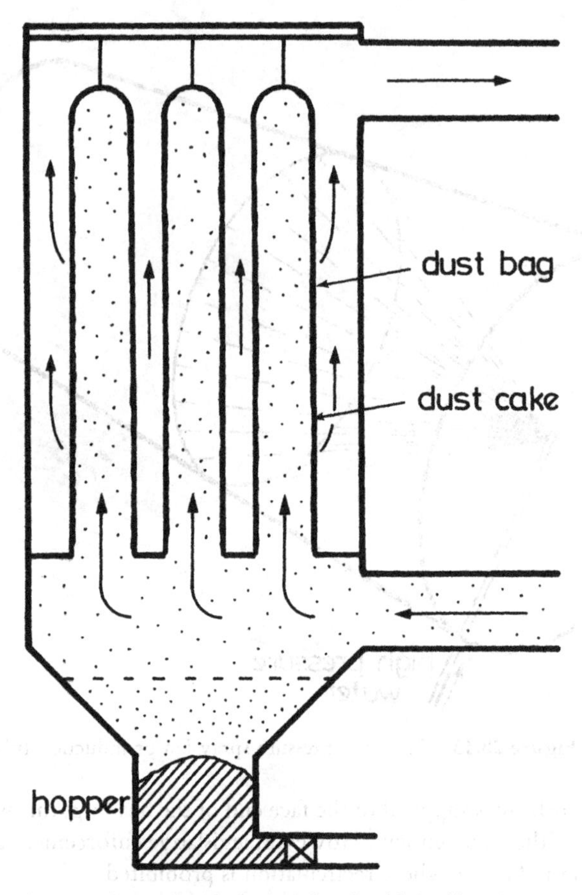

Figure 20.14 Principle of a fabric filter.

combustion. In such circumstances, it may be preferable to employ dry filters to remove airborne dust.

Figure 20.14 illustrates a **fabric filter**. The air passes through a fabric leaving dust particles adhering to the material or to the dust cake that builds up on the high pressure side. Air may flow from the inside to the outside of the bags as illustrated or in the opposite direction, in which case the bags are supported on an internal frame. The dust cake is dislodged at intervals of time by mechanical agitation or a reversed air pulse and falls to be collected in the hopper. The collected dust is either removed dry and bagged by mechanical cleaning system or as a slurry produced by addition of water to the hopper.

The dust cake itself accomplishes most of the filtration and efficiencies of over 99% can be achieved in the submicron range. Airflow through the fine apertures of the dust cake is laminar. Hence, fabric dust collectors tend to follow a linear $p = RQ$ relationship. The overall resistance of the unit arises from the combined effects of the dust-impregnated fabric and the dust cake. It is a straightforward matter to show that

$$R = R(\text{fabric}) + R(\text{cake}) = \frac{\mu}{A}\left(\frac{x}{k_f} + \frac{m}{k_c A \rho}\right) \quad \frac{\text{N s}}{\text{m}^5} \tag{20.55}$$

where μ = dynamic viscosity of air (N s/m^2), A = surface area of filter (m^2), x = thickness of fabric (m), k_f = permeability of dust-impregnated fabric (m^2), k_c = permeability of dust cake (m^2), m = mass of dust in the dust cake (kg) and ρ = density of the dust cake (kg/m^3). (The definition and units of permeability are explained in section 12.3.2. However, manufacturers may assume a standard value of air viscosity and quote filter permeabilities in terms of m^3/s (of flow) through each m^2 of area for unit pressure gradient through the material (Pa/m), i.e. m^3/s per m^2 per Pa/m.)

Equation (20.55) quantifies the increase in resistance as the thickness and, hence, mass of the dust cake builds up. This also increases the capture efficiency of the device. The pressure developed by the unit fan will rise and the air quantity will fall. An excessive pressure may cause rupturing of the filter fabric. The cake must, in any case, be dislodged before the airflow drops to an unacceptably low value. A backward curved (non-overloading) centrifugal fan operating on a steep pressure–quantity portion of its characteristic curve is advisable.

The simplest type of fabric cleaning mechanism is an electromechanical agitator. If operation of the unit can be interrupted every few hours (dependent upon dust loading) then the fan can automatically be switched off and the bags shaken by the agitator. More sophisticated units allow continuous operation by cycling the filtration and cleaning around several separate compartments.

Reverse flow cleaning involves a temporary reversal of air direction. This eliminates the mechanical linkages of the agitator system and is preferred for some types of fabric such as glass cloth where the severe flexing action of mechanical shaking may break the fibres. Pulsed jet reverse flow increases the efficiency of cleaning. Acoustic methods have also been employed to dislodge filter cakes.

The choice of fabric material usually lies between cotton weaves, felted fabrics or a synthetic such as polypropylene. The felted fabrics give an initially higher efficiency but synthetics are preferable where moist conditions or hygroscopic minerals may tend to produce a sticky dust cake. A newly installed bag will have a relatively low resistance. The initial mechanism is that dust particles will become lodged within the material. This increases both the resistance to flow and capture efficiency. Subsequent cleaning cycles will remove the dust cake but will have little effect on dust that has become impregnated in the material (more is dislodged from the smoother fibres of synthetic material). It follows that performance tests on a fabric dust collector should be delayed until dust impregnation of the material has reached steady state.

Two types of **cyclone** have been developed for dust removal, both of which can be operated dry or with the addition of water to improve capture efficiency. The **conical cyclone** operates by the dusty air being constrained into a helical vortex of reducing radius. Figure 19.4 was drawn to illustrate the conical cyclones used in dust samplers. Larger versions can be used as dust collectors. Dust particles are subjected to two opposing forces in a cyclone; the centrifugal force that tends to throw the particles out toward the wall, and drag of the air which tends to pull them inward toward the central air outlet tube. The greater the mass of the particle and the rotational velocity the more efficient the cyclone will become. Hence, the performance is enhanced for larger particles and as the physical size of the cyclone decreases. Cyclones are normally employed in groups for air cleaning. The centrifugal action is improved by arranging for the air to enter tangentially. It is essential to remove the dust from the base continuously in order to avoid re-entrainment. The finer particles that escape in the outlet air may be removed by a second cyclone or other filtration device connected in series.

The **cylindrical cyclone** imparts helical vortices to the airflow by means of turning vanes in a duct. The dust which concentrates and moves in helical fashion along the walls is collected and removed through an annulus formed by a second inner duct. The capital cost of cyclones is relatively low. They have no moving parts and are easy to maintain. However, the power requirements are such that they are constrained to applications of low airflow.

Electrostatic precipitators are used widely as air cleaners in buildings and for surface industrial applications such as the removal of fly ash from power station stacks or capturing aerosols in the chemical and metallurgical industries. Although a well-designed electrostatic precipitator can reach capture efficiencies of over 99% in the submicron range, their need for high voltages prohibits their use in gassy mines and mitigates against their employment in other underground facilities.

The principle of operation of an electrostatic precipitator is that when an aerosol is passed through an electric field produced by a pair of electrodes then the particles will become charged and migrate towards one of those electrodes. For industrial applications, the active electrodes are charged to voltages between 20 and 60 kV while the dust-collecting electrodes are earthed. The electric field is considerably enhanced in regions of sharp curvature on the electrode surfaces. For this reason, the active electrodes are often wires hanging vertically downwards. The wires are usually charged negatively as this gives a more stable performance for heavy duty performance although ozone can be formed. High energy electrons are emitted from the negatively charged wires. Each electron collision with a gas molecule causes the ejection of two further electrons which go on to repeat the process. This escalating process produces an electron avalanche and is often accompanied by a visible glow; hence, the phenomenon is termed a **corona**. The gas molecules that have lost electrons become positive ions and migrate towards the negatively charged wires. However, further away from the active electrodes, the free electrons lose their kinetic energy to the extent that they are no longer capable of dislodging further electrons from gas molecules but are, instead, absorbed into those molecules. The electron avalanche ceases and the edge of the corona is reached.

The gas molecules are then negatively charged, i.e. negative ions, and migrate towards an earthed electrode. During that migration they become attached to dust particles which are also, therefore, drawn towards an earthed electrode and adhere by electrostatic attraction to the surface of that electrode. On contact, the particles begin to leak their charge to the earthed electrode. Other layers of charged particles arrive and build up progressively. They too will gradually give up their charge. However, the outermost layer of dust is always the most heavily charged and will be analogous to a skin compressing the underlying particles and causing the build-up of a dust cake. The dust can be dislodged into an underlying hopper by rapping the earthed electrodes.

In **tube electrostatic precipitators**, a single wire forms the active electrode suspended in a metal cylinder which acts as the grounded electrode. However, for the larger flows found in industrial applications, the **plate electrostatic** precipitator has become more common. This is illustrated in Fig. 20.15. The air passes over the charged wire electrodes which are suspended between a series of grounded plates. The dust collects on the surface of the plates. For some applications, the mechanisms of dislodgement by rapping may be replaced by running a film of liquid down the plate surfaces or by periodically dipping the plates into a liquid bath.

The efficiency of an electrostatic precipitator can be determined by an equation first derived by W. Deutsch in 1922:

$$\eta = 1 - \exp\left(-\frac{Au_e}{Q}\right) \tag{20.56}$$

where A = area of plates (m^2), Q = airflow (m^3) and u_e = electrical (ion) migration velocity (m/s) (see equation (20.28)). The electrical migration velocity depends on

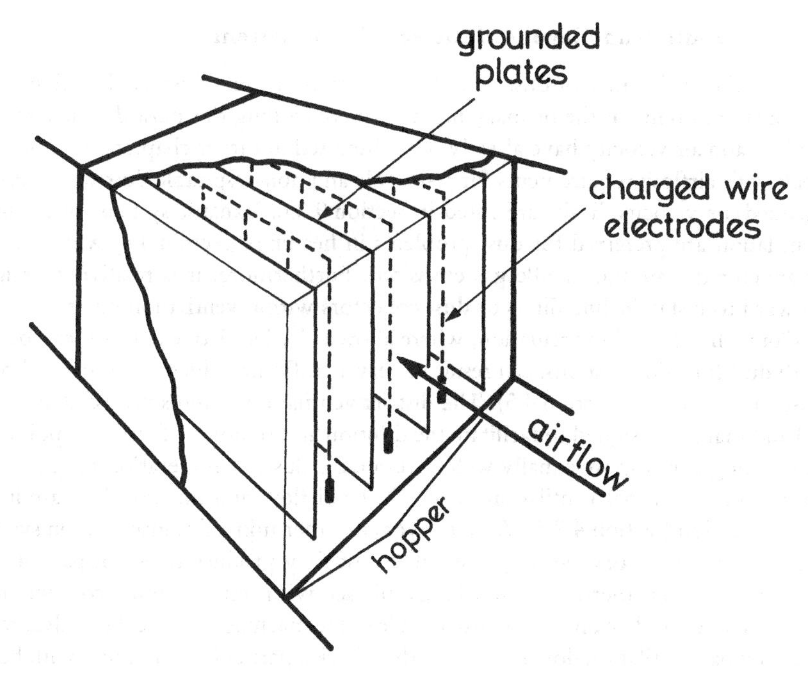

Figure 20.15 A plate electrostatic precipitator.

the type of dust and varies between 0.02 m/s for fly ash to 0.2 m/s for gypsum. Although theoretical procedures have been derived for quantification of the electrical migration velocity, tables of empirical values have, to this time, proved to be more reliable (ASHRAE, 1988).

Personal respirators

Every effort should be made to maintain dust concentrations in subsurface workings within mandatory threshold limits and safe for the health of the workforce. A final line of defence is the personal respirator used to filter inhaled air. Two types are available. The first of these is a mask that fits around the nose and mouth. The filter is necessarily a compromise between dust removal efficiency and resistance. A respirator that requires more than about 150 Pa of pressure difference at normal breathing rates is unlikely to be tolerated by personnel. Furthermore, contact of the mask on the face can be irritating, especially in hot conditions. An improved version, sometimes called an airstream helmet, utilizes a belt-mounted battery to power a small fan. This passes air through a filter and up a tube to the helmet. The cleaned air flows downwards between a transparent visor and the face of the wearer. This device does not rely on breathing effort; nor is there any direct face contact with the visor. It also provides eye protection with less visual impedance than that given by goggles or safety glasses.

20.4.3 Dilution and layout of the ventilation system

Despite the availability of dust collectors, dilution of mine dust by the mine ventilation system remains the primary method of controlling this hazard. The effects of airflow and air velocity have already been discussed in earlier chapters. Section 9.3.3 deals with airflow requirements for respirable and non-respirable dust while recommended air velocity limits are listed in section 9.3.6. Exhaust systems of auxiliary ventilation are preferred for dust problems in headings (section 4.4) while overlap arrangements can also handle gas emissions. Furthermore, it is relatively straightforward to install in-line filters or dust collectors within ventilation ducts.

Controlled partial recirculation, where allowed by legislative authorities, coupled with dust filtration systems, can result in very significant reductions in general body dust concentrations (section 4.5). The district ventilation systems discussed in section 4.3 and that are designed to facilitate the dilution and removal of airborne pollutants in working zones apply equally well to respirable dust. Consideration might also be given to homotropal ventilation in which the airflow and mineral flow are in the same direction (section 4.2.3). As conveyors or other mineral transportation systems are then in return airways, any respirable dust they produce does not pass on to a working area. Furthermore, on a longwall face with unidirectional coal winning, few personnel need be on the downwind side of the machine. Despite these advantages, homotropal ventilation does have some drawbacks, particularly in mines with heavy gas emissions (Stevenson, 1985).

20.4.4 Separation of personnel and dust

In section 20.3.4 we described the re-entry period after blasting in metal mines as a classical example of the separation of personnel from dust concentrations. Several other methods are available to reduce the exposure of individuals or groups to dust. The United States Bureau of Mines has been active in developing this approach, particularly for the protection of the operators of longwall face equipment and continuous miners in room and pillar workings.

Airflow diverters of two types have been fitted to such machines. First, barriers have been added to shearers in order to divide the face airflow before it reaches the location of the cutting drum. This is positioned such that it provides a split of relatively clean air to the shearer operator. A great deal of research has been conducted into the use of spray fans to control the direction and flow of air at continuous miners and longwall shearers. Appropriate location and design of these triangular or cone-shaped sprays not only assists in dust suppression but also ensures that airborne dust is diverted away from operators' positions (National Research Council, 1980).

Air curtains have also been employed to prevent dust clouds from reaching operators' positions, as well as assisting in the ventilation of cutter heads (Ford and Hole, 1984; Froger *et al.*, 1984; James and Browning, 1988). The air curtains may be directed across the top, bottom and sides of the cutting zone. They are produced from tubes of about 10 mm diameter maintained at an air pressure of approximately 1.5 kPa. A 2.5 mm slot runs along the length of the tube with an attached guide plate angled such that air leaves the tube tangentially, clinging to the guide plate (the Coanda effect) until it is deflected into the required direction by a splitter. Entrainment of additional air assists in both the ventilating and dust control effects.

Another development that reduces dust exposure to machine operators has been the growing utilization of remote controls. These allow personnel to stand some distance from the mineral-winning machines while maintaining control by hand-held wireless units. Finally, studies leading to the reorganization of work practices have also promoted reduced dust exposure of face personnel (Tomb *et al.*, 1990).

REFERENCES

Allen, M. D. and Raabe, O. G. (1982) Re-evaluation of Milliken's oil drop data for the motion of small particles in air. *J. Aerosol Sci.* **13**, 537–47.

ASHRAE (1988) *Equipment Handbook*, Chapter 11, pp. 11–2, American Society for Heating, Refrigerating and Air-conditioning Engineers.

Bandopadhyay, S. (1982) Planning with diesel powered equipment in underground mines. *PhD Thesis*, The Pennsylvania State University.

Bhaskar, R. and Ramani, R. V. (1988) Behaviour of dust clouds in mine airways. *Trans. AIME*, **280**, 2051–9.

Bhaskar, R., Gong, R. and Jankowski, R. A. (1991) Studies in underboom dust control to reduce operator exposure to dust. *Proc. 5th US Mine Ventilation Symp., West Virginia*, 197–206.

Booth-Jones, P. A., Annegarn, H. J. and Bluhm, S. J. (1984) Filtration of underground ventilation air by wet dust-scrubbing. *Proc. 3rd Int. Mine Ventilation Congr. Harrogate,* 209–17.

Chung, H. S. (1981) Coagulation processes for fine particles. *PhD Thesis,* The Pennsylvania State University.

Flagan, R. C. and Seinfeld, J. H. (1988) *Fundamentals of Air Pollution Engineering,* 542 pp., Prentice-Hall.

Ford, V. H. W. and Hole, B. J. (1984) Air curtains for reducing exposure of heading machine operators to dust in coal mines, *Ann. Occup. Hyg.* **28**, 93–106.

Froger, C., Courbon, P. and Koniuta, A. (1984) Dust-laden airflow control applied to worker protection: scale model study of a steep seam working. *Proc. 3rd Int. Mine Ventilation Congr. Harrogate,* 215–17.

Gaudin, A. M. and Meloy, T. P. (1962) Model and a comminution distribution equation for repeated fracture, *Trans. AIME* **223**, 243–50.

Hamilton, R. J. and Walton, W. H. (1961) In *The Selective Sampling of Respirable Dust in Inhaled Particles and Vapours* (ed. Davies), Pergamon, Oxford.

Hartman, I. and Grenwald, H. P. (1940) Use of wetting agents for allaying coal dust in mines. *US Bureau of Mines IC-7131,* 12 pp.

Heising, C. and Becker, H. (1980) Dust control in longwall workings. *Proc. 2nd Int. Mine Ventilation Congr., Reno, NV.,* 603–11.

Hood, M. (1991) Personal communication on jet-assisted rock cutting.

Hood, M., Knight, G. C. and Thimons, E. D. (1991) A review of jet-assisted rock cutting. *J. Eng. Ind.,* to be published.

James, G. C. and Browning, E. J. (1988) Extraction techniques for airborne dust control, *Proc. 4th Int. Mine Ventilation Congr. Brisbane,* 539–46.

Jayaraman, N. I., *et al.* (1989) High pressure water-powered scrubbers for continuous miner dust control. *Proc. 4th US Mine Ventilation Symp., Berkeley, CA,* 437–43.

Jones, A. D. (1978a) Experimental and theoretical work on the use of a high pressure water spray to induce airflow in a tube and capture airborne dust. *MRDE Report No.73,* National Coal Board, U.K.

Jones, A. D. (1978b) Optimal design for water-powered dust extraction. *MPhil Thesis,* University of Nottingham.

Knight, G. (1985) Generation and control of mine airborne dust. *Proc. 2nd US Mine Ventilation Symp., Reno, NV,* 139–50.

Marshall, V. C. (1974) *Comminution,* Chameleon Press, London.

National Research Council (1980) Measurement and control of respirable dust in mines. *NMAB-363,* US National Academy of Sciences, Washington, DC.

Owen, P. R. (1969) Dust deposition from a turbulent airstream. In *Aerodynamic Capture of Particles* (ed. Richardson).

Padmanabhan, S. and Mutmansky, J. M. (1989) An analysis of quartz occurrence patterns in airborne coal mine dusts. *Proc. 4th US Mine Ventilation Symp., Berkeley, CA,* 463–74.

Prandtl, L. (1923) *Ergebnisse der aerodynamischen Versuchsanstalt zu Göttingen,* Oldenbourg, Munich, p. 29.

Qin, J. and Ramani, R. V. (1989) Generation and entrainment of coal dust in underground mines. *Proc. 4th US Ventilation Symp., Berkeley, CA,* 454–62.

Ramani, R. V. and Bhaskar, R. (1984) Dust transport in mine airways. *Proc. Coal Mine Dust Conf. W. Virginia University* (ed. Peng), 198–205.

Ramani, R. V., *et al.* (1988) On the relationship between quartz in the coal seam and quartz in the respirable airborne coal dust. *Proc. 4th Int. Mine Ventilation Congr., Brisbane,* 519–26.

Sandys, M. P. J. and Quilliam, J. H. (1982) Sources and methods of dust control. *Environmental Engineering in South African Mines,* Chapter 15, Mine Ventilation Society of South Africa.

Sartaine, J. J. (1985) The use of water-powered scrubbers on NMS Marietta drum miners. *Proc. 2nd US Mine Ventilation Symp., Reno, NV,* 733–40.

Schröder, H. H. E., Runggas, F. M. and Krüss, J. A. L. (1984) Characterization of sonically atomized water-spray plumes, *Proc. 3rd Int. Mine Ventilation Congr. Harrogate,* 219–28.

Seinfeld, J. H. (1986) *Atmospheric Chemistry and Physics of Air Pollution,* Wiley, New York.

Stevenson, J. W. (1985) An operator's experience using antitropal and homotropal longwall face ventilation systems. *Proc. 2nd US Mine Ventilation Symp., Reno, NV,* 551–57.

Stricklin, J. H. (1987) Longwall dust control at Jim Walters Resources. *Proc. 3rd US Mine Ventilation Symp., Penn State,* 558–63.

Taylor, C. D., Kovscek, P. D. and Thimons, E. D. (1988) Dust control on longwall shearers using water-jet-assisted cutting. *Proc. 4th Int. Mine Ventilation Congr., Brisbane,* 547–53.

Tomb, T. F., *et al.* (1990) Evaluation of longwall dust control on longwall mining operations. *Proc. SME Annual Meet. Salt Lake City, UT, February,* 1–10.

Wang, Y. P., *et al.* (1991) Use of surfactants for dust control in mines. *Proc. 5th US Mine Ventilation Symp., West Virginia,* 263–70.

PART SIX

Fires and Explosions

21

Subsurface fires and explosions

21.1 INTRODUCTION

The most feared of hazards in underground mines or other subsurface facilities are those of fires and explosions. Like airplane crashes, these do not occur often but, when they do, have the potential of causing disastrous loss of life and property as well as a temporary or permanent sterilization of mineral reserves. Furthermore, 'near misses' occur all too frequently. The incidence of mine fires appears not to be declining despite greatly improved methods of mine environmental design and hazard control. This is a consequence of two matters; first the growing variety of materials that are imported into modern mine workings, varying from resins and plastics to liquid fuels and hydraulic fluids. A second factor is the continuous increase in the employment of mechanized procedures, many of the machines involving flammable liquids and materials that can produce toxic fumes when overheated. The enormous loss of life due to mine fires and explosions during the eighteenth and nineteenth centuries preoccupied the minds of mining engineers and scientists of the time (Chapter 1). Through the 1980s, mine fires re-emerged as a topic of pressing research need.

The majority of deaths arising from mine fires and explosions are caused, not by burning or blast effects, but by the inhalation of toxic gases, in particular, carbon monoxide. There are two major differences between underground fires and those that occur in surface structures. The first concerns the long distances, often several kilometres, that personnel might be required to travel in passageways that may be smoke filled. Secondly, the ventilation routes are bounded by the confines of the airways and workings, causing closely coupled interactions between the airflows and behaviour of the fire.

It is difficult for anyone who has not had the experience, to comprehend the sensations of complete isolation and disorientation involved in feeling one's way through a long smoke-filled mine airway in zero visibility. It is a cogent exercise to turn off one's caplamp in an unilluminated return airway and to walk just a few steps, even without the trauma of a highly polluted atmosphere.

It is, therefore, a matter of ongoing importance that all personnel involved in the design and operation of underground openings should have some knowledge pertaining to the prevention and detection of subsurface fires, as well as procedures of personnel warning systems, escapeways, firefighting, toxic gases, training, fire drills and the vital need for prompt response to a fire emergency. These are some of the topics that are discussed in this chapter.

21.1.1 The fire triangle and the combustion process

Perhaps the most basic precept in firefighter training is the **fire triangle** shown in Fig. 21.1. This illustrates that the combustion process which we term 'fire' requires three components: fuel, heat and oxygen. Remove any one of these and the fire will be extinguished. The fuel may be solids, liquids or gases. The liquids and gases might be introduced into the mine environment at ambient temperature by natural or mining process, or may be produced by heating solid materials.

Whenever a combustible solid or liquid is heated to a sufficiently high temperature (**flashpoint**), it will produce a vapour that is capable of being ignited by a flame, spark

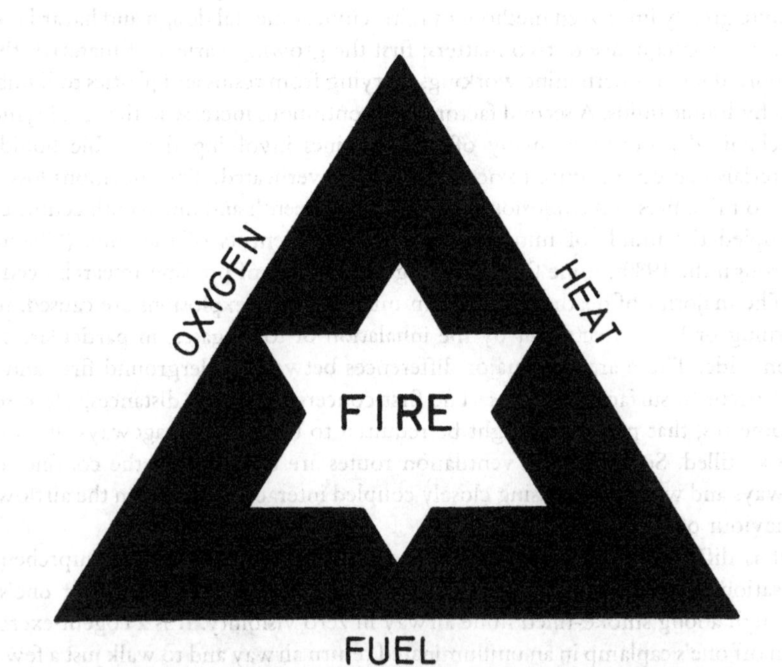

Figure 21.1 The fire triangle.

or hot surface which has the required concentration and duration of thermal energy. Gasoline has a flashpoint of − 45 °C while most commonly available solids require the application of a flame for them to reach flashpoint. The **ignition temperature** of any given substance is the lowest temperature at which sustained combustion is initiated. **Flaming** is the process of rapid oxidation of the vapours accompanied, usually, by the emission of heat and light. In the case of self-sustained burning, that heat is sufficient to raise the temperature of the newly exposed or surrounding areas of surface to flashpoint. However, combustion can continue at a slower rate without flaming through the process we know as **smouldering**. In this case, the oxidation process continues on the surface of the material and produces sufficient heat to be self-sustaining, but not enough to cause the emission of vapours in the quantity required for flaming combustion.

The oxygen which forms the third side of the fire triangle is normally provided by the air. Flammable liquids such as the oil of a flame safety lamp will cease to burn when the oxygen content of the air is reduced to some 16% (section 11.2.2). Flaming combustion of all kinds is extinguished at oxygen contents of 10% to 12% while smouldering is usually terminated at oxygen concentrations below 2%. However, some materials may contain sufficient inherent oxygen for slow combustion to continue at even further reduced levels of atmospheric oxygen. Coupled with the low values of thermal conductivity of crushed material, this can result in 'hot spots' lying dormant in abandoned areas for long periods of time, but capable of re-ignition if a renewed air supply is admitted subsequently.

21.1.2 Classification of mine fires

Fires underground can be classified into two broad groups, **open** and **concealed** fires. Open fires occur in airways, faces and other openings that form part of the active ventilation system of the mine and, hence, affect the quality of the mine airflows quickly and directly. As the term might imply, open fires are often accompanied by flaming combustion because of the availability of oxygen and offer the possibility of direct attack by firefighting teams. Conversely, concealed fires occur in areas that are difficult or impossible to access such as caved or abandoned zones. These are usually, but not necessarily, initiated as a result of spontaneous combustion and can occur in both coal and sulphide ore minerals as well as within any imported organic matter such as paper, discarded fabrics (e.g. oily rags) or timbering in abandoned areas. The degree to which concealed fires propagate and pollute the mine atmosphere depends upon the rate at which air leaks through the areas affected. The matter of spontaneous combustion is discussed in further detail in section 21.4.

21.2 CAUSES OF IGNITIONS

The variety of procedures, processes and materials used in modern mining provides many opportunities for the ignition of flammable materials. However, the most commonly reported causes of fires and explosions in mines are listed in the following subsections.

21.2.1 Mechanized equipment

Machines intended for use underground should be designed to operate with a high degree of safety in a harsh physical environment, and are subject to legal requirements and conditions in most mining countries. It is no surprise, therefore, that the majority of fires attributable to machines arise out of

1. misuse,
2. lack of proper maintenance,
3. removal or bypassing of safety features such as diagnostic devices, environmental monitors or thermal trip switches, and
4. running unattended for long periods of time.

Exhaust systems on diesel equipment should be fitted with scrubbers that not only reduce airborne pollution (section 11.3.2), but also prevent the emission of incandescent particles. Furthermore, hoses, transmission or brake fluids and a variety of components made from synthetic materials on modern diesels may be capable of producing toxic gases when ignited. All vehicles or other diesel equipment should be fitted with on-board fire extinguishers.

It is particularly important that equipment which contains significant quantities of oil, such as large transformers or air compressors, should be safeguarded by thermal trips, pressure relief valves and other devices necessary for automatic cut-off in the event of any abnormal condition. Such devices should be subjected to routine testing and maintenance. Wherever possible (and may be required by law) non-mobile equipment should be located within enclosures with fire-resistant roof, floor and walls, and which are ventilated to a return airway. Again, fire extinguishers and, preferably, an automatic fire suppression system should be available within the chamber. In coal mines, the surroundings in adjacent airways should routinely be coated with stonedust (section 21.8.3).

21.2.2 Electrical apparatus

In addition to the general comments on mechanized equipment made in the preceding subsection, electrical gear can give rise to incendive hazards from sparking and overheating. Switchgear should be sited such that it is not affected by convergence or falls of roof. This is most liable to occur close to mineral winning areas. Furthermore, start switches should be protected against accidental operation by glancing blows from falling debris or passing traffic. Electrical substations and battery charging chambers should be equipped with non-aqueous fire extinguishers.

Cables in airways should be hung in catenary fashion on cradles suspended from the roof. They should be located such that they will not be pinched by convergence or the yielding of roof supports, nor be impacted by vehicles. The insulation and type of sheathing must be suitable for the electrical load and rigours of the underground environment. All such cables should be inspected routinely for evidence of physical damage.

Electrical failures should result in immediate isolation of the power by means of overload and earth leakage protective devices. In gassy mines, all electrical motors and heavy current devices should be enclosed within flameproof casings so that any ignition of methane is contained within the equipment. Signalling or other light current apparatus should be certified as intrinsically safe, i.e. incapable of producing sparks of sufficient energy to ignite a methane air:mixture.

During non-working shifts, the electrical power supplies to each area of the mine should be isolated at the appropriate highest level control centre or substation. Precautions should be taken against power surges caused by lightning strikes on surface power lines, transformers or substations. Similarly, particular care should be taken against electrical leakage in the vicinity of explosives or fuel storage areas.

21.2.3 Conveyors

Conveyor fires have been subjected to particular study because of the rapidity of fire propagation along the early rubber-based types of conveyor belting. Modern conveyor belting for underground use must be subjected to fire propagation tests (e.g. Verakis, 1991). Three types of materials are used for mine conveyor belts, namely, styrene-butadiene rubber (SBR), neoprene (NP) and polyvinyl chloride (PVC). Composites of these materials are also employed. Following ignition of the belt material and removal of the igniting source, the fire should preferably fail to propagate or, if it does, move at a slow rate. However, it should be noted that heated belt material may produce hazardous fumes.

Numerous tests have indicated that fire propagation rates along conveyor belting are influenced by airspeed (e.g. Hwang et al., 1991). At a relative velocity of 1.5 m/s between the belt surface and the nearby airstream, a phenomenon known as **flash over** attains its maximum effect. This occurs when a flame front from the burning belt reaches forward over an unburned surface with an optimum angle and length such that the radiant effect on that surface reaches a maximum. This can cause flaming of the top layer of belting and a significant increase in flame propagation rate along the surface of the belt. Deeper layers in the weave of the material may or may not be ignited. The effect appears to be most pronounced in SBR belting (Verakis and Dalzell, 1988). Flashover involves a serious hazard as belt surface propagation rates may reach some 10 m/min. The spread of fire along mine conveyors is influenced strongly by the turbulence of the airflow. Hence, laboratory tests of small samples of belting can give misleading results. Large-scale gallery tests are more reliable.

Conveyor fires are most likely to be initiated by friction. If the belt becomes staked at any point along its length and the drive rollers continue to turn, then high temperatures will be generated at the drive head. Temperatures monitors or belt tension transducers can sense this condition. Such devices should be wired to isolate electrical power from the conveyor drive when an alarm condition is detected.

Similarly, a seized idler or return roller can become red hot from the friction of a belt moving over or around it. Conveyors should be patrolled during operation in order to detect the development of faulty rollers. Worn bearings will often be noisy and may also be detected by the smell of heated surfaces. A further frictional hazard can occur if the conveyor becomes misaligned to the extent that the belt rubs against surrounding surfaces such as the conveyor structure or airway sides.

In all of these cases, a fire may be initiated when lubricants, coal dust or flammable debris reach their ignition points. It follows that dust or spillage should not be allowed to accumulate round and, particularly, underneath conveyors. A clean conveyor road is more likely to be a safe one.

21.2.4 Other frictional ignitions

The main cause of methane ignitions on the working faces of coal mines is frictional sparking at the pick points of coal winning machinery. This occurs particularly when the machine cuts through sandstone or pyritic material. Two approaches have been taken to reduce this hazard. One is to ensure that there is sufficient ventilation around the cutting drum to provide rapid dilution of the methane as soon as it is emitted. It is, of course, important that the overall airflow at the working face is adequate to prevent methane layering (section 12.4.2) and that the layout of the system minimizes flushes of methane from worked-out areas (section 4.3.2). A number of devices have been employed to enhance air movement across the pick points of shearers and continuous miners (e.g. Browning, 1988). Unfortunately, these may exacerbate the dust problem unless combined with a wet scrubber (section 20.4.2).

The second approach to the incendive streak of sparks that sometimes trails behind a cutter pick is to quench it with water. This technique combines the suppression of both dust and methane ignitions. It is achieved by pick face flushing and, even more efficiently, by jet-assisted cutting (section 20.4.1).

Rope haulage systems have been the cause of some mine fires. Care should be taken that all pulleys and return wheels are routinely serviced and lubricated. Ropes should not be allowed to rub against solid surfaces such as the roof, sides or floor of airways and, particularly, timber supports. If haulage ropes must pass through holes in stoppings then, again, the ropes should not contact the sides of the orifices. Fluid couplings and enclosed gearings or direct drives are preferred to mechanical clutches, belts or V-drives for the transmissions of mining machinery. However, where the latter are employed then, again, regular inspections and maintenance are required to ensure their continued safe operation. Similarly, mechanical braking systems should be well looked after.

21.2.5 Explosives

The initiation of fires from explosives or igniter cord remains a danger in non-gassy mines. Incandescent particles from blasting operations may contain sufficient heat energy to ignite dry wood or combustible waste material. Igniter cord should never be hung on timber supports. A strict record should be maintained on all explosives and detonating devices at the times of issue and and return to the stores. The relevant national or state legislation should be consulted for the conditions under which explosives may be stored or transported underground.

21.2.6 Welding

All welding operations that are permitted underground should be carried out under well-controlled conditions. Where there is any possibility of methane or other flammable gases being present then testing for those gases should be carried out before and, at intervals, during the welding operations. Hot slag and sparks from welding are easily capable of igniting combustible materials such as coal, wood, paper and waste rags. Wherever possible, such materials should be removed from the vicinity of welding operations and the remainder wetted down or coated by stonedust. Molten metal should not be allowed to drop on the floor. Slag pans should be used to capture hot run-off. This is particularly important in coal-mines, in shafts and near timber supports. Fire extinguishers must be available at the sites of all welding operations.

Gas containers employed in oxy-acetylene cutting should be stored and used in a secure upright position. Gas bottles must never be used in the vicinity of explosives or concentrations of flammable liquids.

21.2.7 Smoking and flame safety lamps

It is sad fact that the use of smoking materials has been suspected as the cause of some fires and explosions in mines. In those mines that have been classified as gassy, carrying such materials (often known as **contraband**) into the subsurface is illegal. This law should be enforced with the utmost rigour. Furthermore, through well chosen examples during training and refresher classes a workforce will, themselves, ensure compliance with non-smoking regulations.

In subsurface openings where smoking is permitted then, again, education, posters and warning signs should be employed as ongoing reminders of the possible disastrous consequences of careless disposal of smoking materials.

Damaged flame safety lamps have also been suspected of igniting a methane:air mixture. Where these devices remain in use, they should be treated with care and subjected to inspection after each shift. When a high concentration of methane is detected by a blue flame spiralling rapidly within a flame safety lamp (section 11.4.2) then the lamp should be lowered gently and, if necessary, smothered inside one's clothing. Familiarity with the procedure should be gained through training and will counter the natural reaction of the untrained person to drop the lamp or to throw it away in panic.

21.3 OPEN FIRES

Fires that occur in mine airways usually commence from a single point of ignition. The initial fire is often quite small and, indeed, most fires are extinguished rapidly by prompt local action. Speed is of the essence. An energetic ignition that remains undetected, even for only a few minutes, can develop into a conflagration that becomes difficult or impossible to deal with. Sealing off the district or mine may then become inevitable.

The rate at which an open fire develops depends, initially, on the heat produced from the igniting source. A fine spray of burning oil from a damaged air compressor can be like a flamethrower and ignite nearby combustibles within seconds. On the other hand, an earth leakage from a faulty cable may cause several hours of smouldering before flames appear. The further propagation of the fire depends on the availability of fuel and oxygen (Fig. 21.1). A machine fire in an untimbered metal mine airway will remain localized if there is little else to burn in the vicinity. Conversely, an airway that is heavily timbered or with coal surfaces in the roof, floor or sides will provide a ready path for speedy development and propagation of a fire.

When an open fire has developed to the extent of causing a measurable change in the temperature of the airflow then it can affect the magnitudes and distributions of flow within the mine ventilation system. Conversely, the availability of oxygen to the fire site controls the development of the fire. This section discusses the coupled interaction between fire propagation and ventilation, and the means by which open fires in mines may be fought.

21.3.1 Oxygen-rich and fuel-rich fires

At the start of most open fires in ventilated areas, there is a plentiful supply of oxygen—more than sufficient for combustion of the burning material. Indeed, if the air velocity is brisk then heat may be removed at a rate greater than that at which it is produced. The heat side of the fire triangle is removed and the fire is 'blown out'.These are examples of **oxygen-rich** fires. Assuming that the fire continues to proliferate, it will consume increasing amounts of oxygen and, at the same time, produce greater volumes of distilled gases and vapours. The point may be reached when the heat of combustion produces temperatures that continue to remain high enough to distil gases and vapours from the coal, timber or other available fuels but with insufficient oxygen to burn those gases and vapours completely. The fire has then become **fuel rich**.

The development of an oxygen-rich into a fuel-rich fire is a serious progression and produces a much more dangerous situation for firefighters. When flammable gases at temperatures exceeding their ignition point meet relatively fresh air then they will ignite along the gas–air interfaces. The added turbulence may produce intimate mixing of air and unburned gases resulting in explosions. These phenomena can occur downstream from an open fire if air leaks into the firepath from adjacent airways. Firefighters are then faced with a difficult decision. Leakage of air from adjacent airways must be into the firepath in order to prevent spread of the fire into those adjacent airways, yet the admittance of that air may cause explosions and propagation of the fire at a rate much greater than that allowed by burning of the solid material itself.

A similar effect occurs when buoyancy of the hot gases causes **roll-back** of smoke at roof level against the ventilating current (section 21.3.2). This can occur over the heads of workers who are fighting the fire from an upstream position. Again, burning of the gases along the air interface can occur, igniting coal or timber in the roof and producing the danger of explosion. Personnel involved in fighting a fuel-rich fire may become aware of pressure pulses or rapid fluctuations in the movement of the air. These are caused by rolling flames and 'soft' explosions as gases ignite along gas:air mixing zones. The same phenomena can be observed following ignitions of methane (section 1.2). Such pulsations may be a precursor to a larger and more violent explosion.

It follows that every attempt should be made to prevent an oxygen-rich fire from developing into a fuel-rich fire. This underlines the need for early detection and prompt action. An intuitive reaction to a fire may be to restrict the air supply and, hence, to remove the oxygen leg of the fire triangle. This can be accomplished by building stoppings or erecting brattice cloths upstream from an airway fire. However, consideration of the dangers inherent in fuel-rich fires indicates that restricting the airflow may be inadvisable. Analyses of gases downstream from fires can be interpreted to indicate whether a fire is oxygen-rich or fuel-rich (section 21.7).

21.3.2 Effects of fires on ventilation

An open fire causes a sharp increase in the temperature of the air. The resulting expansion of the air produces two distinct effects. First the expansion attempts to take place in both directions along the airway. The tendency to expand against the prevailing direction produces a reduction in the airflow. This is known as the **choke** or **throttle** effect. Secondly, the decreased density results in the heated air becoming more buoyant and causes local effects as well as changes in the magnitudes of natural ventilating energy.

The choke effect

Consider an airway before it is affected by a fire. Air flows along it at a mass flowrate of M(kg/s) and doing work against friction at a rate of F(J/kg). The airpower dissipated against friction, P_{ow}, is the product of the two

$$P_{ow} = FM \quad \frac{J}{kg}\frac{kg}{s} = W \tag{21.1}$$

The effect of a fire in the airway on P_{ow} depends on the reactions of fans, natural ventilating pressures and ventilation controls throughout the system. However, if no deliberate action is taken to change these factors, it is reasonable to estimate that it remains sensibly constant.

Equation (7.10) gave us

$$F = \frac{p}{\rho} = R_t Q^2 \quad \frac{J}{kg} \tag{21.2}$$

where p = frictional pressure drop (Pa), ρ = mean density of air (kg/m^3), R_t = rational turbulent resistance of the airway (m^{-4}) and Q = mean value of airflow (m^3/s). Combining equations (21.1) and (21.2) gives

$$P_{ow} = FM = R_t MQ^2$$

or

$$P_{ow} = R_t \frac{M^3}{\rho^2} \quad \text{W} \tag{21.3}$$

as $Q = M/\rho$. Equation (21.3) may be rewritten as

$$M = \left(\frac{P_{ow}}{R_t}\right)^{1/3} \rho^{2/3} \quad \frac{\text{kg}}{\text{s}} \tag{21.4}$$

As P_{ow} and R_t are constants,

$$M \propto \rho^{2/3} \tag{21.5}$$

where $\propto$ means 'proportional to'. Hence, as the fire causes the density of the air to decrease, the mass flow of air will also decrease for the same energy dissipation. This phenomenon produces the choke effect. It should be noted, however, that the volume flow exiting the airway has increased. As $M = \rho Q$, proportionality (21.5) can be written as

$$Q \propto \frac{1}{\rho^{1/3}} \quad \frac{\text{m}^3}{\text{s}} \tag{21.6}$$

Note, also, from $P_{ow} = FM$ = constant, that as M decreases, the work done against friction per kg of air, F, must increase—a result of the increased volume flow and, hence, turbulence.

The choke effect is analogous to increasing the resistance of the airway. For the purposes of ventilation network analyses based on a standard value of air density, the raised value of this 'pseudo-resistance', R_t', can be estimated in terms of the air temperature as follows.

From equation (21.3)

$$R_t = P_{ow} \frac{\rho^2}{M^3} \quad (\text{m}^{-4})$$

Hence, for any standard (fixed) value of density and constant air power loss,

$$R_t' \propto \frac{1}{M^3}$$

However, combining with proportionality (21.5) which represents the actual reduction in mass flow,

$$R_t' \propto \frac{1}{\rho^2}$$

The general gas law (section 3.3.1) gives

$$\rho \propto \frac{1}{T}$$

where T = absolute temperature (K). Hence,

$$R_t' \propto T^2 \qquad (21.7)$$

The value of the pseudo-resistance, R_t', increases with the square of the absolute temperature. However, it should be recalled that this somewhat artificial device is required only to represent the choke effect in an incompressible flow analysis.

Litton *et al.* (1987) have also produced an estimate of the increased resistance in terms of the carbon dioxide evolved from a fire.

The buoyancy (natural draft) effect

The most immediate effect of heat on the ventilating airstream is a very local one. The reduced density causes the mixture of hot air and products of combustion to rise and flow preferentially along the roof of the airway. The pronounced buoyancy effect causes smoke and hot gases to form a layer along the roof and, in a level or descentional airway, will back up against the direction of airflow. The layering effect can be estimated using the method given in section 12.4.2.

This phenomenon of **roll-back** creates considerable difficulties for firefighters upstream from the fire, particularly if the conflagration has become fuel rich. The roll-back is visually obvious because of the smoke. However, it is likely to contain hidden but high concentrations of carbon monoxide. Furthermore, the temperatures of the roll-back may initiate roof fires of any combustible material above the heads of firefighters. The most critical danger is that tidal flames or a local explosion may occur throughout the roll-back, engulfing firefighters in burning gases.

One method of reducing roll-back is to increase the airflow in the airway. This, however, will increase the rate of propagation of the fire. Another method is to advance with hurdle cloths covering the lower 60% to 80% of the airway (section 12.4.2). The increased air velocity at roof level will help to control the roll-back and allow firefighters to approach closer to the fire. However, this technique may also cause the roll-back gases to mix with the air and to produce an explosive mixture on the forward side of the hurdle cloth. Furthermore, the added resistance of the hurdle cloth might reduce the total airflow to the extent that a fuel-rich situation is promoted. The behaviour of open fires is very sensitive to modifications to the airflow. Hence, any such changes should be made slowly, in small increments, and the effects observed carefully.

A third method of combating roll-back is to direct fog sprays towards the roof. In addition to wetting roof material, the air induction effects of the sprays will assist in promoting airflow in the correct direction at roof level.

A more widespread effect of reductions in air density is the influence they exert in shafts or inclined airways. This was handled in detail under the name of natural ventilation in section 8.3.1. The effect is most pronounced when the fire itself is in the shaft or inclined airway, promoting airflow if the ventilation is ascentional and opposing the flow in descentional airways. Indeed, in the latter case, the flow may be reversed and can result in uncontrolled recirculation of toxic atmospheres.

If the air temperatures can be estimated for paths downstream of the fire then the methods given in section 8.3.1 may be employed to determine the modified natural ventilating pressures. Those temperatures vary with respect to

1. size and intensity of the fire,
2. distance from the fire,
3. time,
4. leakage of cool air into the airways affected, and
5. heat transfer characteristics between the air and the surrounding strata.

At any given time, air temperatures tend to fall exponentially with respect to distance downstream from a fire. Climatic simulation models (Chapter 16) may also be employed to track the time-transient behaviour of air temperatures downstream from a fire. However, in that case, two matters should be checked. One is that the limits of application of the program may be exceeded for the high temperatures that are involved. Secondly, the transient heat flux between the air and strata will be much quicker than for normal climatic variations. Hence, the virgin rock temperature (VRT) in the simulation input should be replaced by a 'surrounding rock temperature' (SRT), this being an estimate of the mean temperature of the immediate envelope of rock around the airway before the fire occurs.

Having determined air temperatures in all paths downstream from the fire, the revised natural ventilation pressures for the mine can be determined. These may then be utilized in network analysis exercises to predict the changes in flow and direction that will be caused by a fire of given thermal output. A number of fire simulation packages have been developed to allow numerical modelling of mine fires (e.g. Greuer, 1984, 1988; Dziurzynski *et al.*, 1988; Deliac *et al.*, 1985; Stefanov *et al.*, 1984).

21.3.3 Methods of fighting open fires

The majority of open fires can be extinguished quickly if prompt action is taken. This underlines the importance of fire detection systems, training, a well-designed firefighting system and the ready availability of fully operational firefighting equipment. Fire extinguishers of an appropriate type should be available on vehicles and on the upstream side of all zones of increased fire hazard. These include storage areas and fixed locations of equipment such as electrical or compressor stations and conveyor gearheads.

Neither water nor foam should be used where electricity is involved until it is certain that the power has been switched off. Fire extinguishers that employ carbon dioxide or dry powders are suitable for electrical fires or those involving flammable liquids.

Deluge and sprinkler systems can be very effective in areas of fixed equipment, stores and over conveyors. These should be activated by thermal sensors rather than smoke or gas detectors in order to ensure that they are operated only when open combustion occurs in the near vicinity.

The two direct methods of firefighting introduced in this section involve the application of water and high expansion foam. The additional or complementary means of fire management by adjustment of ventilation controls and the injection of an inert gas are discussed in sections 21.3.4 and 21.6 respectively.

Firefighting with water

Except where electricity or flammable liquids are involved, water is the most common medium of firefighting. When applied to a burning surface, water helps to remove two sides of the fire triangle. The latent heat of the water as it vaporizes and the subsequent thermal capacity of the water vapour assist in removing heat from the burning material. Furthermore, the displacement of air by water vapour and the liquid coating on cooler surfaces help to isolate oxygen from the fire.

Water is normally applied by hosepipes upstream from the fire. A difficulty is the limited reach of water jets imposed by the height of the airway. This underlines the vital need for water to be available at adequate pressure and quantity in the firefighting range. In order for a water jet to reach some 30 m in a typical coal mine entry, water pressures should be in the range 800 to 1400 kPa (Mitchell, 1990) and capable of supplying up to five hoses from a manifold connected to a single hydrant. In practice, the range of water jets in mine airways may often be no greater than 10 m. The nozzles should preferentially be of the adjustable type to give either a jet or a fog spray.

Hard-won lessons indicate the need for careful forethought in designing a mine firefighting network. The air and the water should flow in the same direction so that firefighters do not become dependent on a water supply that passes through the fire before it reaches them. Hydrants should be located at strategic points with respect to areas of increased fire hazard, at intervals along airways and at cross-cuts with access doors. All fittings for hydrants and range components should be standardized throughout any given mine. Hydrant outlets should be protected against damage and corrosion by non-metallic caps. However, these must always be removable by hand and without undue force. Supplies at firefighting stations should be inspected at set intervals to ensure their operational efficiency at all times. Range fittings should include tee-pieces, blank-off caps and manifolds. It is particularly important that hosepipes be unrolled and examined for deterioration on a planned maintenance schedule and that they should be stored according to manufacturers' recommendations.

If access can be gained to an airway that runs parallel to a fire then fog sprays can be directed through doors or holed stoppings into the path of the fire. This can be effective if the sprays are employed at an early stage and immediately downstream from the fire front. However, for a large conflagration or where the fire has become fuel rich, it is likely to lose its effectiveness.

The locations of pumps and configuration of their power supplies should be considered carefully with respect to the layout of the mine. The pumps and routes of their cables should be chosen such that they are least likely to be disrupted by a fire. Dual power supplies via alternative routes may be considered. Furthermore, power for firefighting pumps should be capable of being maintained when electricity to working sections of the mine has to be isolated. Underground sumps can provide valuable water capacity. However, the firefighting system should also allow water to be supplied in adequate quantities from surface locations.

High expansion foam

Large volumes of water-based foam provide a valuable tool for fighting fires in enclosed spaces such as the basements of buildings or in the holds of ships. It has been employed for mine fires since at least 1956 (Eisner and Smith, 1956). The method is employed on large fires and, although it has had somewhat limited success in extinguishing mine fires, it can play a valuable role in cooling and quenching an area to an extent that allows firefighters with hoses to approach closer to the firefront. Even when sealing an area has become inevitable, valuable time for rescue operations can be bought by employing high expansion foam.

The bubbles are generated by a fan which blows air through a fabric net stretched across a diffuser. The net is sprayed continuously with a mixture of water and foaming agent. Bubbles can be produced at a rate of several cubic metres per second (Strang and MacKenzie-Wood, 1985). Compounds such as ammonium lauryl sulphate may be employed as the foaming agent while the addition of carboxymethylcellulose improves the stability of the bubbles (Grieg *et al.*, 1975).

The objective is to form a plug of high expansion foam which fills the airway and is advanced on to the fire by the ventilating pressure. The ratio of air to water within the foam may be in the range 100:1 to 1000:1. As the foam advances, bubbles break around the perimeter when they touch a dry surface. However, the liquid that is released wets that surface and allows advancement of the following bubbles. Shrinkage of the foam occurs continuously at the leading edges and accelerates because of radiant effects as it approaches the burning material.

Control of the combustion process is achieved by two primary mechanisms. First, vaporization of the water removes heat from the site and, secondly, the increased concentration of water vapour may produce an extinguishing atmosphere. As the air within the bubbles is heated to 100 °C it will expand by some 30%. However, the vaporization of liquid water to a gas involves an expansion of about 1700:1. Assuming an air:water mix in the foam of 1000:1 then, following evaporation of the water, 1000 l of air has expanded to 1300 l while 1 l of water has become 1700 l of water vapour giving 3000 l of mixture. If the air originally had an oxygen content of 21% then the evaporation of water will reduce that to

$$21 \times \frac{1300}{3000} = 9.1\%$$

which will extinguish flaming combustion.

Despite these mechanisms, high expansion foam does have some drawbacks. First, it may be quite difficult to generate a foam plug that fills the airway completely. As the plug builds up, the air velocity will increase through the narrowing channel between the plug and the roof, tending to maintain the gap open. Judicious employment of brattice cloths may assist in forming a complete plug of foam. It is important to control the path of the foam and, in multi-entry systems, this can be quite difficult. The natural direction of movement of the foam is dictated by the ventilating pressure. Here again, brattice cloths or stoppings in cross-cuts to adjacent parallel entries can assist in controlling the direction of the foam. Major obstructions caused by roof falls are quite liable to occur during a large underground fire. A foam plug may not be able to climb over such obstructions with the ventilating pressure available.

However, the greatest danger of foam plugs is that the reduction in airflow may promote a fuel-rich fire with the attendant danger of explosion. Downstream gases should be monitored for the development of this condition. Both increases and decreases in combustible gases have been reported in differing fires when high expansion foam has been employed. The reduction in airflow will tend to raise the concentration of combustible gases. However, as the inert mixture of air and water progresses downstream, condensation of the water occurs, allowing the air fraction to increase and, hence, modifying the combustible gas concentrations.

After the application of high expansion foam has been initiated, it is important to maintain it in operation during fire-fighting as intermittent production of foam can exacerbate the development of an explosive atmosphere. This underlines the need for good training so that operators are familiar with the equipment and procedure. Furthermore, care should be taken that sufficient supplies of foaming agent are available before the operation is started (Timko *et al.*, 1988).

21.3.4 Control by ventilation

When contemplating changes to airflows and applied pressure differentials during a fire emergency, there are four types of effects that must be considered most carefully.

1. *The effect on the combustion process.* The importance of avoiding the progression of an oxygen-rich fire into a fuel-rich fire has already been stressed in section 21.3.1.
2. *The effect on direction and rate of propagation of the fire.* Every attempt should normally be made to prevent an open fire from spreading into other airways. However, exceptions from this general rule may become necessary to guide products of combustion away from trapped personnel. An example may be the deliberate destruction of a stopping or air crossing to divert or short circuit a fire path from an intake airway into an adjacent return. Again, any modifications of the airflow passing through the fire zone must seek to achieve a balance between speed of propagation and control of the combustion process.
3. *Effects on the distributions of products of combustion.* This becomes a critical issue when personnel have become trapped inbye the fire, particularly if their exact whereabouts are unknown. However, any steps that will improve atmospheric conditions in escapeways require to be investigated.

4. *Effects on airflow distributions in other parts of the mine.* While the consequences of ventilation changes in the zone affected by the fire are of immediate concern, the effects of such changes throughout the rest of the mine should not be overlooked, particularly in a gassy mine or when personnel may still be evacuating other areas.

If a computer model of the mine has been maintained up to date then this will prove invaluable in investigating the predicted effects of proposed changes to the ventilation system. With a modern network analysis package (section 7.4), a personal computer or terminal in the emergency control centre can produce such predictions within seconds. Nevertheless, the uncertainties inherent in a fire situation demand that actual changes to the airflow system be made incrementally while observing the reactions on distributions and gas concentrations. The following subsections discuss the practical strategies that may be employed to control a fire by ventilation.

Pressure control

Airways that are parallel and adjacent to the fire path will remain unpolluted provided that they are maintained at a higher atmospheric pressure. These will allow access for escape, the building or strengthening of stoppings in cross-cuts, or the application of water sprays into the fire path. In multi-entry workings, control of such pressure differentials can be achieved by the erection of brattice cloths in the adjacent airway as illustrated in Fig. 21.2. Even if the pressure differential in the desired direction is not completely achieved, the reduced rate of toxic leakage may allow time for personnel to escape. If necessary, the brattice cloths may be advanced pillar by pillar to remove smoke sequentially from the adjacent airway. Devices such as the 'parachute stopping' or 'inflatable seal' have been developed to replace brattice cloths in such circumstances. These can be erected quickly and give improved seals around the perimeter of the airway (Kissell and Timko, 1991).

A consequence of this technique is that the airflow over the fire will be increased to an extent that depends on the configuration and resistances of the local airways. Pressure differentials between airways can also be modified by the use of a temporary fan instead of a restriction in the adjacent airway. In this case, airflow over the fire

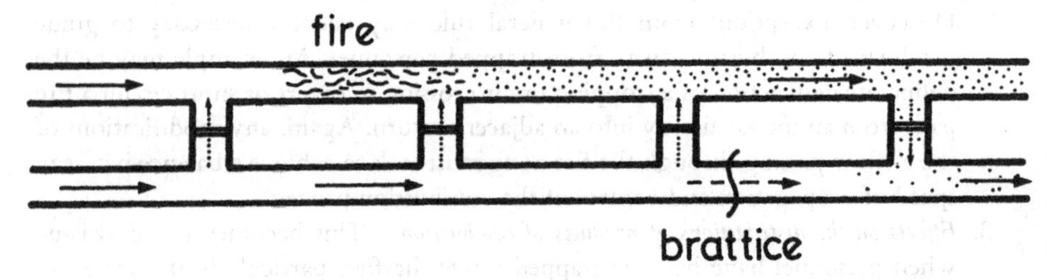

Figure 21.2 A brattice in the adjoining airway clears smoke from that airway by promoting leakage into the firepath.

will be reduced. The location and pressure developed by the fan must be selected with care in order to avoid recirculation of products of combustion. Where pressure differentials are small, even the few pascals developed by a free-standing auxiliary fan can induce the desired effect (section 4.4.3).

Airflow reversal

Many mines operate under a legislative requirement that the airflow provided by main fans must be capable of being reversed promptly. The background to such laws is the fear of a fire or other inundation of airborne pollutants occurring within a downcast shaft or main intake airway. Noxious and, possibly, flammable gases could then contaminate all, or most, of the ventilation system including working areas and return escapeways. If the fire is detected at an early stage then pollution of the complete system may be prevented by prompt reversal of the airflow. Even where contamination of the total network has occurred, air reversal may allow clearance of pollutants from return airways to the extent that a fresh air route may be established between surface and refuge chambers where personnel may be trapped.

The decision to reverse a complete mine ventilation system is fraught with difficulties and has very seldom been taken in practice.

There are essentially three methods of achieving the reversal of airflow in a ventilation system. Where axial impellers are used on the main fans, then changing the direction of rotation will cause reversal of the airflow. This can be implemented electrically at the fan motor. However, axial fans operate efficiently in one direction only. The aerofoil section of each impeller blade is designed to give aerodynamic stability of flow through the fan. When operating in reverse, breakway of the boundary layers over the blades occurs, resulting in high shock losses. The 'lift' of the blades and, hence, the throughflow of air is greatly reduced (section 10.3.2). Similarly, fixed guide vanes, fan casings and evasees, all designed for a forward direction, will produce high shock losses when the airflow is reversed. The reversed air quantity may be reduced to less than 50% of the normal forward flow (Dunn *et al.* 1982).

In the case of centrifugal fans, airflow reversal can be achieved only by means of reversal doors. The flow direction through the fan itself remains unchanged. For an exhausting centrifugal fan located at the mine surface, hydraulic or pneumatic activation of the reversal doors opens the fan inlet to the outside atmosphere and, simultaneously, diverts the fan exhaust into the mine shaft or slope. The opposite occurs for a forcing centrifugal fan. Where reversal doors are fitted as part of a surface fan installation, their operation should be checked routinely as part of a planned maintenance procedure.

Although the flow direction through a centrifugal fan remains unchanged, the shock losses incurred when air reversal doors are activated results in a reduced flow. The amount of the reduction is site specific and depends entirely on the design and siting of the reversal doors, and the configuration of the fan with respect to the inlet and outlet duct arrangements.

The possibility of requiring air reversal should be considered when designing the layout of airways and ventilation doors around an underground main fan. Such reversal should be attainable rapidly by opening or closing those doors.

During the course of ventilation network planning exercises (Chapter 9), it is often possible to design systems that allow rapid reversal of airflow in one section of the mine, or in a single airway, without total reversal at the main fans. This may be achieved by the strategic location of doors that can be opened or closed to permit airflow in either direction. Means of such local reversal might be considered, for example, for a conveyor route that is also to serve as an intake during normal operations.

Mandating the provision of air reversal facilities for an underground mine appears to be a reasonable safeguard. However, except in a clear-cut case such as a fire in or very close to a downcast shaft, the potential risks associated with reversing the airflow may be greater than those of maintaining the normal direction of flow, particularly in the short time period that may be available for making critical decisions.

The reasons that mine managements have very rarely decided to reverse ventilation during an emergency are both practical, and also because of possible litigation should lives be lost as a consequency of the reversal. During the trauma of a major emergency involving changing conditions in air quality and possibly disruptions of ventilation structures and communications, it may be impossible to know with certainty the locations, movements and dispersal of the workforce. Reversal of the airflow could then result in smoke and toxic gases being drawn over personnel who had assembled in a previously unpolluted zone. It may be expected that people who work routinely in a section of the mine will be familiar with the local ventilation system and, in case of an emergency, will act in accordance with that knowledge. Reversing the airflow could create additional uncertainty and confusion in their actions.

The majority of doors in airlocks or access paths between intakes and returns are designed to be self-closing, assisted by the mine ventilating pressure. This may be a legislative requirement. In the event of airflow reversal, those doors will be blown open and create short circuits unless they are provided with self-locking devices. Even in the latter situation or where powered doors are employed, the proportion of air leakage must be expected to increase when the pressure differential across the door reverses. Hence, coupled with the diminution in overall flow caused by the reversal procedures, the reversed ventilation reaching the working areas must be expected to be much lower than the normal forward flow.

High temperatures usually prevent firefighting rescue teams from approaching a fire from the downstream side. If the airflow is reversed over a fire in an intake, fire-fighting teams must transport their equipment and materials to a fresh air base inbye the fire.

The expansion of gases held in old workings or other voidage, and resulting from a drop in barometric pressure, is discussed in section 4.2.2. In the case of a forcing fan being reversed to create an exhausting system, the rapid fall in barometric pressure throughout the system may cause large emissions of voidage gas. If these contain high concentrations of methane, passing it into the fire zone could result in a series of explosions propagating far back into the mine. Furthermore, during the actual process of reversal, flammable gases from the strata or produced by incomplete combustion or volatilization of hydrocarbons may be drawn back over the fire, again leading to the possibility of explosions.

Although there may be several hundred tonnes of moving air in a major subsurface structure, the braking effect of viscous shear and turbulence causes it to be a well-damped system. Hence, when a main fan stops, the effect is noticeable almost immediately at all places underground. In the majority of cases, natural (thermal) ventilating effects will maintain movement in the normal direction.

In the situation of forced reversal, the transient effects will exist for much longer time periods than for simple stoppage of a main fan. This is because redistributions of the natural ventilating effects will not be completed until a new equilibrium of heat transfer has been established between the strata and the airflow. This may take several hours or even days. Furthermore, the fire itself will create thermally induced airflows with, perhaps, local reversals and recirculations. There is, therefore, some uncertainty concerning the speed at which reversal can be attained throughout the system and the stability of the reversed airflows. Research in Poland has indicated the efficacy of accelerating reversal by the use of intensive water sprays directed into the top of an upcast shaft (Trutwin, 1975).

21.4 SPONTANEOUS COMBUSTION

When air is allowed to percolate through many organic materials including coal then there will be a measurable rise in temperature. The same phenomenon can be observed in crushed sulphide ores and is caused by a progressive series of adsorptive, absorptive and chemical processes. These produce heat and the observable elevation in temperature. The percolating airflow will, therefore, remove that heat increasingly as the temperature of the material rises. If the leakage airflow is sufficiently high then a balanced equilibrium will be reached at which the rate of heat removal is equal to the rate at which heat is produced; the temperature will stabilize. The process will also reach an air-constrained equilibrium if the airflow is sufficiently low to inhibit the oxidation processes. However, between these two limits there is a dangerous range of percolating airflows that will encourage spontaneous heating.

Each material that is liable to spontaneous combustion has a critical temperature known as the minimum **self-heating temperature** (SHT). This is the lowest temperature that will produce a sustained exothermic reaction or **thermal runaway**. Hence, if the temperature reaches the SHT before thermal equilibrium is attained then the oxidation process will accelerate. The temperature will escalate rapidly, encouraging even higher rates of oxidation until the material becomes incandescent. At this stage, smoke and gaseous products of combustion appear in the subsurface ventilation system. The mine then has a **concealed** fire. The primary dangers of such occurrences are the evolution of carbon monoxide, the ignition of methane and combustion progressing into airways to produce open fires.

The phenomenon of spontaneous combustion has been recognized since at least the seventeenth century (PD–NCB Consultants, 1978). An early theory postulated that oxidation of pyritic material within coal provided centres of enhanced activity. Bacterial action is certainly a factor in the initial natural heating of hay and other foodstocks. This can play a part in the spontaneous combustion of timber or organic waste material underground but is unlikely to contribute significantly to the self-heating of coal or other minerals. Similarly, while increases in the temperature of materials can be observed in ore passes (section 15.3.6), the gravitational energy of collapsing waste areas in mines produces temperature rises that are insufficient to promote spontaneous combustion. Current concepts of the initiation of self-heatings are discussed in the following subsection.

21.4.1 The mechanisms of spontaneous combustion in minerals

Although spontaneous combustion can occur in crushed or caved sulphide minerals and in heavily timbered areas within metal mines, the problem is most common in coal mines. Research in this area has concentrated on the spontaneous combustion of coals.

The development of self-heating requires the large surface area of crushed material, combined with a slow migration of air through that material. Hence, the problem arises in goaf (gob) areas, caved zones, crushed pillar edges, fractured coal bands in roof or floor strata, stockpiles and tips on surface, and within abandoned sections of mines. The progressive stages of spontaneous combustion appear to be complex and not yet fully understood. Here, we shall examine the effects of oxygen, elapsed time and water.

The phases of oxidation

The oxidation processes of coal occur in four stages (Banerjee, 1985).

1. Physical adsorption (section 12.2.1) of oxygen commences at a temperature of about −80 °C and is reversible but diminishes rapidly as the temperature increases to become negligible beyond 30 to 50 °C. The process of adsorption produces heat as a byproduct of the modified surface energy of the material. This causes the initial rise in temperature.
2. Chemical adsorption (known also as chemisorption or activated sorption) becomes significant at about 5 °C. This progressively causes the formation of unstable compounds of hydrocarbons and oxygen known as peroxy complexes.
3. At a temperature which appears to approximate to the self-heating temperature (SHT) of the coal, the peroxy complexes decompose at an accelerating rate to provide additional oxygen for the further stages of oxidation. This occurs within the range of approximately 50 to 120 °C with a typical value of 70 °C. At higher temperatures, the peroxy complexes decompose at a greater rate than they are formed (Chakravorty, 1960) and the gaseous products of chemical reaction appear—in particular, carbon monoxide, carbon dioxide, water vapour, and the oxalic acids, aromatic acids and unsaturated hydrocarbons that give the characteristic odour of 'gobstink' (section 21.4.4).

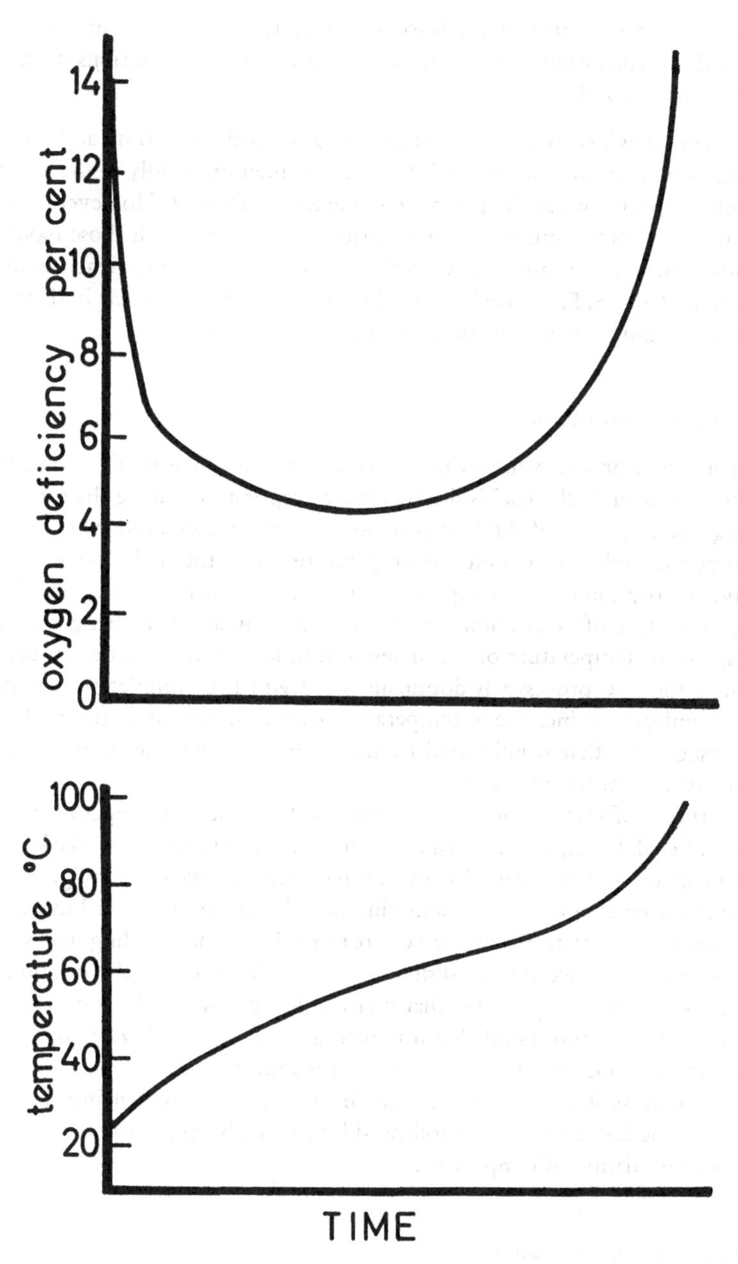

Figure 21.3 Examples of changes in oxidation rate and temperature of coal with respect to time. (Developed from Muzyczuk, cited by Banerjee (1985).) The time scale is omitted as this can vary from a few hours to many days, depending on the type and fineness of the coal, and the flowrate and psychrometric condition of the air.

4. When the temperature exceeds some 150 °C, the combustion process accelerates rapidly. Incineration of the coal occurs with escalating emissions of the gaseous products of combustion.

The rate at which oxygen is consumed varies both with time and the phase of oxidation as illustrated in Fig. 21.3. Oxygen is taken up rapidly in the earlier stages of chemisorption and as the peroxy complexes are formed. However, this reduces with time and as the surfaces become coated (weathered) with those oxygen compounds. The temperature curve tends to level off and may reach equilibrium. However, if the SHT is reached, as illustrated in Fig. 21.3, then both the rate of oxygen consumption and the temperature escalate rapidly.

The effects of water vapour

There are two processes involving water that act in opposite directions. First, the moisture content of the coal is driven off by evaporation during the early stages of heating. Hence, some of the heat is removed in the water vapour as latent heat of evaporation, tending to inhibit the temperature rise of the coal. The second process involves adsorption of water vapour from the air by the coal (Hodges and Hinsley, 1964). The heat of adsorption (sometimes called **heat of wetting**) produces an increase in the temperature of the material. It follows that the net effect depends on which of the two processes is dominant. Coal which is completely saturated with water is unlikely to increase in temperature owing to heat of wetting. However, a dry crushed coal that is infiltrated by moist air can exhibit an initial rapid rise in temperature from this phenomenon.

Adsorption of water vapour adds considerably to the early stages of spontaneous heating of coal. It is significant that coal mines in dry climates tend to be less troubled by spontaneous combustion. However, it has been observed that re-ignition of coal frequently occurs on re-opening a mining area that has been flooded to extinguish a fire. This is thought to occur for two reasons. First, the flooding and subsequent drainage may produce further disintegration of the coal and the creation of new surfaces; secondly, it is probable that many points of higher elevation or entrapped gases may not be fully flooded and remain at a sufficiently high temperature (hot spots) to inhibit adsorption. When cooler air is admitted subsequently it will pick up water vapour as it progresses through the wet conditions. An initial short-lived cooling of the hot spots may be followed by rapid adsorption of water vapour and a renewed escalation of temperature.

The path of spontaneous heating

The processes of oxidation and adsorption do not occur uniformly throughout a mass of crushed combustible material. The rate and directions of air migration and the air:surface contact area depend on the geometry of the zone, compaction from overlying strata and the fineness of the crushed material. Hence, the rapid escalation of temperature that characterizes the later stages of a concealed fire occurs first at discrete foci or 'hot spots'. The synergistic effects of iron pyrites are now thought to be caused by differential rates of expansion during the early stages of heating. The coal around the pyrites becomes more finely crushed and produces additional area for oxidation. Furthermore, the pyrite itself becomes oxidized, adding to the escalation in temperature. Green crystals of ferrous sulphate are formed which oxidize further to the yellow hydrated ferric oxide, a characteristic feature of zones in coal mines that have been involved in spontaneous combustion.

The migration of the heating again depends on the rate and direction of air leakage. However, in contrast to open fires, the tendency is for a spontaneous heating to propagate through crushed material against the airflow, i.e. towards the intake airways.

21.4.2 Susceptibility to spontaneous combustion

A large number of tests and indices have been devised as suggested measures of the liability of differing coals and other materials to spontaneous combustion. These have involved

1. coal petrology and rank, younger and lower ranks of coal being more susceptible,
2. rates of oxygen consumption or temperature rise at specified phases of the oxidation process,
3. self-heating temperatures (SHT) or other temperatures at specified stages of the heating process, and
4. rates of heat production during isothermal or adiabatic tests.

No such test or index has been found to have universal application. The difficulty is that susceptibility to spontaneous combustion depends not only on the material but also its physical state as well as the psychrometric condition and migration paths of the leakage airflows, the latter depending, in turn, on the mining methods and layout. The matters of additional relevance are as follows:

1. amount and degree of comminution of crushed material left in the goaf (gob) area, these depending on
   - friability of coal,
   - type of coal winning machine,
   - efficiency of coal clearance,
   - roof, floor or pillar coal left (percentage extraction),
   - thickness of seams and need for multilift or caving mining systems,
   - depth and pre-stressing (microfracturing) of coal, and
   - geological disturbances;
2. methods of stowing and sealing of roadsides;
3. gradient of the seam and proximity of other seams;
4. length of face;
5. rate of face advance or retreat;

6. roof/floor stability and strata stresses in the vicinity of pillar edges, stoppings, air crossings and ventilation doors;
7. the air pressure differential across the affected area;
8. the layout and resistances of surrounding airways and faces, including obstructions;
9. degree of consolidation and, hence, resistance of the caved areas;
10. the relative moisture contents of the coal and air;
11. the reduction of oxygen content in the goaf (gob) areas by the emission of methane or other gases.

With this number and variety of factors it is not too surprising that efforts to characterize the susceptibility of coals to spontaneous combustion purely on the basis of laboratory tests have met with rather limited success. An improved approach involves allocating weighted credit to each of the factors listed in an attempt to develop a site-specific indication of liability to spontaneous combustion (e.g. Banerjee, 1985; PD–NCB, 1978).

21.4.3 Precautions against spontaneous combustion

As with most potential hazards in the subsurface environment, precautionary measures against spontaneous commence at the time of planning and design of the mine. Core samples of the seam or ore should be subjected to susceptibility tests discussed in the previous section. The layout of the ventilation network should be designed to minimize pressure differentials between adjoining airways and across caved areas. This might be arranged, for example, by favouring through-flow rather than U-tube arrangements (section 4.3.1). While design airflows must be sufficient to deal with gases or other airborne pollutants, consideration should be given to means of reducing those airflows such as methane drainage. Booster fans, where allowed by law, provide a powerful means of air pressure management and, coupled with the techniques of network analysis to investigate locations and fan duties, are most valuable in reducing incidences of spontaneous combustion. Branch resistances in the surrounding ventilation network should be kept as low as practicable by means of larger cross-sections or driving parallel entries. Furthermore, obstructions in those airways should be avoided.

The probability of spontaneous combustion can be reduced by minimizing the amount of coal, timber, paper, oily rags or other combustible materials that are left in gob areas. This may be inevitable if top coal must be left for the purposes of roof control. Nevertheless, efficient clearance of the fragmented coal from the face and good housekeeping should be practiced in mines that have a history of spontaneous combustion.

It is important for the mine ventilation engineer to be conscious of the zones in which spontaneous combustion is most likely to occur. Recalling that some leakage takes place through the strata around stoppings and doors, spontaneous heating may occur in coal that exists within that strata, whether it be in the roof, floor or sides. Good strata control and the liberal application of roadway sealants can help in such circumstances.

Pillars in coal mines should be designed large enough to minimize crushing at the edges and corners. Side bolts can help to maintain the integrity of pillars while the injection of low viscosity grouts might be used as a last resort. Here again, the application of surface sealants assists in preventing ingress of air. However, the most difficult types of spontaneous fires occur within caved zones and, in particular, in the goaf (gob) areas of coal mines.

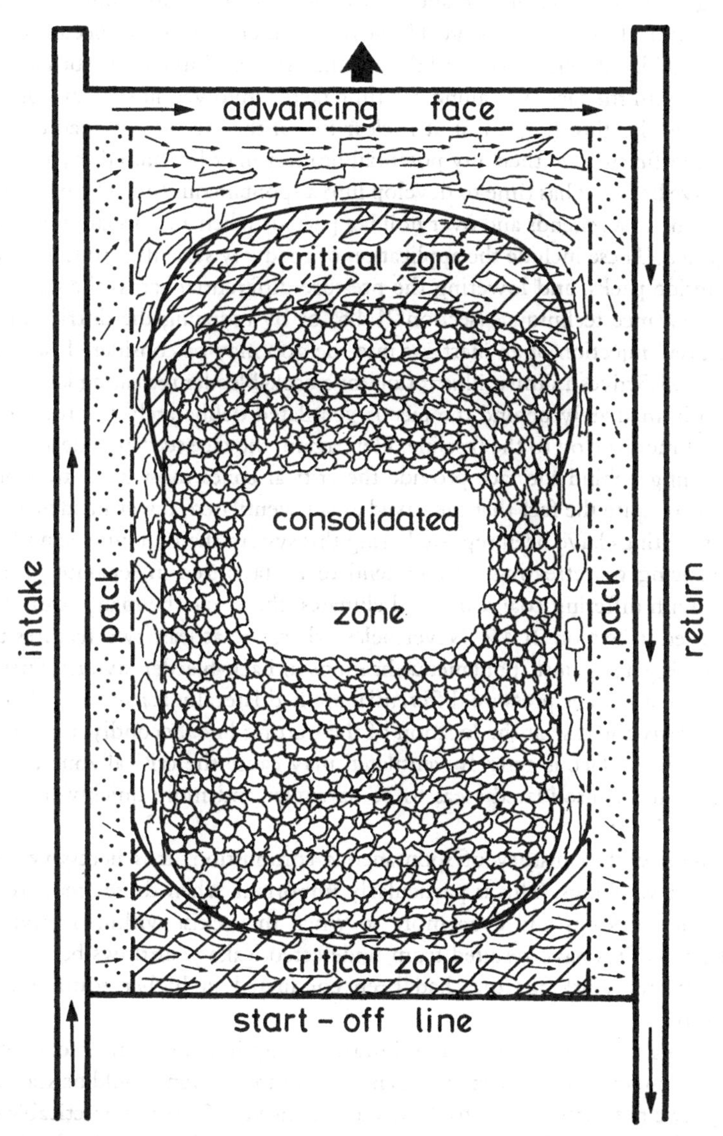

Figure 21.4 Zones most liable to spontaneous combustion in the caved area behind a single-entry advancing longwall face.

Figure 21.4 illustrates the air migration paths and the zones most liable to spontaneous heating in the caved area of an advancing longwall. Those critical zones occur where the leakage airflows lie within that range which provides sufficient oxygen to promote continued oxidation of combustible material, but not enough to remove heat at the rate at which it is generated. There are two distinct zones. One of these lies along the original starting line of the face where incomplete consolidation allows a leakage path between intake and return. Within the central portion of the gob, consolidation allows little leakage. However, another critical zone occurs between the fully consolidated central core and the advancing face. This zone is not stationary but advances with the face. Recalling the time factor involved in the development of a spontaneous heating (Fig. 21.3), it is clear that if the face is advanced continuously and at a sufficient rate then any potential heating may be 'buried' by consolidation of the cave before it has time to develop into a spontaneous fire. The most hazardous times occur at weekends and over holiday periods when additional precautions may be required. These include the application of roadway sealants, grout injections into the roadside packs, and lowering the pressure differential across the gob, either by pressure balance techniques (section 21.4.5) or by reducing the district airflows. In severe cases, injection of an inert gas into the gob may be considered (section 21.6).

Figure 21.5 gives a similar illustration for a back-bleeder retreating longwall where the gob immediately behind the face is ventilated deliberately in order to prevent flushes of methane on to the faceline. Here again, the face start-up line and a moving zone trailing behind the face provide the critical zones most liable to spontaneous heatings. Despite the fact that the caved area is ventilated, fewer incidences of spontaneous heatings have been reported using this system than for advancing longwalls. As advancing or retreating systems tend to be favoured in separate geographical regions with differing coal seams and climates, there may be many reasons for this, as discussed in section 21.4.2. Nevertheless, there are at least two features that may tend to mitigate against spontaneous heatings in back-bleeder layouts. First, there is likely to be a smaller pressure differential applied laterally across the gob area. This is particularly the case at the location of the starting line. Secondly, the fact that the area behind the face is actively ventilated may cause the critical zone between the ventilated and consolidated areas to be narrower and more quickly buried by the caving roof strata.

In mines with a history of spontaneous combustion, it is necessary to seal all abandoned workings. This is particularly important when those areas are adjacent to current workings and even more so when they exist within overlying strata. Spontaneous fires have resulted from undetected leakage airflows between current gob areas and workings that had been conducted in higher seams many years previously.

Following the completion and withdrawal of equipment from a section of a mine liable to spontaneous combustion, all entries into the section should be sealed and the atmospheric pressures applied to those seals balanced as far as it is practicable to do so. This may be accomplished simply by a re-arrangements of doors or by the dismantling of stoppings and air crossings. Active pressure balancing techniques may be employed

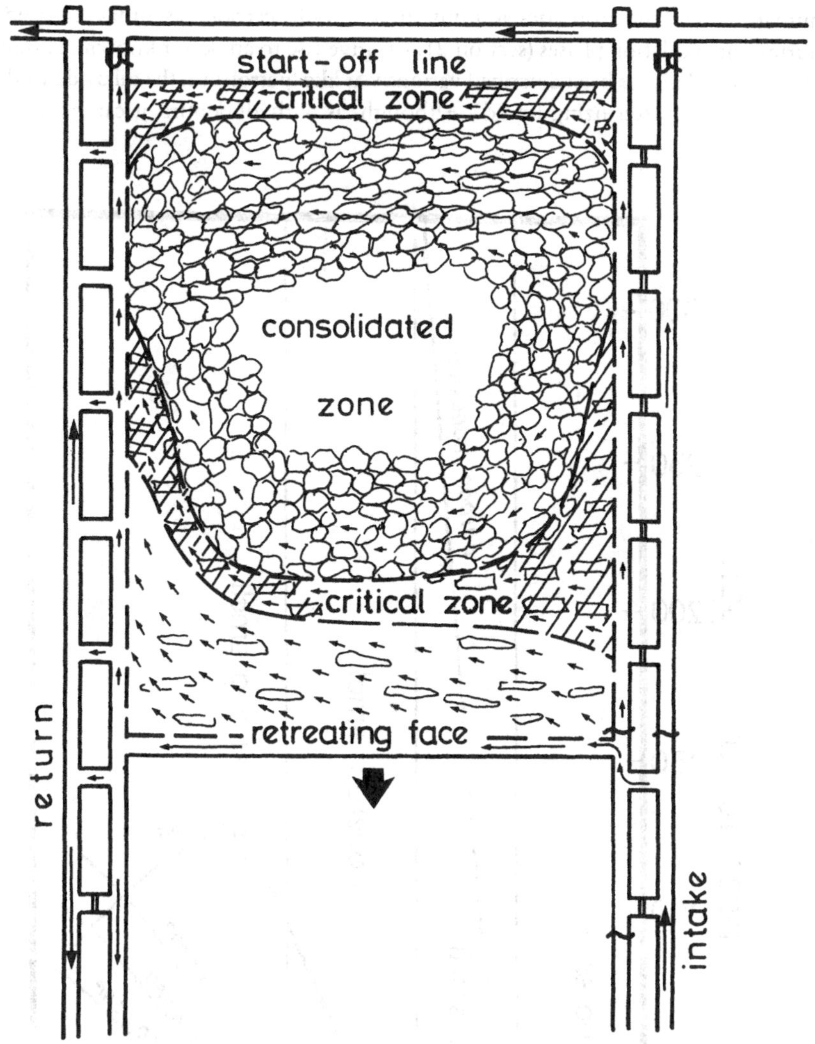

Figure 21.5 Zones most liable to spontaneous combustion in a multiple-entry retreating longwall with back bleeders.

as discussed in section 21.5.5. The sites of seals should be prepared at strategic control points during the development of a section and should involve channels being excavated into the roof and sides, and providing a nearby stock of building materials. This will facilitate sealing the district rapidly in the event of an uncontrollable fire.

21.4.4 Detection of a spontaneous heating

There are essentially three classes of detection for incipient or active spontaneous heatings. The oldest and, many would argue, still the most reliable is through the human senses. The aromatics and unsaturated hydrocarbon gases that are produced during the oxidation phases (section 21.4.1) give rise to an odour known colloquially as gobstink. This has been described variously as like 'petroleum, the oil used in a flame-safety lamp and sterilizing liquid'. Although, as with all odours, it can be described

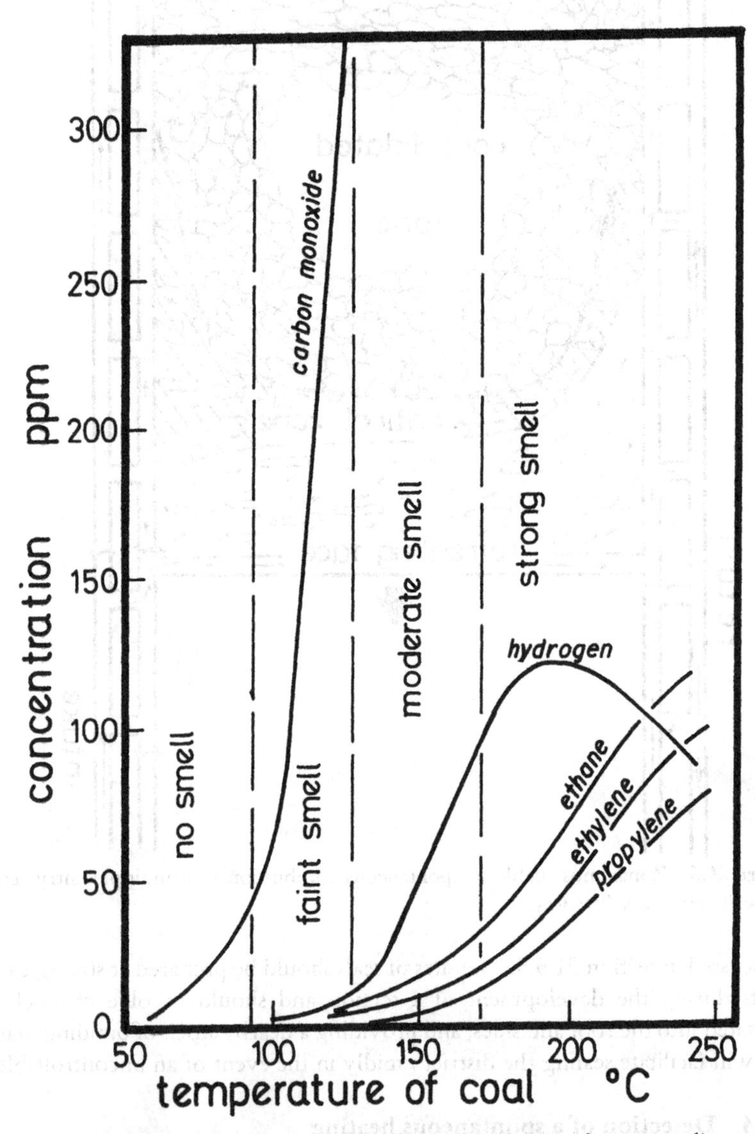

Figure 21.6 Example of the development of odour and gases from a medium-volatile bituminous coal. The relative positions of the gas curves will remain the same for other coals but the actual concentrations will vary with type of coal and magnitude of leakage airflows.

only by analogy it is, nevertheless, very distinctive and unlikely to be forgotten. Figure 21.6 illustrates the increasing strength of the smell with respect to gas concentrations and temperature of the oxidizing coal.

The odour is likely to be the first indication of a heating that can be detected by the human senses. The output of water vapour may also be observed either as a haze in the air or from beads of condensation on supports or other surfaces in locations where such condensation is unusual. At a later stage, the matter is put beyond doubt by the appearance of smoke.

The second class of detectors are thermal devices used to determine increases in temperature. Infrared scans of roadway sides have been employed to identify the emission points of warm gases into airways and are useful for localized heatings in pillars or around stoppings (Chakravorty and Woolf, 1980). They are less useful as a permanent and continuous means of warning. Thermocouples or thermistors have been left in gob areas. However, they too have met with very little success to the present time. First, their wiring is unlikely to withstand the mechanical stresses of an active caving zone, even when sheathed. Secondly, the thermal conductivity of crushed rock is low. Hence, the temperature even within a metre of an active centre of heating may indicate no abnormal condition.

The most widespread method of detecting the onset and development of a spontaneous heating is by monitoring gas concentrations in return airways. The gases that are evolved are indicated on Fig. 21.6. The interpretation of the relative values and trends of these concentrations is discussed in detail in section 21.7.

21.4.5 Dealing with a spontaneous heating

There is a procedure that should be followed when detection systems indicate that an active spontaneous heating is developing in a mine. First, a gas monitoring station should be set up downstream from the affected area and air samples taken at intervals of not more than 30 min. If an air monitoring system already exists in the mine then even greater detail of gas concentration trends can be followed. Personnel should be evacuated from all return airways affected and, if the condition is developing rapidly, also from the rest of the mine—except for those involved in dealing with the situation.

Simultaneously, steps should be undertaken to identify the location of the fire. This may be obvious if smoke appears from discrete places in a close vicinity such as from a fire in leakage paths around a stopping or in the crushed corner of a pillar. The location of the fire is more difficult to detect if it occurs in a caved zone or inaccessible old workings. Whether or not smoke is present, it is useful to conduct a carbon monoxide survey. Measurements can be made by hand-held instruments or by stain tubes (section 11.4.2). Indicating the results on a mine map can assist in selecting the most probable sites of the fire.

Having located the fire, the next step is to decide how to control and, if possible, to extinguish it. A variety of methods exist for this purpose. The injection of an inert gas as a practical and powerful method of dealing with both open and concealed fires in mines was developed through the 1980s and is discussed separately in section 21.6. Other techniques of fighting concealed fires are introduced in the following subsections.

Excavating the fire

If the fire is known to be within a few metres of an airway then it may be possible to dig out, cool and remove the incandescent material. This can be the case for fires in pillars, around stoppings or air crossings, or for some gob fires. It may be necessary to drill holes into the zone in order to determine more accurately the site and extent of the fire. The excavation should commence from the upwind side of the fire to minimize the exposure of personnel to smoke and carbon monoxide. The airway should be wetted and/or coated in stonedust for a distance of some 10 m on either side of the excavation. Care should be taken to ensure control of the roof which may have become weakened by heat. Readings of methane and carbon monoxide concentrations should be taken frequently downstream from the site. As the fire is approached, it may be necessary to spray water on the workers to cool them. When the seat of the heating is exposed, it should be cooled by water jets applied around the periphery. Spraying water into the heart of a glowing carboniferous mass may result in the formation of water gas (section 11.2.6). The hot material should be loaded into metal conveyances and dampened thoroughly before transporting it out of the mine. When all of the incinerated material has been removed, the void should be cooled and, after some 24 h to ensure no re-ignition, filled with an inert material such as limestone dust or gypsum-based wet fillers.

If it is impracticable to remove the fire physically then it becomes necessary to prevent ingress of air to the fire location. This leads us into the remaining techniques of dealing with a concealed fire.

Burying the fire

In some cases, it is possible to prevent or reduce access of air to the fire location by burying it under collapsed roof strata. Localized leakage through a roadside pack can be reduced by bringing the roof down in the airway, employing shotfiring if necessary, removing only a portion of the debris and compacting the remainder over the area to be sealed. The excavated roof must, of course, be well supported leaving the airway with an anticline.

If the heating occurs in the critical moving zone behind a longwall face (Fig. 21.4 and 21.5) and is detected sufficiently early, then it is sometimes possible to bury it under consolidated caved material by a temporary increase in the rate of face advance (or retreat). This is likely to be successful only if the incipient heating has been detected by an early warning gas detection system and well before smoke appears.

Sealants

A variety of sealants have been employed on stoppings, airway surfaces, roadside packs and pillars in order to increase their resistance and, hence, reduce the access of leakage air. These may be applied on external surfaces or injected as grouts into the strata or packed material. While sealants that include resins or gels produce the lowest permeabilities and allow a degree of flexibility, the choice must often be made quickly and on the basis of local availability. Concrete and gypsum plasters can be sprayed quickly and effectively onto airway surfaces, as well as being useful for grouting and sealant infills between stoppings. Water-based slurries using mill tailings or other waste material have also been employed as grouts or for injecting into fire zones. These may be applied through boreholes drilled either from an underground location or from the surface. The injection of sodium silicate into coal pillars has been found to be effective (Banerjee, 1985).

While all known leakage paths connecting to the fire zone through roadsides should be sealed, it appears to be particularly advantageous to seal on the inlet or high pressure side. It is, however, often difficult to locate the relevant points of inward leakage on the inlet side. One way of doing this, where allowed by law, is to employ the techniques of pressure management (following subsection) to reverse intentionally and temporarily the direction of leakage across the fire area. The appearance of smoke or elevated concentrations of carbon monoxide in the intake will identify the normal inlet points after which the flow reversal should be terminated or, better still, the leakage reduced to zero. This method should be employed with caution and only when adequate gas monitoring facilities are available in order to avoid unknown recirculation. The approval of governmental agencies should be sought if necessary.

Localized pressure balancing

If no differential pressure exists across a level permeable zone then there can be no air leakage through it. Pressure balancing involves raising the pressure on the return side or decreasing the pressure in the intake until the leakage flow is reduced to near zero. This principle can be applied to complete sections of abandoned workings as described in section, 21.5.5 or, in a more localized manner, to gain control of a gob fire without sealing the district.

Figure 21.7 shows the principle of this technique which can be applied to a variety of situations. In the example illustrated, a fire has commenced along the starting line of an advancing longwall. The pressure differential between intake and return in that vicinity has been reduced to near zero by the installation of a fan and regulator in the return airway. In cases of low normal pressure differentials, an induction fan without any surrounding brattice may be sufficient (section 4.4.3). The hydraulic gradients shown on the same figure illustrate the pressure differentials with and without the pressure balance. Where applicable, the method can be applied quickly and at low cost to arrest the development of a spontaneous heating and to bring about an immediate reduction in carbon monoxide emissions. A pressure survey (section 6.3) may be run around the district to determine the effective fan pressure required. If there is convenient access, then a length of pressure tubing can be installed between the relevant points in the intake and return airways including an in-line pressure gauge. The regulator in the return airway can be adjusted until the pressure balance is

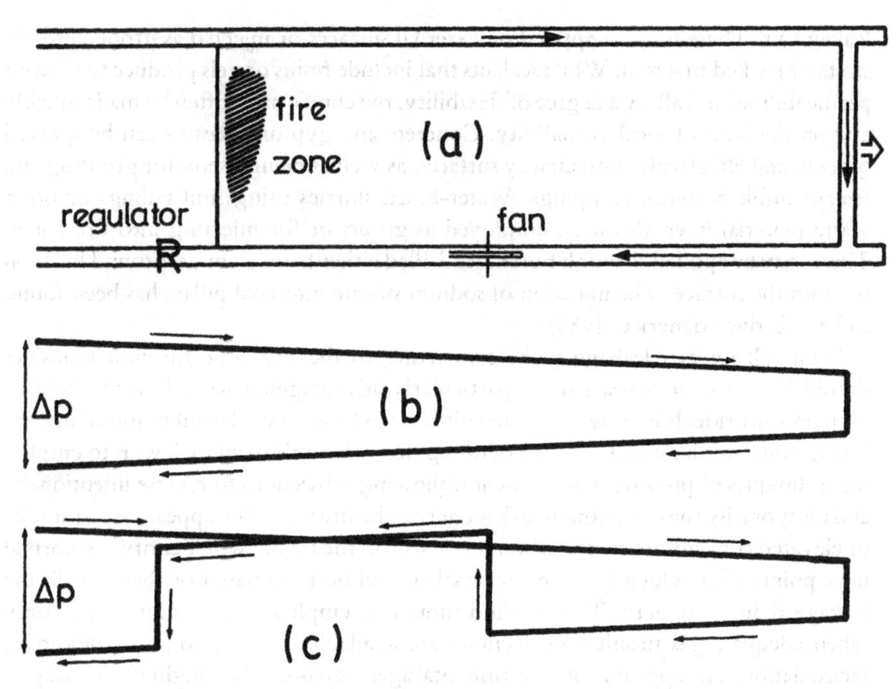

Figure 21.7 Example of localized pressure balancing to control a spontaneous heating: (a) plan of the district; (b) hydraulic gradient without the pressure balance; (c) hydraulic gradient with the pressure balance in place.

achieved. Furthermore, an adjustable orifice within the short length of fan ducting can be used to modify the effective fan pressure. The locations of the fan and regulator may be changed in order to achieve finer control of the zone in which the pressure balance is applied.

The method of localized pressure balancing is very flexible and can achieve spectacular and speedy success in many scenarios. However, it requires skilled personnel to devise and control each installation. If applied inexpertly, it can result in partial recirculation of products of combustion. In any case, it is a prudent precaution to employ a carbon monoxide monitor in the intake to detect any reversal of leakage flows.

Flooding and sealing off

Flooding can be used in two ways to extinguish concealed fires. First, if the affected zone lies to the dip of current workings then the fire itself may be flooded. This should be done in a controlled manner in order to be able to handle the products of combustion and other gases displaced by the water. Furthermore, if a previously flooded area is re-opened then it may rapidly re-ignite (section 21.4.1). The second use of flooding is to provide very effective seals in airways that have low-lying sections. The water penetrates fissures in the surrounding strata and provides a near-perfect barrier against air leakage.

This brings us to the larger matter of building seals to control concealed fires in inbye areas. While sealing a section of a mine is often considered as a last resort, it is important to recognize those situations in which it will become inevitable and then to seal quickly. This allows an inert atmosphere to build up and to extinguish active combustion. Re-entry and a resumption of mining may then be possible within a relatively short time. If, however, sealing-off has been delayed unduly, then the fire may have developed into such a large conflagration that the area could be sterilized indefinitely. The one situation in which sealing off must be delayed is while there is the slightest possibility of rescuing trapped personnel. The procedures involved in constructing seals during a fire emergency are discussed in section 21.5.

21.5 STOPPINGS, SEALS AND SECTION PRESSURE BALANCES

There is considerable divergence within differing geographical areas on the meaning of the terms **stoppings** and **seals**. In some regions, the two are regarded as synonymous while, in others, they are interpreted as quite different types of structures. To prevent confusion, we shall use the terms here according to the following definitions.

1. *Temporary stopping.* A light structure erected from brattice cloths, other fabrics or boarding but which will not withstand any significant physical loading.
2. *Stopping.* A single- or double-walled structure constructed from blocks, bricks, sandbags or from substantial boarding attached to tight roof supports, but not intended to be explosion proof.
3. *Seal.* Two or more barriers covering the full cross-section of the airway, 5 to 10 m apart, with the intervening space totally filled with an inert material. Steel girders may be interlaced to add structural strength. A seal is designed to withstand explosions.

Temporary stoppings, often in the form of brattice cloths or quickly erected alternatives are employed during the fighting of open fires in order to regulate the airflow over the fire, change the air pressure in nearbye airways (section 21.3.4) or reroute airflows to assist in rescue operations.

Stoppings and seals are used when alternative means of controlling a fire or spontaneous heating have been exhausted and rescue operations have been terminated. The purpose is to stop the flow of air and to allow an inert atmosphere to build up within the affected zone. The decision on when to erect stoppings or seals should be contingent on the particular circumstances. In the case of a deeply seated concealed fire and when there is no perceived imminent danger of an explosion, stoppings may be erected in all entries to the area at a fairly early stage and before the heating reaches an open airway. The district may subsequently be re-entered for salvage of equipment or resumption of mining either after a cool-down period or when further arrangements have been made to control the heating. Sealing or stopping-off an airway fire should, again, be carried out as soon as it becomes clear that the fire is out of control and when any rescue operations have been completed.

In this section, we shall discuss the further decisions and procedures involved in terminating airflows through a subsurface fire zone.

21.5.1 Site selection

It is important that all openings into a zone to be stopped or sealed should be closed, except for sampling pipes. Such openings may include boreholes and connections into old workings. In the shallower mines, there may be fractures extending through to the surface. The potential effects on surface structures and their inhabitants should be considered.

The stoppings or seals should be close to the fire area. However, if there is a possibility of explosion then a sufficiently long expansion zone should be left to reduce the magnitude of the pressure pulse on the barriers. The sites should be selected to minimize the number of stoppings or seals required. The precise locations should be in well-supported areas with solid roof, floor and sides. In the case of stoppings, there should be a sufficient length of airway to allow the erection of a second stopping outbye and conversion into a seal, should that prove to be necessary.

A well-organized mine will have pre-prepared sites for stoppings at control points in each district and with a stock of the appropriate materials nearbye (IME, 1985). In any case, those sites should allow ready access for the supply of further materials and, also, for the provision of ventilation to personnel involved in building the stopping or seal. The latter may necessitate a temporary duct and auxiliary fan. Existing regulators, doors or door frames can be very useful as locations for stoppings. If the site is polluted by products of combustion then the construction must be undertaken by teams fitted with breathing apparatus. Where there is danger of an explosion or outrush of gases during construction, sites should be chosen that permit rapid escape of the personnel.

It is unlikely that locations will be found which satisfy all of those requirements. Site selection for stoppings and seals invariably requires compromises between optimum positions and practical considerations.

21.5.2 Sequence of building seals and stoppings

In the case of concealed fires and where there is no danger of explosion, it is preferable to complete intake stoppings first in order to terminate airflow into the fire zone, and then to follow up quickly with closure of the return airways. However, in the case of an open fire, or where monitored gas concentrations indicate that an explosive atmosphere may develop then all stoppings or seals should be completed simultaneously and personnel evacuated from the mine for a period of 24 h.

21.5.3 Construction of seals and stoppings

It is unusual for coal dust explosions in mines to generate pressure peaks of more than some 350 kPa on the faces of seals although considerably higher pressures have been measured in experimental explosions (Strang and MacKenzie-Wood, 1985). Hence, seals which (by our definition) should be explosion proof must be able to withstand such dynamic pressure pulses. Figure 21.8 illustrates a typical seal. The end walls may be constructed from sandbags, masonry, any form of blocks, boards or even stiffened brattice secured firmly to chocks (cribs) or props. It is, however, preferable that the inner barrier be flame proof. The strength of the seal should be provided by the infill and any girders or other supports that may be employed in the intervening space. The length of seal is shown as 5 to 10 m and should be chosen with reference to the size of the airway, the condition of the surrounding strata and the type of infill available.

Construction of a seal in a coal mine should commence by applying stonedust liberally to the airway inbye the seal. Stonedust barriers may also be erected inbye the seal to arrest an explosion before it reaches the face of the seal (section 21.8.3). Figure 21.8 illustrates the end walls recessed into the roof, floor and sides. This is considered to be good practice although it should not be necessary provided that the length of seal is adequate and a modern (quick-setting) wet infill is available. Conveyor structure should be dismantled and cleared from the area. Similarly, pipes and cables should be dislocated and removed.

While dry material such as sand, stonedust or fly ash may be used for the infill, a better seal is obtained by employing a gypsum-based plaster. This is pumped in as a liquid which penetrates immediate fractures in the surrounding strata and then sets to a compact solid material.

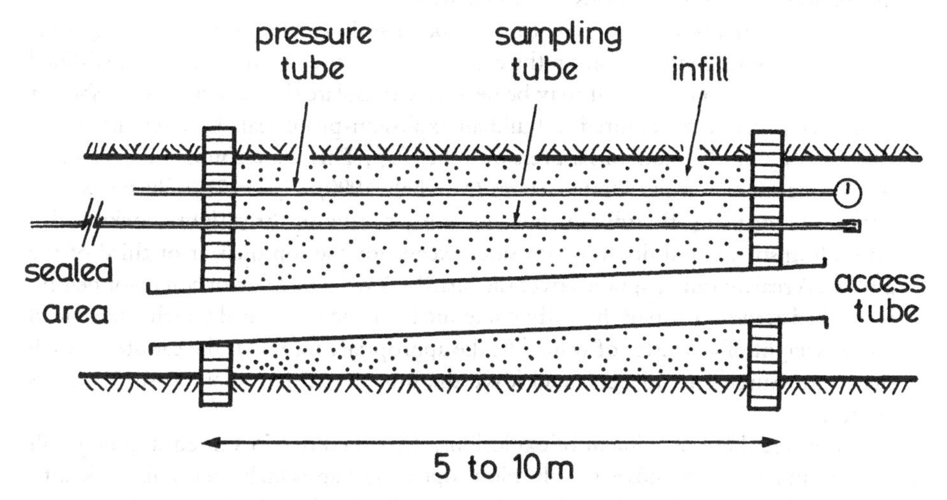

Figure 21.8 Example of a section through a seal (explosion proof).

The gas sampling pipe should extend at least 30 m inbye the seal to allow for 'breathing' induced by fluctuations in barometric pressure. Three flexible tubes may be threaded through the metal sampling pipe in order to draw independent samples from roof, floor and middle of the sealed airway.

The steel access tube shown on Fig. 21.8 has a number of purposes. First, it allows ventilation to be supplied to the fire while construction of the seal is in process. This tends to retard the development of a fuel-rich situation and, hence, reduces the risk of explosion whilst workers are still at the site. Despite the access tube, workers may be subjected to roll-back of smoke during construction of an upstream seal. When each of the seals has been completed then both ends of all the access tubes should be blanked off simultaneously by strong steel plates. All personnel must leave the mine for a period of 24 h or until monitored signals of gas concentration indicate that an inert atmosphere has been attained.

Following the sudden cessation of airflow, significant changes take place within the sealed zone. The concentration of combustible gases increases while the oxygen content decreases. In the case of open fires, it is probable that the mixture of gases will pass through an explosive range (section 21.7). Meanwhile, convection effects will create movements of the changing atmosphere and, perhaps, rolling or 'tidal' flames. The combination of these effects is likely to produce a series of explosions. These can be monitored by pressure transducers connected to the pressure tubes on the seals. An older but simpler method is to leave a glass U-tube containing mercury attached to a pressure tube. If the mercury is subsequently found to have been blown out then an explosion has occurred within the sealed zone.

A second reason for the access tube is to allow subsequent re-entry inspection by rescue teams equipped with breathing apparatus. The outbye end of the tube should be at a convenient height for this purpose. Should it become necessary, the access tube can be filled with infill material. This is the reason for the slight downward inclination of the tube towards the sealed area.

This is one major disadvantage to the type of seal we have been discussing—the time taken for its construction. In the case of an open fire in a timbered or coal-lined airway, rapid action is vital. It may be necessary to isolate the zone in a much shorter time period than that required to build an explosion-proof seal. An alternative is to erect a stopping that contains a pressure-relief flap. This is a technique that has met with some success in the United States (Mitchell, 1990). Figure 21.9 illustrates such a **vented stopping.** Boards are attached firmly to two or three tight chocks (cribs) on both upstream and downstream sides except for the top quarter or third of the airway. A sealant can be sprayed over the boards. A weighted flap of conveyor belting is attached across the top of the outbye side and held open by a cord which is tensioned by the weight of a canister of water. At the appropriate moment, the canister at each site is punctured and all personnel leave the mine. The flaps close as the canisters empty.

Except for the most violent of explosions, the structure of a vented stopping will remain intact. Lesser pulses simply blow open the flap which then falls back into place leaving the integrity of the stopping secure. Although the vented stopping

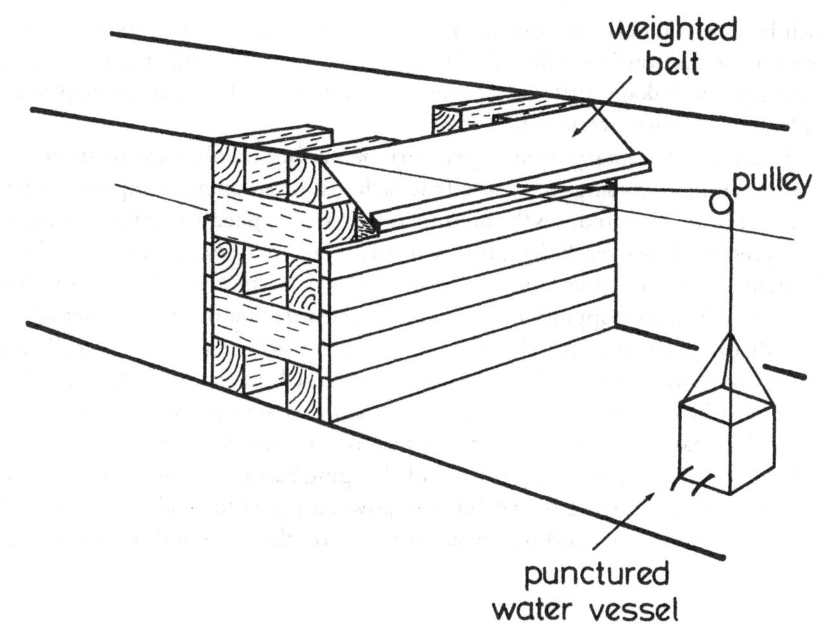

Figure 21.9 A vented emergency stopping.

does not have the structural strength or resistance of an explosion-proof seal, it provides an excellent temporary expedient until permanent seals can be established, should those prove to be necessary.

21.5.4 Re-opening a sealed area

The simple method of breeching stoppings or seals and re-establishing an air circuit suddenly is not one that can be recommended when a mine or section of a mine has been sealed because of a fire. Even when gas samples indicate that the fire is no longer active, hot spots may still remain that could result in re-ignition when supplied with oxygen. The preferred technique is first to send in rescue teams equipped with breathing apparatus to make a thorough inspection of the area and to take additional air samples. If the sealed area is extensive and, particularly, if the atmosphere will pass through an explosive range when mixed with fresh air, it may be necessary to reventilate in stages, building additional seals or stoppings further inbye and closer to the original fire zone.

21.5.5 Section pressure balances

No stopping or seal in the subsurface has infinite resistance. Even if the stopping or seal itself were perfect, the potential for leakage still occurs in the surrounding strata.

Such leakage can delay the extinction of a fire or, if serious, may maintain the fire indefinitely. Section 21.4.5 described how air pressure management can be employed to control the leakage airflows feeding a concealed fire. The same principle can be applied to complete sections of a mine.

The most rudimentary form of **pressure balancing** involves the re-arrangement of doors and air-crossings to ensure that each seal or stopping is exposed to return airway pressure. This reduces the pressure differential applied across the sealed district to the pressure drop over the length(s) of airway between the stoppings. Figure 21.10(a) illustrates an improved version. An additional wall is constructed some 4 to 5 m in front of each main stopping or seal to form **pressure chambers**. A duct of about 0.5 m diameter connects the chambers and, hence, equalizes their pressures. There is then no pressure differential applied across the sealed district, nor can there be any continuous flow around it. These are examples of **passive** pressure balancing.

The drawback to passive pressure balances is that they do not prevent 'breathing' of the seals or stoppings during periods of changing barometric pressure. This can be overcome by employing an **active** (or powered) pressure balance system. This provides a means of equalizing the air pressures on the inner and outer faces of the

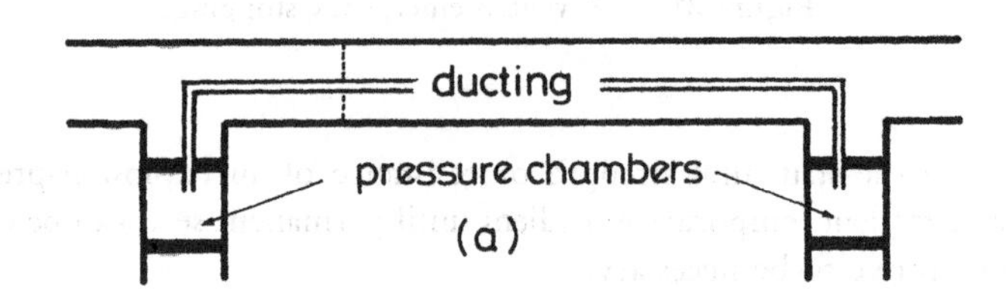

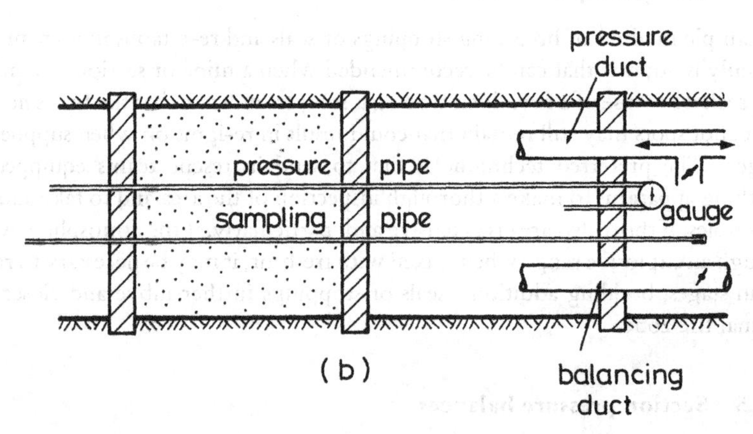

Figure 21.10 (a) A passive pressure balance. (b) An active pressure balance chamber.

same seal or stopping and is illustrated on Figure 21.10(b). Two ducts pass through the outer wall into the pressure chamber. One of these (the pressure duct) is supplied by a ventilating pressure either by a small fan or by laying the duct through a nearbye door or stopping in order to utilize the mine ventilating pressure. The pressure chamber will be pressurized positively if the seal is on the return side of the main ventilation network and negatively if it faces an intake airway. The second (balancing) duct simply passes through the outer wall. Both ducts are fitted with variable dampers to give a flexible means of fine adjustment to the pressure within the chamber.

Two tubes are employed to monitor air pressure, one passing completely through the seal and the other into the chamber. A gauge across these two tubes will indicate zero when there is no pressure differential between the chamber and the inbye side of the seal. The duct dampers can be adjusted manually when the pressure gauge deviates from zero. The arrangement can be made automatic by employing an electronic pressure gauge which transmits an amplified signal to servo-motors on one or both of the duct dampers.

A number of special-purpose adaptations to the principle of pressure chambers have been devised. One example may occur in a nuclear waste repository to ensure that leakage through an airlock takes place in one consistent direction at all times, irrespective of the direction of pressure differential across the airlock (Brunner *et al.*, 1991). Another example, where permitted by legislation, is the use of compressed air to provide positive pressurization of the pressure chamber.

21.6 THE USE OF INERT GASES

The injection of inert gases to assist in the control of subsurface fires has been undertaken since, at least, the 1950s (Herbert, 1988). However, from 1974, significant developments in the deployment of nitrogen took place in Germany. The technique has become commonplace in coal mining areas where spontaneous combustion occurs frequently (Both, 1981). The overall purpose of injecting an inert gas is to reduce the oxygen content in order to prevent or inhibit combustion. The objectives may further be classified as follows:

1. to prevent concealed heatings in zones that are highly susceptible to spontaneous combustion;
2. to reduce the risk of explosions during sealing or stopping-off procedures;
3. to accelerate the development of an inert atmosphere in a newly sealed zone and to prevent the creation of an explosive mixture when it is re-opened.;
4. to control the propagation of an open fire during rescue, firefighting and sealing operations.

Three types of gases have been used in the procedure for which the term **inertization** has been coined; carbon dioxide, products of combustion and nitrogen. In this section we shall discuss the employment of these gases in addition to methods of application and control.

21.6.1 Carbon dioxide

Carbon dioxide has a density of 1.52 relative to air (Table 11.1). This makes it particularly useful for the treatment of fires in low-lying areas such as dip workings or inclined drifts (Froger, 1985). However, the same property can render it difficult to control in horizontal workings. A 20 t tanker of liquid carbon dioxide will produce some 9000 m^3 of cool gas. The liquid form may be piped into the area where it is required and expanded through an orifice or, indeed, injected directly into a localized heating. In both cases, the gas removes heat from the fire as well as promoting an inert atmosphere. However, piping the liquid carbon dioxide can give rise to freezing problems as well as difficulties in handling the pipes.

The use of carbon dioxide as an inerting gas has several other disadvantages. It is quite soluble in water and can suffer some loss in wet conditions. More significant perhaps, is the fact that it adsorbs readily on to coal and coked surfaces, even more so than methane (Fig. 12.2(b)). When exposed to incandescent carboniferous surfaces it may be reduced to carbon monoxide. Furthermore, it is considerably more expensive than nitrogen.

21.6.2 Combustion gases

Following the sealing of a fire zone, gases produced by the combustion processes, combined with the consumption of oxygen, will produce an extinguishing atmosphere. However, it may be rich in combustible gases and become explosive if air is subsequently re-admitted (section 21.7.3). The products of full combustion, primarily mixtures of carbon dioxide, nitrogen and water vapour, have been employed as an injected inert gas. Flue gas from burning coal has been used in China (Sun, 1963) while modified jet engines have been employed in Poland, Russia and Czechoslovakia (Strang and MacKenzie–Wood, 1985) The latter method involves burning kerosene at rates of some 0.7 kg/s to produce 30 m^3/s of inert exhaust gases. These are cooled by large quantities of water and admitted into the fire zone. The engine produces a power output of about 30 MW which can be usefully employed. Where the law allows the underground use of a jet engine or where it can be employed on surface for a drift mine, then the large output of inert exhaust gases makes it attractive. However, it nullifies the employment of gas analysis as a means of following the progression of the fire, the capital cost is high and a highly specialized team is required to operate and maintain it.

21.6.3 Nitrogen

Liquid nitrogen is the basis for the majority of inertization schemes now employed for subsurface fires. Again, the liquid gas is supplied in tankers of, typically, 20 t capacity giving about 16 500 m^3 of gas. For continuous operation throughout a period of gas injection, the tankers may unload into a bulk storage vessel of up to 40 t capacity and which has been brought to the mine surface.

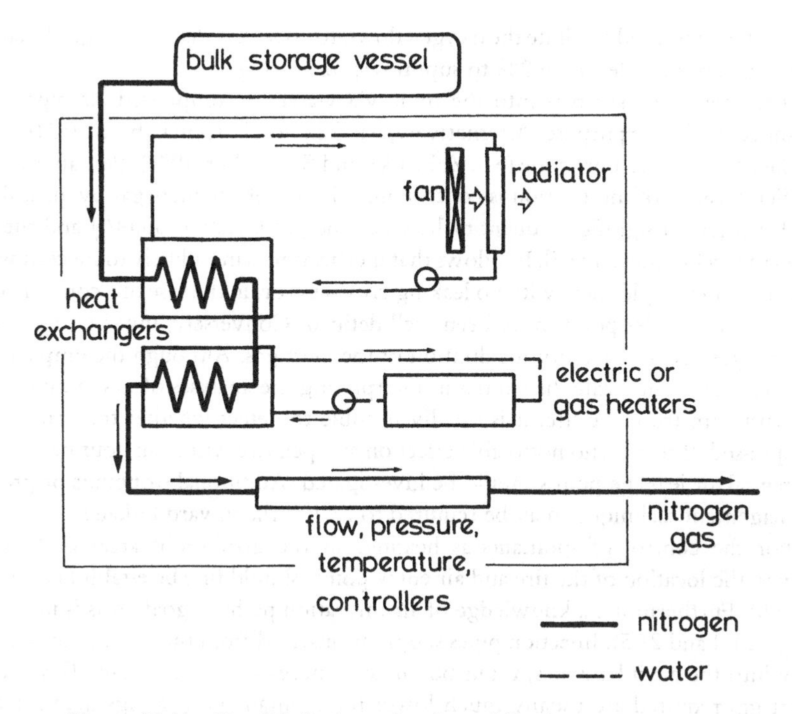

Figure 21.11 Simplified schematic of a mobile nitrogen evaporator.

Because of the low boiling temperature of nitrogen, the liquid must be evaporated before piping it into the mine. Figure 21.11 indicates the principle of a mobile evaporator which, again, has been brought to the mine for the emergency period. Typically, two water circuits are employed; a primary circuit using atmospheric heat and secondary heaters powered by electricity or liquid–gas fuels. The gaseous nitrogen passes through a bank of controllers before entering the mine pipeline. A subsidiary nitrogen line provides a feedback to maintain a constant pressure in the storage vessel. The maximum gas feed rate into the mine depends on the duty of the evaporator but may, typically, be within the range 1 to 6 m^3/s.

Liquid nitrogen is a by-product of the commercial production of oxygen and is much less expensive than liquid carbon dioxide. Furthermore, it is not so soluble as the latter, does not adsorb so readily on carbon surfaces and with a density approximating that of air, mixes readily without stratification.

21.6.4 Methods of application and control

In order to assess the volume flow of inert gas required, the rate of oxygen supply to the fire may be determined from the inlet oxygen concentration and a measured or estimated airflow. It is then a straightforward calculation to determine the flowrate of inert gas required to dilute the oxygen down to 10% in order to extinguish flaming combustion, or to less than 2% to suppress smouldering.

The inert gas may pass into the mine via water or compressed air pipes commandeered for the purpose. Alternatively, gas feed boreholes may be drilled from the surface to intersect a fire zone (e.g. Zabrosky and Klinefelter, 1988). Perhaps the most difficult aspect of inertization is controlling dilution of the inert gas by air leakage. If this enters in significant quantity between the gas injection point(s) and the fire, then the technique may fail. It follows that inertization is most likely to succeed where the fire is in a single entry with no leaking cross-cuts or, in the case of a concealed fire, where the air inlet points have been well defined. Conversely, multi-entry systems offer a greater opportunity for dilution of the inert gas. Although the employment of inert gases can create difficulties in interpreting the analyses of gas samples taken downstream from the fire, it is usually possible to detect whether the fire is being suppressed. If there is no noticeable effect on an open fire within an hour or two then potential air leakage points should be investigated. Additional stoppings or pressure management techniques may be required to reduce the inward leakage.

For the control of spontaneous heatings in the goaf (gob) areas of longwall mines, the location of the fire and air entry points should first be established (section 21.4.5). Furthermore, a knowledge of air migration paths in goaf areas is invaluable (Figs. 21.4 and 21.5). Injection pipes should be inserted from the airways or working face into the air inlet zones, using boreholes if necessary. The volume flowrates of inert gas required are usually much lower for spontaneous heatings than for open fires. However, the reaction may be slower. Indeed, the monitored concentrations of carbon monoxide and methane may increase for up to 36 h as the inert gas displaces those gases from the fire zone. A steady nerve is useful at such times. When the carbon monoxide concentration begins to fall, that can be used as a controlling guide to the required injection rate of the inert gas. Oxygen concentration in the return airways downstream from goaf (gob) inertization should be monitored to ensure that it remains above the relevant mandatory limits (19 to 19.5%).

For coal mines with a history of recurring spontaneous combustion, the trend is towards establishing a permanent nitrogen 'fixing' plant on the surface. The lower grade of nitrogen produce by this method is of little consequence for inertization. A permanent network of nitrogen pipelines throughout the working sections of the mine then allows the gas to be fed at a relatively low rate but continuously from each longwall face into the caved zone. Properly designed, this creates an inert atmosphere throughout the critical zones (Figs. 21.4 and 21.5). As a further projection to future developments, the infrastructure is then in place for complete inertization of the working face when the techniques of automation and remote control make that cost effective.

Where inertization is having a beneficial effect it is important that it be maintained for as long as required. While a deeply seated fire can be controlled by an inert gas, it will seldom be cooled sufficiently to be extinguished. Hence, premature cessation of the operation may result in a rapid escalation of the fire. Similarly, injection into a sealed area should be continued until the oxygen content falls below 2%.

21.7 FIRE GASES AND THEIR INTERPRETATION

In section 11.3.3 we introduced briefly the gases that are produced in the majority of underground fires. Other than use of the human senses, monitoring the quality of the air in a mine is the dominant method of detecting a fire or spontaneous heating. Sampling the air downstream from a fire or from within a newly sealed area and plotting the trends is the primary method of tracking the behaviour of the fire and the development of atmospheres that are, or may become, explosive. However, as the gases emitted vary with the phases of oxidation, time and temperature, it is necessary to employ skilled interpretation of those trends.

21.7.1 The processes of burning and the gases produced.

When coal or timber are burning, three processes are in progress.

1. Gases are distilled from the solid material.
2. The solid material on its surface is oxidized with the emission of heat and light (this is why the surface glows more brightly when fanned with fresh air).
3. Flaming combustion—the burning of combustible gases produced by the first two processes—takes place. Again heat and light are produced. Some of the heat passes back to the surface by radiation and convection to assist in the promotion of further distillation.

The fire gases and their relative proportions depend on the contributions of each of these three processes. The gases of distillation from coal are carbon monoxide, carbon dioxide, hydrogen and water vapour. Methane is also desorbed as the temperature increases. Timber distils the same gases although the amount of hydrogen may be negligible. When flaming combustion occurs, the combustible gases burn to a degree that is governed by the availability of oxygen in the air. The final mixture leaving the fire zone is, therefore, a result of the gases of distillation together with the extent to which the fire has become fuel rich.

The complexity of the processes involved can be illustrated by considering just a few of the ways in which methane can burn:

$$CH_4 + 2O_2 \rightarrow 2H_2O + CO_2$$
$$2CH_4 + 3O_2 \rightarrow 4H_2O + 2CO$$
$$3CH_4 + 5O_2 \rightarrow 6H_2O + 2CO + CO_2$$

Furthermore, secondary reactions may produce water gas and reduction of carbon dioxide to carbon monoxide.

21.7.2 The detection and trend analysis of fire gases

For many purposes of analysis the sampled atmosphere is considered as composed of air, combustibles and inerts (excess nitrogen and carbon dioxide). Figure 21.6 illustrates the combustible gases emitted as coal is heated in a limited air supply such as a spontaneous heating. It is clear that carbon monoxide is a leading indicator of the early stages of such a fire. However, the saturated hydrocarbons, in particular ethylene, are useful indicators of burning coal. In general, as the fire develops into open combustion, the major gaseous product, carbon dioxide, forms an increasing proportion of the pollutant emission.

During any fire incident, running graphs should be maintained of the concentrations of the gases that are monitored or gained from the analysis of samples. Transmitting transducers installed downstream from the fire give a stream of near continuous data. However, in the majority of cases, samples must still be obtained manually by means of evacuated chambers or hand pumps attached to sample containers. The sample vessels should be clean, dry and free from any lubricants that may contaminate the gas sample. A mobile laboratory should be available at the mine surface during any major incident to provide rapid analysis of samples. The rate of sampling must be dictated by the urgency of the situation and the speed at which conditions are changing.

Since the beginning of the twentieth century a number of ratios and composites of gas concentrations have been suggested to assist in the interpretation of fire gases. Table 21.1 indicates some of these.

A feature of several of the ratios is the **oxygen deficiency** ΔO_2. This is a measure of the oxygen that has been consumed and is based on two assumptions; first, that the air has been supplied with 20.93% oxygen and 79.04% inert gases (excepting 0.03% carbon dioxide) That 79.04% contains traces of other gases but is referred to simply as nitrogen. Secondly, it is assumed that no nitrogen has been consumed or added (except from the air) through the area under consideration. If no oxygen is consumed,

Table 21.1 Gas ratios used in interpreting trends of gas concentrations produced by mine fires

Ratio	*Name*
$\frac{[CO]}{\Delta O_2}$	Graham's ratio or index for carbon monoxide (ICO)
$\frac{[CO_2]}{\Delta O_2}$	Young's ratio
$\frac{CO}{\text{excess } N_2 + CO_2 + \text{combustibles}}$	Willett's ratio
$\frac{[CO_2] + 0.75[CO] - 0.25[H_2]}{\Delta O_2}$	Jones and Trickett ratio
$\frac{[CO]}{[CO_2]}$	Oxides of carbon ratio

then the $O_2:N_2$ ratio would remain at $20.93/79.04 = 0.2648$ irrespective of the addition of other gases. For any measured values of O_2 and N_2, the concentration of oxygen that was originally in place can be calculated as

$$\frac{20.93}{79.04}[N_2]$$

Hence, the amount of oxygen that has been consumed, or oxygen deficiency is given as

$$\Delta O_2 = \frac{20.93}{79.04}[N_2] - [O_2] \quad \% \qquad (21.8)$$

The oxidation of coal and the corresponding production of gases were studied by Dr. J. S. Haldane and others at the beginning of the twentieth century (e.g. Haldane and Meachen, 1898). By 1914, Ivon Graham had recognized the importance of carbon monoxide as an early indicator of the spontaneous heating of coal and the equally vital influence of the oxygen that was consumed. He first suggested using the index $[CO]/\Delta O_2$, now known as **Graham's ratio** or the index for carbon monoxide (ICO).

Graham's ratio is the most widely used indicator of an incipient heating in coal mines and has often given warnings several weeks before any odour could be detected. It has the significant advantage that it is almost independent of dilution by leakage of air as this affects both numerator and denominator equally. However, none of the indices listed in Table 21.1 is infallible and Graham's ratio does have some drawbacks. First, its accuracy becomes suspect if very little oxygen has been consumed, i.e. Graham's ratio is unreliable if the oxygen deficiency, ΔO_2, is less than 0.3%. This is a weakness shared by the other indices that involve oxygen deficiency. Secondly, it will be affected by sources of carbon monoxide other than the fire including the use of diesel equipment, or if the air supplied to the fire is not fresh. The latter can occur if the fire is fed, partially, by air that has migrated through old workings and contains blackdamp (de-oxygenated air). Again, like other trace gases or indices, a normal range of Graham's ratio should be established for any given mine. This will usually be less than 0.5%. Any consistently rising values in excess of 0.5% are indicative of a heating.

Example An air sample taken from a return airway yields the following analysis:

nitrogen,	$[N_2]$	$= 79.22\%$
oxygen,	$[O_2]$	$= 20.05\%$
carbon monoxide,	$[CO]$	$= 18 \text{ ppm} = 0.0018\%$
oxygen deficiency,	$[\Delta O_2]$	$= \frac{20.93}{79.04} \times 79.22 - 20.05$

$$= 20.98 - 20.05$$

$$= 0.93\%$$

$$\frac{[CO]}{\Delta O_2} = \frac{0.0018}{0.93} \times 100 = 0.19\%$$

The $[CO_2]/\Delta O_2$, or Young's ratio is, again, nearly independent of dilution by fresh air. Carbon dioxide is the most prolific of the gases produced in mine fires. Hence, the values of $[CO_2]/\Delta O_2$ will be much higher than $[CO]/\Delta O_2$. As a fire progresses from smouldering to open flame, the burning of carbon monoxide will produce an increase in carbon dioxide. Hence a simultaneous rise in $[CO_2]/\Delta O_2$ and fall in $[CO]/\Delta O_2$ indicates further development of the fire. However, as both ratios have the same denominator, the straightward plots of carbon monoxide and carbon dioxide show the same trends. Young's ratio suffers from similar limitations to Graham's ratio. Additionally, the concentration of carbon dioxide may have been influenced by adsorption, its solubility in water, strata emissions of the gas and other chemical reactions.

Willett's ratio was introduced by Dr. H. L. Willett in 1951 with specific reference to situations where there is a higher than usual evolution of carbon monoxide by ongoing low temperature oxidation. In these cases, the gradual extinction of a fire in a sealed area may not be reflected well by the carbon monoxide trend alone but as a percentage of the air-free content of the sample.

The Trickett or Jones–Trickett ratio is used as a measure of reliability of sample analysis and also as an indicator of the type of fuel involved. It can be used for the gaseous products of both fires and explosions. Typical values are shown on Table 21.2. Dilution by fresh air has no effect on the Jones–Trickett ratio. However, it is subject to the limitations of oxygen deficiency.

Table 21.2 Typical values of the Jones–Trickett ratio (after Strang and MacKenzie-Wood, 1985)

Fuel	*Jones–Trickett ratio*
Fires	
Methane	0.4 to 0.5
Coal, oil, conveyor belting, insulation and polyurethanes	0.5 to 1.0
Timber	0.9 to 1.6
Explosions	
Methane	0.5
Coal dust	0.87
Methane and coal dust	0.5 to 0.87
No combustion process	<0.4
Impossible mixture (reliability check)	>1.6

The oxides of carbon ratio, $[CO]/[CO_2]$ is a useful pointer to the progression of the fire, rising during the early stages and tending to remain constant during flaming combustion. However, $[CO]/[CO_2]$ rises rapidly again as a fire becomes fuel rich and is an excellent indicator of this condition. This ratio may also be favoured because it is unaffected by inflows of air, methane or injected nitrogen (Mitchell, 1990). It is, however, subject to variations in carbon monoxide and carbon dioxide that are not caused by the fire.

It is clear that each gas and composite ratio as a warning of impending fire or indicator of fire development has both strengths and weaknesses. During any fire, it is recommended that spreadsheet and graphics software should be utilized to store all monitored data and sample analyses on a desk computer. This should be located in, or close to, the emergency control centre. All gas concentrations and preferred ratios should be made available as screen and hard-copy graphs plotted against time. These should be reviewed repeatedly as new data or analyses become available. The trends are more informative than the absolute values. Hence, repeated samples from a few set stations are preferred to samples taken at many different locations. The injection of inert gases will affect all of the gas analyses with the exception of the $[CO]/[CO_2]$ ratio and where nitrogen is used as the injecting gas.

Gas concentrations and indices given pointers to the average intensity but not the size of the fire. In some cases, where there is limited leakage between the fire and sampling points, an estimate of the extent of the fire may be made from the flowrate of each of the gases. This is given as the product of the air flowrate and the relevant gas concentration.

The principles of gas detection and methods of sampling are discussed in section 11.4. Ionization smoke detectors have been developed in which smoke particles are drawn through a chamber where they are charged by a radioactive source such as americium-241 or krypton-85. An ion-collecting grid and amplifier produce an electrical output that is a function of the smoke concentration (e.g. Pomroy, 1988). Versions of this have been produced that can distinguish between smoke from a fire and diesel particulate matter (Litton, 1988).

The improvements that have been made in electrochemical methods of detecting carbon monoxide (section 11.4.2) have led to this being a common method of fire detection as part of a mine environmental monitoring system. The employment of computer analysis allows filtering of the false alarms from diesels or blasting operations that gave earlier versions a doubtful reputation (e.g. Eicker and Kartenberg, 1984). The filter program may be fairly simple such as ignoring short-term peaks and giving audiovisual alarms only when a significant upward trend is indicated. More sophisticated systems take into account the time-variant generation of carbon monoxide between successive sensors distributed along an airflow path (Boulton, 1991).

The combination of computer-controlled monitoring systems and ventilation network analysis programs allows the possibility of not only detecting the existence of a fire at an early stage, but also its probable location. This is facilitated by strategic location of the sensors (Pomroy and Laage, 1988). The employment of **tube bundle**

sampling is particularly useful for detecting the early and slow development stages of a spontaneous heating. This technique is described in section 11.4.3.

Temperature monitors, often known rather loosely as 'heat sensors' have a application when mounted above fixed equipment and, particularly, when used to activate deluge or sprinkler systems (section 21.3.3). They are of somewhat limited use as fire detectors in airways as the air temperature drops rapidly downstream from a fire. Such sensors may be subjected to greater variations in air temperature from normal operations (such as passage of a diesel vehicle) than from the early stages of a fire. If made sufficiently sensitive, temperature sensors may lose credibility because of an excessive recurrence of false alarms.

21.7.3 Explosibility diagrams

When a fire becomes fuel rich, there is a danger that explosive mixtures of gases will propagate away from the immediate fire zone. Furthermore, following the stopping-off or sealing of a fire area, it frequently occurs that the rising concentration of combustible gases and falling concentration of oxygen pass through a range that is explosive. Similarly, the dilution of combustible gases that occurs when a sealed area is re-opened may, again, result in passing through an explosive range. In order to be able to predict and control such circumstances, it is necessary to have an understanding of the flammability limits of gases and gas mixtures.

This subject was introduced in section 11.2.4 and illustrated by the Coward diagram for methane shown in Fig. 11.1. That diagram should be reviewed, if necessary, as a reminder of the concepts of upper and lower flammability and how these converge to a nose limit as inert gases are added.

In many situations, the large majority of combustible gas is methane and Fig. 11.1 may be employed to determine whether the mixture of oxygen, methane and inerts lies within the explosive triangle or is likely to do so in the near future. However, in other cases, the presence of carbon monoxide and/or hydrogen may cause significant changes in the Coward diagram. Figure 21.12 shows the individual explosive triangles for all three of these combustible gases. The corresponding coordinate points are given in Table 21.3. If the method of the Coward diagram is to be useful for mixtures involving more than one combustible gas then we must be able to quantify the explosive triangle for those composite mixtures. Let us attempt to do that.

A basic precept in the world of science and engineering is that if anything is done to upset the equilibrium of a system, then that system will react in an attempt to reach a new equilibrium with the minimum adjustment of the component parts (Le Chatelier's principle). Let us adopt subscripts 1, 2 and 3 for the three combustible gases and call their percentage concentrations p_1, p_2 and p_3 respectively. If they do not react chemically with each other then the mixture will have a total combustible concentration of

$$p_t = p_1 + p_2 + p_3 \quad \% \tag{21.9}$$

Furthermore, Le Chatelier's principle leads to the prediction that for gas flammability

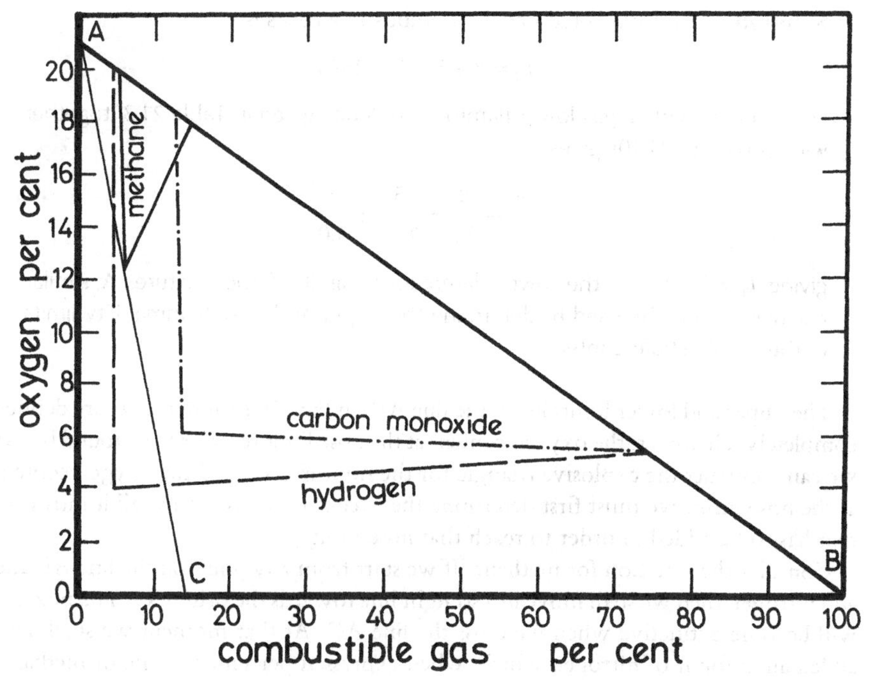

Figure 21.12 Coward diagrams for methane, carbon monoxide and hydrogen.

Table 21.3 Vertices of explosive triangles (percentages)

Gas	*Flammable limits*		*Nose limits*	
	Lower	*Upper*	*Gas*	*Oxygen*
Methane	5.0	14.0	5.9	12.2
Carbon monoxide	12.5	74.2	13.8	6.1
Hydrogen	4.0	74.2	4.3	5.1

limits, L_1, L_2 and L_3 (where these can be upper, lower or nose limits), the corresponding gas flammability limit of the mixture, L_t, will be given by

$$\frac{p_t}{L_t} = \frac{p_1}{L_1} + \frac{p_2}{L_2} + \frac{p_3}{L_3} \tag{21.10}$$

Example An air sample produces the following analysis:

CH_4	8%
CO	5%
H_2	3%

Determine the lower flammability limit of this mixture.

Solution The total percentage of combustible gases is

$$p_t = 8 + 5 + 3 = 16\%$$

Using the individual gas lower flammability limits given in Table 21.3, together with equation (21.10), gives

$$\frac{16}{L_t} = \frac{8}{5.0} + \frac{5}{12.5} + \frac{3}{4.0}$$

giving $L_t = 5.82\%$ as the lower flammability limit of the mixture. A similar calculation may be used to determine the upper and nose flammability limits of the combustible content.

The upper and lower limits lie on the line AB on Fig. 21.12 and, hence, are defined completely. However, the oxygen content at the nose limit remains to be found before we can construct the explosive triangle for the mixture. To find the oxygen content at the nose limit, we must first determine the excess inert gas (let us call it nitrogen) that has to be added in order to reach that nose limit.

Consider the situation for methane. If we start from any point on the line AB and add nitrogen then we shall move in a straight line towards the origin, O. The mixture will become extinctive when we cross the line AC. At that moment we shall have added an amount of nitrogen which, when expressed per unit volume of methane, is a constant. (This follows from the fact that both AB and AC are straight lines.) As we can commence at any position of AB, let us choose point B. On adding nitrogen and moving towards O, we shall cross the extinction line at point C where the methane concentration is 14.14%. The remaining $100 - 14.14 = 85.86\%$ is nitrogen. We have, therefore, added $85.86/14.14 = 6.07\ m^3$ of nitrogen for each cubic metre of methane. A similar exercise can be carried out for carbon monoxide and hydrogen to give the values in Table 21.4.

This table gives the excess nitrogen to be added, N^+, if the combustible content consisted of one gas only. For a mixture of combustible gases the excess nitrogen

Table 21.4 Volumes of excess nitrogen to be added, N^+, in order to make flammable gases extinctive

Combustible gas	*Nitrogen to be added to make mixture extinctive* N^+ *(m^3 (nitrogen) for each m^3 (combustible gas))*
Methane	6.07
Carbon monoxide	4.13
Hydrogen	16.59

required, N_{ex}, is given as

$$N_{ex} = \frac{L_n}{p_t}(N_1^+ p_1 + N_2^+ p_2 + N_3^+ p_3) \quad \% \qquad (21.11)$$

where L_n = percentage of combustible mixture already found from equation (21.10). The required oxygen content at the mixture nose limit is then simply 20.93% of the air fraction, i.e.

$$\text{oxygen (nose limit)} = 0.2093(100 - N_{ex} - L_n) \quad \% \qquad (21.12)$$

Example A sample taken from a sealed area yields the following analysis:

methane	8%	
carbon monoxide	5%	$p_t = 16\%$
hydrogen	3%	
oxygen	6%	
inerts	78%	

Construct the Coward diagram for this condition.

Solution

1. Using equation (21.10) and Table 21.3 gives the following. The lower flammability limit, L_{low}, is obtained from

$$\frac{16}{L_{low}} = \frac{8}{5} + \frac{5}{12.5} + \frac{3}{4}$$

giving L_{low} = 5.82% combustible. The upper flammability limit, L_{up}, is obtained from

$$\frac{16}{L_{up}} = \frac{8}{14.0} + \frac{5}{74.2} + \frac{3}{74.2}$$

giving L_{up} = 23.56% combustible. The nose flammability limit, L_n, is obtained from

$$\frac{16}{L_n} = \frac{8}{5.9} + \frac{5}{13.8} + \frac{3}{4.3}$$

giving L_n = 6.62% combustible.

2. Equation (21.11) and Table 21.4 give the excess nitrogen required for an extinctive atmosphere to be

$$N_{ex} = \frac{6.62}{16}(6.07 \times 8 + 4.13 \times 5 + 16.59 \times 3)$$

$$= 49.24\%$$

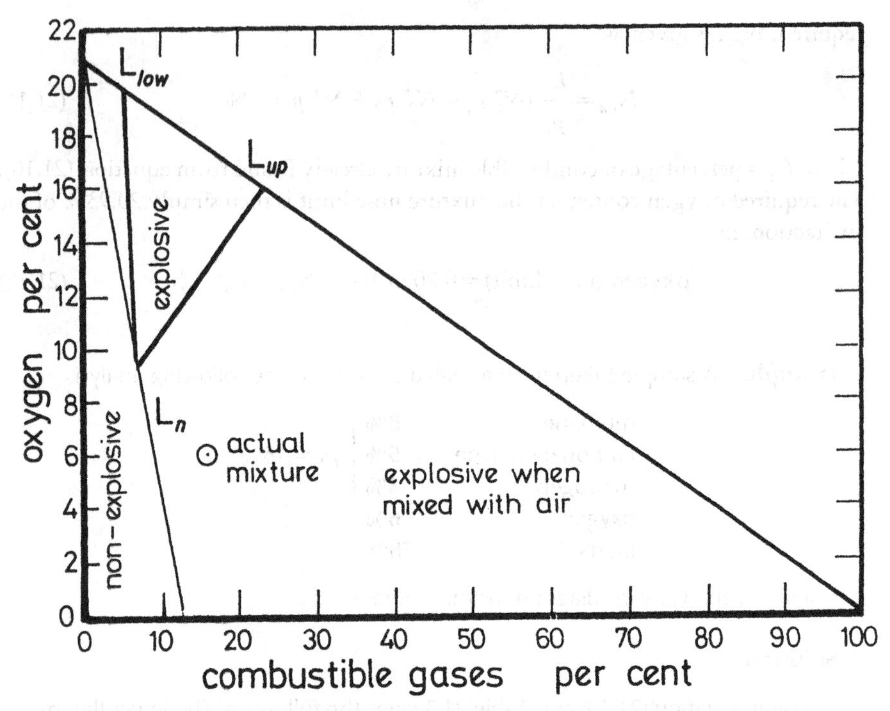

Figure 21.13 Coward diagram for the mixture given in the example.

and, finally, oxygen content at the nose limit is given by equation (21.12) as

$$0.2093(100 - 49.24 - 6.62) = 9.24\%$$

The explosive triangle has now been defined completely and has been constructed as Fig. 21.13. The actual mixture point has also been entered on the diagram and illustrates that although the mixture is not explosive, it will become so if air is allowed to enter the area.

Coward diagrams are most useful in tracking trend directions of gas mixtures. They do, however, have a drawback. As each new sample analysis becomes available, the mixture point on the Coward diagram moves—but the explosive triangle also changes its shape and position. It is analogous to shooting at a moving target. Fortunately, the calculations are simple and a new triangle can be developed for each sample in order to follow actual trends. The complete process can readily be programmed for automatic appearance of the updated Coward diagram and mixture point on a computer screen. Running through a set of consecutive data for any given sampling location produces a dynamic picture of explosive triangles and mixture points moving over the screen. This creates a strong visual impact of time-transient trends.

An older method of dealing with variations in gas mixtures is the US Bureau of Mines composite diagram illustrated on Fig. 21.14 and founded on earlier work by

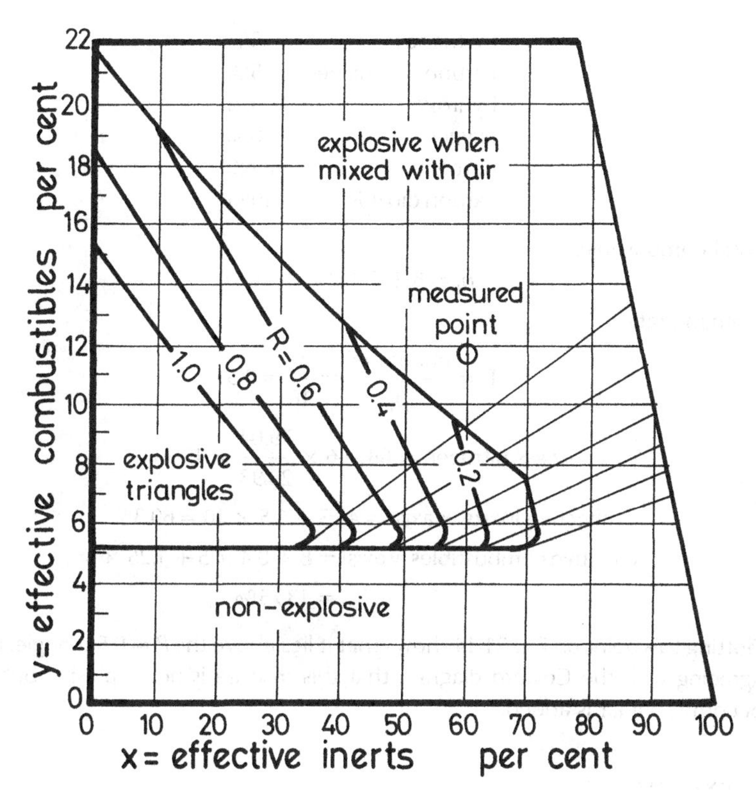

Figure 21.14 US Bureau of Mines diagram. Total combustible, p_t = methane + carbon monoxide + hydrogen; R = methane/p_t; x = excess nitrogen + 1.5 × carbon dioxide; y = methane + 0.4 × carbon monoxide + 1.25 × hydrogen. All gases expressed in percentage by volume.

Zabetakis *et al.* (1959). In this diagram, the y axis is an 'effective combustible' defined by a weighted combination of the volumetric percentages of the three combustible gases. The weighting takes account of the explosibility of carbon monoxide and hydrogen compared with that of methane. The x axis is a combination of the excess nitrogen and 1.5 times the concentration of carbon dioxide, the 1.5 allowing for the greater extinctive power of carbon dioxide. The 'excess nitrogen' required here is the percentage of nitrogen in excess of that justified by the oxygen present:

$$\text{excess nitrogen} = \text{actual } N_2 - [O_2] \times \frac{79.04}{20.93}$$

The set of explosive triangles are approximations based on methane but adjusted for incremental combined additions of carbon monoxide and hydrogen.

Example Using a more detailed analysis of the sample used in the previous example gives

methane	8%
carbon monoxide	5%
hydrogen	3%
oxygen	6%
nitrogen	68%
carbon dioxide	10%

Total combustibles,

$$p_t = 8 + 5 + 3 = 16\%$$

Methane ratio,

$$R = \frac{[\text{methane}]}{p_t} = \frac{8}{16} = 0.5$$

$$\text{excess nitrogen} = 68 - 6 \times \frac{79.04}{20.93} = 45.3\%$$

$$\text{effective inerts } (x \text{ axis}) = 45.3 + 1.5 \times 10 = 60.3\%$$

$$\text{effective combustibles } (y \text{ axis}) = 8 + 0.4 \times 5 + 1.25 \times 3 = 13.75\%$$

Plotting this point on Fig. 21.14 shows that it lies above the $R = 0.5$ triangle, thus agreeing with the Coward diagram that this mixture is not explosive but will become so if air is added.

21.8 EXPLOSIONS

There has scarcely been a major mining industry that has not been traumatized by underground explosions of gases, dusts and mixtures of the two. The potential for disastrous loss of life when such an explosion takes place is very high. Fatality counts have all too often been in the hundreds from a single incident. Fatalities and injuries produced by explosions arise from blast effects, burning and, primarily, from the carbon monoxide content of the afterdamp—the mixture of gases produced by the explosion. In the dust explosion at Courriéres coal mine in France (1906), 1099 men lost their lives. One of the most catastrophic explosions on record occurred at Honkieko, Manchuria (1942) when over 1500 miners died. The carnage that took place in coal mines during the Industrial Revolution (section 1.2) was caused primarily by underground explosions and resulted in the start of legislation governing the operation of mines.

For an explosion to occur, the same three components of the fire triangle must exist simultaneously; namely, a combustible material, oxygen and a source of ignition. However, there is a further condition; the combustible material must be a gas or finely divided dust mixed intimately with the air and in concentrations that lie between lower and upper flammability limits.

The majority of explosions in mines have been initiated by the ignition of methane. This, in itself, is a very dangerous occurrence. However, it becomes much worse when the shock wave raises combustible dust into the air such that it can be ignited by the flame of the burning methane. The resulting dust explosion is likely to be much more violent than the initial methane blast. Indeed, the majority of methane ignitions result in blue flames flickering backwards and forwards along the gas–air interface without developing into an explosion. It is only when the turbulence of the airflow or thermal effects produce a mixture within the explosive triangle (section 21.7.3) that the combustion accelerates into an explosion.

Finely divided particles of any combustible solid can become explosive, including metallic dusts, sulphide ores and most organic materials. Precautions must be taken against explosions in the manufacturing, processing and silo storage of many foodstuffs including grain. However, in this section, we shall confine ourselves to the igniting sources, mechanisms and suppression of explosions that occur in subsurface ventilation systems.

21.8.1 Initiation of explosions

In section 21.2 we classified the major igniting sources of fires and explosions in mines. Unfortunately, explosible mixtures of gases can be ignited by electrical sparks of energy levels as low as 0.3 mJ, as illustrated by the tiny spark that ignites gas in a piezoelectric or induction coil cigarette lighter. It is, therefore, prudent to give a little further attention to the initiation of explosions by sparking phenomena in addition to those sources of ignition discussed in section 21.2.

Incendive sparking arises from heat produced from a chemical reaction. The **thermite** process involves the reaction of aluminium powder with iron oxide:

$$2Al + Fe_2O_3 \rightarrow Al_2O_3 + 2Fe \tag{21.13}$$

This process produces so much heat that it has been used on small-scale welding operations. A similar effect occurs when a surface consisting of aluminium, magnesium or their alloys collides with a rusted steel or iron surface. Incendive sparks are produced that are well capable of igniting a methane:air mixture. It is prudent to prohibit the importation of any light alloys into gassy mines and, indeed, this is enforced by law in a number of countries.

Frictional sparking has been the cause of the increased incidence of methane ignitions on coal faces and which have accompanied the proliferation of mechanized mineral winning. This is discussed in section 21.2.4.

In addition to sparking caused by the misuse of or damage to electrical equipment, electrostatic sparks may also be capable of igniting a flammable mixture of gases. Electrostatic charges are built up on non-conducting (or poorly conducting) surfaces as a regular feature of many everyday operations and, particularly, at pointed or sharply curved regions on those surfaces. Electrical potentials of 10 000 V are commonly generated. This phenomenon may occur, for example, where belts run over pulleys, at the nozzles of compressed air jets (particularly, if liquid or solid particles are entrained) and within non-conducting ventilation ducts. Even the charge that builds up on the human body in dry conditions can produce dangerous sparks (Strang and McKenzie-Wood, 1985). In such conditions, workers should not wear rubber-soled footwear.

All machines with moving parts should be adequately earthed against the build-up of electrostatic charges. Other devices that are liable to this phenomenon should be similarly protected. 'Anti-static' materials are available for ducting, belts and pipes.

21.8.2 Mechanisms of explosions

An explosion may be defined as a process in which the rates of heat generation, temperature rise and pressure increase become very great due to the rapidity of combustion through the mixture. In a typical methane: air explosion the temperature rises to some 2000 °C, i.e. by a factor of about seven. The speed of the process is so great that it is essentially adiabatic. The result is that the pressure in the immediate vicinity increases to a peak value very rapidly and is relieved by expansion of the air. This produces a shock wave that propagates in all available directions. In a mine opening the expansion is constrained by the airway walls, giving rise to high velocities of propagation. If the explosion occurs at the face of a heading then the expansion can take place in one direction only. This leads to the most violent of mine explosions. An ignition of a 10% methane:air mixture in a heading may produce flame speeds of 660 m/s (twice the speed of sound).

The shock wave is essentially the moving boundary between the normal mine atmosphere and the elevated pressure of the expanding explosion. Hence, there is a highly concentrated pressure gradient across the shock wave. It is this that causes 'blast' effects and can inflict extremely high dynamic loads on objects that lie in the path of the explosion. The shock wave produces another effect. It causes intimate mixing of any accumulations of methane that it passes through, producing a continuous explosive path. Even worse, it raises dust into the air to produce concentrations which, if the dust is combustible, will themselves be explosive—resulting in even more favourable conditions for enriching the fuel path of the explosion.

The shock wave is followed by a front of burning gas and/or dust. This **flame front** continues to provide the expansion of gases that drives the shock wave forward. Except in the most violent of explosions (discussed in a subsequent section on coal dust explosions), the flame front travels more slowly than the shock wave. Between the two, the air is pulsed forward at a velocity of about 85% of the flame speed.

These are the basic mechanisms that drive explosions through subsurface ventilation networks. Bends, junctions, obstructions and availability of fuel all affect the rate of propagation. A mine explosion is a highly dynamic phenomenon and seldom reaches steady state even for a few milliseconds. For most of its life it is either proliferating or subsiding.

Gas explosions

Explosions of methane from coal are by no means confined to mines. Ignitions have too frequently occurred in storage silos and in coal cargo ships, causing great damage and loss of life. Methane concentrations of 40% have been measured in such facilities. Layering phenomena and the production of explosive mixtures with air make unloading operations particularly hazardous. Precautions against ignitions in these circumstances include good ventilation of the facility, monitoring for methane and the gaseous products of spontaneous combustion, ready availability of nitrogen or carbon dioxide cylinders for inertization, strict control of ignition sources and prohibition of smoking in the vicinity. Bacteria have also been used to convert the methane into carbon dioxide (Kolada and Chakravorty, 1987).

Incidences of hydrogen explosions at battery charging stations and ignitions of oil vapours from machines have been reported. However, the vast majority of gas explosions in mines have involved methane. Modern systems and standards of mine ventilation are successful in maintaining general body concentrations of methane at safe levels throughout the ventilated areas. However, strata emissions of gas are often at methane concentrations of over 90%. It follows that, between the points of emission and the general body of the airstream, zones of an explosive mixture must exist. The ventilation arrangements should ensure that these zones are as small as possible and, wherever practicable, are maintained free from igniting sources. For example, the danger of methane, coal dust, air and sparks existing simultaneously at the pick-point of a coal winning machine can be reduced by pick-face flushing or jet-assisted cutting (section 20.4.1).

Dangerous accumulations of methane in roof cavities should be avoided by filling those cavities with an inert material. Air velocities should be sufficient to prevent methane layering (section 12.4.2) and longwall systems should be designed to maintian the gas–air interface within the goaf (gob) and without flushing on to the faceline. In appropriate circumstances, methane drainage will reduce the amount of gas that enters the mine ventilation system (section 12.5).

Coal dust explosions

The danger of coal dust being explosible in air depends on a number of factors including the following.

1. *Concentration of the dust and presence of methane.* The flame of a dust explosion is propagated by the combustion of each discrete particle in turn. If the particles are sufficiently far apart then the heat produced by a burning particle will be insufficient to ignite its closest neighbour. The flame will not propagate. The lower flammable limit is attained when the particles become sufficiently close to allow sequential ignition. At the other extreme, if the dust particles are too close together, there will not be enough oxygen to allow complete combustion. The system will become internally fuel rich and the rate of heat generation will fall. The upper flammability limit is attained when the temperature is reduced below the ignition point of the dust.

   The lower flammability limit may be of the order of 50 g/m^3 with maximum explosibility at 150 to 350 g/m^3 depending on the volatile content of the coal

(Holding, 1982). The upper flammability limit might be as high as 5000 g/m^3 (Strang and McKenzie-Wood, 1985). To put these flammability limits into perspective, it should be recalled that threshold limit values for respirable coal dust concentrations can be as low as 2 mg/m^3 (Table 19.2). A concentration of 50 g/m^3 produces a suffocating atmosphere. It is unlikely that such concentrations of airborne dust will exist under normal conditions in the active branches of a mine ventilation network. Nevertheless, in the absence of dust suppression, explosible dust concentrations might occur around the cutting heads of rock-breaking machines. Furthermore, a very thin coating of dust on the surfaces of airways or a conveyor can produce an explosive atmosphere when disturbed by a shock wave.

Coal dust and methane are synergistic in their explosibility characteristics. The flammability limits of each of them are widened in the presence of the other. The relationship between the two varies with the volatile content of the coal. A typical example, based on results reported for an Australian coal, is shown on Fig 21.15.

2. *Fineness of the dust.* There appears to have been very little work done in establishing definitive relationships between explosibility and the size of dust particles. However, the increased surface area and more intimate mixing with the air given by the finer particles suggest that explosibility will increase as the average

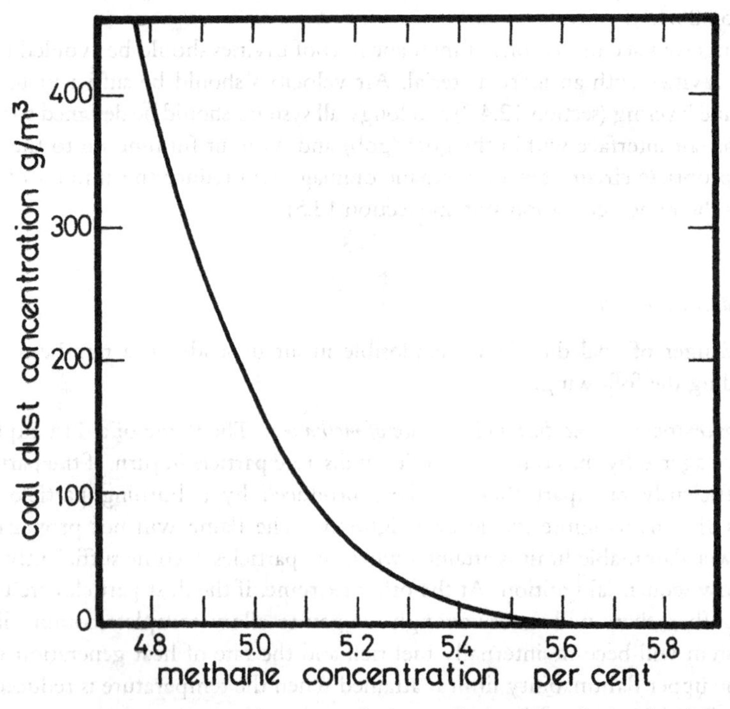

Figure 21.15 Example of the reduction in the lower flammability limit of coal dust in the presence of methane. Other curves will vary according to the rank of the coal. (Based on work reported by Jensen *et al.* (1989).)

particle size decreases. An accepted assumption is that any combustible dust which can be raised into the air is explosible. This encompasses all particles of less than some 250 μm diameter.

3. *Type of coal.* There is considerable evidence to show that coal dust explosibility increases with volatile content (e.g. Holding, 1982). Hence, low rank coals are more prone to dust ignitions. Anthracite dust is normally considered to be non-explosible in mining conditions. However, with a sufficiently high energy of initiation, low volatile coal dusts can produce violent explosions. The disastrous explosion in Manchuria (1942) occurred while mining a coal of volatile content 15% to 19%. The explosibility of coal dust decreases with respect to ash and moisture content as well as with age. The latter factor is considered to be due to the coating of a partially oxidized layer on older dust particles. It follows that coal dust on, or close to, a working face is more liable to ignitions.
4. *Strength of initiating source.* The energy level of the initial ignition plays a large role in governing the rate of propagation and growth in power level of a dust explosion. The majority of coal dust explosions in mines have been initiated by a methane ignition, this providing the starting shock wave and flame front. However, any other igniting source of sufficient power can result in a dust explosion in the absence of a flammable gas. The Courriéres dust explosion (1906) is thought to have been caused in a methane-free atmosphere by a case of prohibited explosives (Cybulska, 1981).

To comprehend the power levels that can be attained by coal dust explosions, let us return to the matter of the spearheading shock wave driven by a more slowly advancing flame front. This type of explosion is known as a **deflagration**. A flame velocity of 50 m/s will produce a shock wave velocity of some 375 m/s. (A lower flame velocity is unlikely to propagate.) However, Fig. 21.16 shows that if the dust concentration favours the growth of the explosion then the flame velocity will increase at a greater rate than that of the shock wave until both exceed some 1100 m/s. Beyond that, the process may develop further into a **detonation** when the adiabatic compression in the shock wave produces temperatures that exceed the ignition point of the dust. Dust particles no longer have to wait until they are ignited by a burning neighbour. They ignite spontaneously. The flame front and shock wave advance in unison and may reach speeds of over 2000 m/s (six times the speed of sound). It is little wonder that coal dust explosions are so feared throughout the world of mining.

Figure 21.16 also illustrates the dynamic pressure developed across the shock wave. Even at the lower end of this curve the pressures are sufficient to disrupt doors, stoppings and air-crossings, while the upper end explains the devastation that can be caused by a well-developed coal dust explosion.

During post-explosion inspections, the dominant path of the explosion may be deduced from the twisting of steel supports, the distortion of track rails, the direction of failure of strong stoppings and the dislocation of heavy equipment. The displacement of less sturdy objects may be misleading and can be caused by secondary deflections of the shock wave, particularly at bends or junctions. Furthermore, an

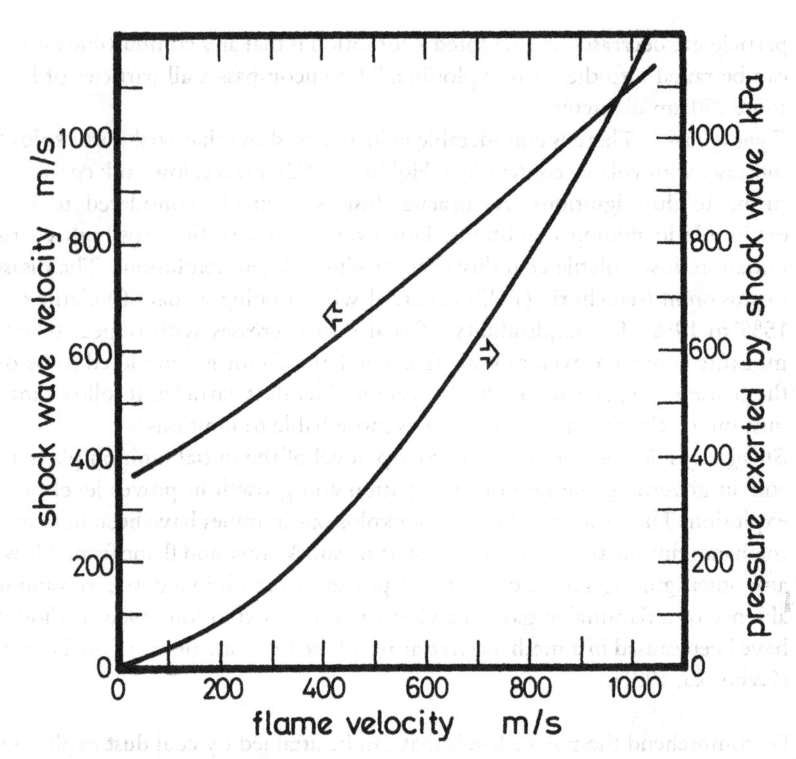

Figure 21.16 Variation of shock wave velocity and dynamic pressure with respect to flame velocity. (Developed from data given by Strang and MacKenzie-Wood (1985).)

explosion is often followed rapidly by an implosion of fresh air due to rapid cooling and contraction of the afterdamp gases. If open burning is in progress then the fresh air can initiate repeated explosions. These phenomena can also leave a confused picture for subsequent investigators.

Coked dust may be found on the sides of the airways, particularly if the explosion has tended to be fuel rich. This condition may also leave a legacy of carbon filaments projecting from the roof and sides. On the other hand, a high velocity flame may leave little evidence of coking. The passage of an explosive flame is unlikely to ignite solid coal. However, paper, clothing and, especially, pockets of methane may be left burning if there is sufficient oxygen remaining in the area. This is most probable in the fringe areas of the explosion.

Sulphide dust explosions

Although explosions of sulphide ore dust are a hazard in some metal mines, they do not have the destructive power of coal dust explosions and have attracted much less attention by researchers (Holding, 1975). These explosions are initiated, primarily, during blasting operations of ores that contain more than 40% sulphur. The main danger to life is production of the highly toxic sulphur dioxide gas which may be retained for several hours, particularly if ventilation controls have been disrupted. Furthermore, the gas reacts with water vapour causing corrosion problems from the sulphuric acid fumes that are produced. A number of precautionary measures have been suggested including hosing dust depositions from the walls of stopes prior to blasting and the use of water atomizers before and during blasting (section 20.3.4). Other techniques involve stemming the blast holes with water ampoules or powdered limestone and inserting explosive charges with zero-time detonators into bags of hydrated limestone powder suspended in the stopes. The latter technique fills the air with inert and suppressant dust particles immediately prior to detonation of the main round (Hall *et al.*, 1989).

Explosibility tests

Explosibility tests for mine dusts vary considerably in the equipment and procedures employed. The essential sequence consists of producing a dust:air or dust:oxygen mixture and applying an igniting source, usually an electrical spark of constant energy level. The parameters that are varied include the dust concentration, the type of dust, added amounts of stonedust (usually powdered limestone) and methane concentration. The results may be recorded simply as 'ignition' or 'no ignition', or, in the more sophisticated tests, the rates of increase of pressure and temperature and the peak values of those variables (Jensen *et al.*, 1989).

Considerable variations occur in the reported explosibilities of similar dusts. These may be attributable, in part, to differences in test conditions. The results are influenced, for example, by variations in the uniformity of the dispersed dust, means of measuring dust concentration, strength of the igniting source and turbulence within the test chamber. This latter factor also indicates that whilst laboratory test results may provide valuable comparisons between differing dusts, they are not necessarily indicative of actual behaviour within the mine environment. Full-scale galleries have been constructed in several countries for investigations into mine fires and explosions.

Indices of dust explosibility have been defined and referred to specific equipment and procedures in differing countries (e.g., Holding, 1982). While these produce useful relative information, caution should be applied when comparing one index with another.

21.8.3 Suppression of mine explosions

The primary safeguards against mine explosions are, once again, a well-designed and operated ventilation system, planned maintenance of equipment and good housekeeping to control accumulations of combustible dust and to ensure adequate applications of stonedust. This brings us to the most widely used method of suppressing the propagation of mine explosions.

Stonedust and water barriers

Finely divided limestone dust serves at least two purposes when mixed intimately with coal dust in a mine airway. First, it will act as an inert dilutent when the mixture becomes airborne, serving to increase the distance between combustible dust particles. Secondly, particles of stonedust will absorb heat and reduce the ability of airborne coal dust to propagate a flame. Provided that there is sufficient stonedust present, this technique provides an efficient means of suppressing the full development of an explosion.

It is important that the stonedust be mixed well with the coal dust. Tests have shown that strips of coal dust interspersed with strips of stonedust will sustain an explosion. Furthermore, combustible dust on a conveyor may propagate an explosion even when the airway is adequately stonedusted. Precautions should, therefore, be taken to ensure that the stonedust is spread on roof, floor and sides often and consistently. An efficient means of doing this is to employ **trickle dusters**. These are devices that emit stonedust into the air at a controlled rate. The stonedust mixes with the airborne coal dust and settles with it to produce a uniformly mixed deposition on all surfaces.

An additional means of using stonedust, particularly applicable to single-entry longwall systems, involves **stonedust barriers**. These are boards supported on pivots across the airway, usually near the roof, and loaded with stonedust. Dust loadings vary from 30 to 60 kg per metre length of board. A number of boards are located close together within a short length of airway to form the complete stonedust barrier. The intention is that the boards and their contents will be dislodged by the shock wave of an explosion to produce a high concentration of airborne stonedust at the time the flame front arrives and, hence, to prevent its further propagation.

The locations of stonedust barriers should be considered carefully. If they are too close to the seat of the explosion then the flame front may have passed before the stonedust has become adequately distributed; if too far away, then the stonedust will become too dispersed by the time the flame front arrives. A sensible arrangement is to have lightly loaded (30 kg/m) barrier that can more easily be displaced, located within some 200 m of face operations and more heavily loaded barriers further outbye. While stonedust barriers may be sited in all district intakes and returns, conveyor entries are particularly recommended to be fitted with these devices.

Stonedust barriers do have some disadvantages. They must be inspected regularly to ensure that they remain capable of being dislodged—yet this renders them prone to accidental or mischievous disturbance. In damp conditions, moisture-proofed stonedust should be employed to ensure that the dust will be dispersed adequately by a shock wave. Stonedust barriers may fail to suppress methane explosions. Furthermore, if an explosion is allowed to develop into a detonation then it is unlikely to be halted by a stonedust barrier. Indeed, the objective of barriers is to suppress an explosion at an early stage of its development.

A water barrier is intended to serve the same purpose as a stonedust barrier. In this case, troughs holding some 40 to 90 l of water are positioned across the airway and attached to roof supports. The troughs themselves are constructed from polystyrene foam or a similarly weak material so that they will disintegrate when subjected to a shock wave travelling at 100 m/s. The cross-section becomes filled with water droplets. Suppression of the flame is achieved by the cooling effect of evaporation and displacement of oxygen.

Triggered barriers and explosion detectors

Stonedust and normal water barriers both suffer from the disadvantage that they rely on the shock wave to disperse the stonedust or water. Triggered barriers are designed to incorporate an internal power source. A typical design consists of an enclosed water tank connected to a concentrated array of nozzles along a short length of airway. A containment diaphragm prevents flow from the tank to the nozzles under normal operating conditions. A bottle of compressed nitrogen or carbon dioxide is located inside the water tank. The gas container is also fitted with a rupture disk.

On activation, a heater within the gas bottle causes the gas to expand and puncture the rupture disk. This causes very rapid pressurization of the water, breakage of the containment diaphragm and delivery of water at high pressure to the nozzles. Several hundred litres of water can be dispersed in less than 1 s. Some designs utilize a powdered flame suppressant in place of the water and a soft explosive instead of the gas container.

Activation of a triggered barrier is initiated by an electrical signal from a detector device located closer to the working area, where an explosion is more likely to commence. Infrared, ultraviolet, temperature and pressure pulse sensors have all been employed as detection devices.

In addition to the active power source to disperse the fire suppressant, an advantage of the triggered barrier is that an optimum distance can be selected between the sensor and barrier. This ensures that the barrier will be activated at the correct moment with respect to the approaching flame front.

21.9 PROTECTION OF PERSONNEL

21.9.1 Training and preparedness

In all underground emergencies, the first priority is the safety of personnel. The surest protection against the initiation and hazards of mine fires and explosions is training and practice of safety procedures. Classes and practical sessions should be held, not only for new recruits, but at regular intervals of time for all employees. These sessions should include discussions on the causes of fires and explosions and how they propagate. The elements of subsurface firefighting and operation of fire-suppression equipment should be outlined. It is prudent to ensure that supervisors and other selected members of the workforce should receive additional instruction and practice in firefighting. As in mine rescue work, competitions between teams can do much to foster interest. This further training may include the use of temporary stoppings, airflow control and air pressure management. However, training of all underground personnel should concentrate particularly on

1. warning systems and location of trapped persons,
2. self-rescuers,
3. escapeways, and
4. refuge chambers.

These four items are discussed in the following subsections.

Mine management plays a crucial role in minimizing the risks of fires and explosions, and in responding correctly when they do occur. In addition to matters of training, the planned layout of services dictates the efficiency of emergency response (section 21.10). Communications, routes of water pipes, cables, airflows and escapeways, availability and effectiveness of firefighting equipment, reliable environmental monitors and an integrated fire-protection policy (Sheer, 1988) can make the difference between a localized ignition that is quickly extinguished and disastrous loss of life. Ventilation and safety engineers should engage in mock scenarios and computer simulations of emergency situations. Such exercises assist in raising questions related to response strategies and mine preparedness.

All underground workers should be required to engage in fire drills and emergency evacuation procedures at least once in each three months. These should involve leaving the mine by escapeways other than the normal travelling routes. Such drills should be accompanied by discussions on choices of escape routes and ventilation systems, and should also encourage familiarity with the layout of airways within the mine.

It is a useful exercise, as part of emergency preparedness, for mine management to establish prior friendly relationships with local police, fire, hospital and news agencies. This can be further facilitated by arranging occasional tours of the mine for such people.

21.9.2 Methods of warning and locating personnel

The rates at which mine fires can propagate and products of combustion spread throughout an underground network of airways makes it vital that all personnel in the mine be warned of the emergency as quickly as possible. This can be difficult because of the distances involved and wide dispersal of the workforce, particularly in metal mines.

Many mining industries employ 'firebosses, firemasters' (or similar titles) whose task is to patrol the mine before and, perhaps, during each working shift and following blasting operations to search for any indications of a fire. To ensure adequate and timely coverage of the mine, tags are left at strategic points showing the time and date of the last visit. Fire patrols should follow the direction of airflow in order to locate more easily the probable source of any smoke or fire odour.

Modern paging-type mine telephone systems enable voice warnings to be transmitted to main work areas and along major transport routes. These are particularly suitable for concentrated mining methods such as the longwall system. However, the extension of paging techniques along the several hundred kilometres of active airways in a large metal mine would be cost prohibitive. Two main methods have been devised to transmit fire warnings to a widespread workforce in under-ground openings:

1. stench warning systems, and
2. ultralow frequency radio signals.

Following receipt of any form of fire warning, mine workers should immediately engage in the emergency evacuation or personnel protection procedures in which they should have been trained and practiced (section 21.9.1).

Stench warning systems

Versions of this technique have been employed in large metal mines since the 1930s. The procedure involves releasing a gas having a very distinctive odour into intake airways or compressed air systems. Having been exposed to low concentrations of the stench gas during training sessions, miners recognize the odour and can initiate emergency procedures. The gases used resemble those introduced into natural gas distribution systems. Ethyl mercaptan mixed into freon has been commonly employed. However, this can be toxic, corrosive and produce an unbearable stench at high concentrations. Furthermore, it may lose its potency when transported through steel pipes. A safer and healthier alternative is tetrahydrothiaphene (Ouderkirk *et al.*, 1985).

A basic disadvantage of the stench warning technique is that it relies on the mine ventilation system. Transmission times of the stench depend on the large range of air velocities that exist within a mine ventilation network. In openings of large cross-sectional area, travel times can become dangerously high to serve as a fire warning system. Furthermore, the concentration of the odour will also vary from very high, close to the emission point, to barely detectable in remote areas of a complex ventilation network. To add to the difficulties, the ventilation system may have been modified or disrupted by the fire (section 21.3.2).

These drawbacks can be minimized by selecting multiple injection points in locations reasonably close to major work areas, in addition to downcast shafts or other primary inlets. Stench gas injector mechanisms should incorporate pressure balance arrangements and controlled release rates to prolong the warning and to avoid short peaks of excessively high concentration. Inbye injectors should be capable of remote activation. The choice of injector sites and release rates should be made carefully. The objective is to carry a clearly detectable odour as rapidly as possible to all parts of the mine where personnel may be working. Ventilation network analysis packages that incorporate gas distribution modules are very helpful in selecting optimum locations and release rates (section 7.4.4).

Ultralow frequency radio signals

Many investigations have been conducted into the use of radio waves for warning mine personnel of an emergency condition and, also, for locating persons who have become trapped. High or medium frequency radio signals are limited to line-of-sight transmission or depend on the existence of metal conductors such as pipes, cables or conveyor structure. For personnel warning or location, the radio signals must be capable of transmission through rock and of being detected by lightweight and low-powered personal receiver units.

The most promising developments in this area have involved ultralow frequency devices. At frequencies in the range 630 to 2000 Hz and employing transmitting powers of 1 kW, rock penetration distances of over 1600 m have been achieved (Hjelmstad and Pomroy, 1990). This indicates that transmitters located on surface will be suitable for most mines, while underground transmission sites will serve for the deeper workings. The signal strength decays inversely with the cube of the distance through the rock.

The transmitter consists of a radio (electromagnetic) signal generator, amplifier and antenna. The latter may consists of ten coils of insulated copper wire coiled into a loop of some 30 m diameter. The personal receiver units are small, intrinsically safe and may be powered by the wearer's caplamp battery. The pencil-size receiver antenna is formed from a high permeability ferrite core wound with copper wire and is very efficient in detecting electromagnetic radiation. More highly powered units can be fitted to vehicles or other equipment. The transmitter and receivers are tuned to a common resonant frequency to discriminate against electromagnetic noise or stray signals. The voltage produced by a receiver unit is amplified and may then be used to generate an audio-visual warning. Incorporation into the caplamp unit can produce on–off blinking of the light until nullified by operation of a switch. Research continues into the use of rock-penetrating radio waves for voice communication. Again, this will serve for both warning and personnel location purposes.

Locating trapped persons

The situation has all too often arisen where miners have become trapped in locations that are inaccessible because of accumulations of gases, products of combustion, water or falls of roof. To facilitate rescue operations, it is necessary to identify the locations of the trapped personnel.

Here again, work is in progress to develop two-way radio communication. However, a very old method that has often led to life-saving operations is the seismic technique. An explosive charge or heavy hammer (pile driver) blow on the mine surface may be heard or felt by personnel in relatively shallow mines. In deeper operations, similar seismic signals may be generated from underground locations. On hearing these signals, the trapped personnel respond by hammering on pipes, rails or, simply, the floor or walls. If this response is detected by the ears or geophones of rescue personnel then mine maps are consulted, if necessary, to select the probable location(s) of the trapped persons. Pilot boreholes may be drilled to those locations to supply food and air while rescue operations proceed.

21.9.3 Self-rescuers

An estimated 80% to 90% of fatalities in mine fires or explosions are caused by carbon monoxide poisoning. Self-rescuers are compact versions of breathing apparatus and are intended to allow the wearers to pass through atmospheres that are contaminated by carbon monoxide. In many countries, it is mandatory for all persons who enter an underground mine to carry a self-rescuer. The lighter versions can be worn on the belt of each person who ventures underground while caches of the heavier self-rescuers may be kept close to work areas and on vehicles. The purpose of self-rescuers is to allow the wearers to escape from or through contaminated atmospheres. Indeed, they are often known as escape breathing apparatus (EBAs). They are not intended for rescue operations or other type of work and should not be confused with the more specialized breathing equipment used by trained rescue teams.

The two essential features of self-rescuers are the types of atmospheres in which they are effective and the time they allow for escape. Their operational duration depends not only on the type of self-rescuer but also on

1. manual effort (e.g. speed of walking),
2. breathing habits, experience, physical condition and mental state of the wearer, and
3. environmental conditions.

Although many lives have been saved by self-rescuers, others have been lost even when such devices were available. In most of these cases. the error has been in failing to don the apparatus at the appropriate time. This matter should be stressed heavily during training sessions. In particular, self-rescuers should be put on as soon as it is suspected that the air is, or may become, contaminated, i.e. by means of telephone messages, personnel warning systems (section 21.9.2), shock waves or strong pulses in the ventilation, unusual odours or the appearance of smoke. Furthermore, during training, personnel should be required to practice unpacking and donning the apparatus. Training versions of the units are available for this purpose. Hard-won experience has indicated that simply observing a demonstration is insufficient.

There are basically two types of self-rescuer units, the filter self-rescuer (FSR) and the self contained self-rescuer (SCSR).

Filter self-rescuers (FSRs)

Each of these small and compact units fits inside a hermetically sealed plastic or stainless steel case which is worn on the belt. To use the device, the seal is broken by a lever arrangement, the mouthpiece is inserted inside the lips and gripped by the teeth, and nose clips are put on. The complete unit is held close to the face and chin by head straps. Air is drawn through three types of filters before passing through a heat exchanger to the mouthpiece. Exhaled air also traverses the heat exchanger and through a one-way value to the external atmosphere.

The first level of filtration simply removes dust particles. The second filter is a drying agent—typically 9% lithium chloride and 91% calcium bromide impregnated into activated charcoal (Strang and MacKenzie-Wood, 1985). The activated charcoal assists in removing sulphur dioxide, hydrogen sulphide and oxides of nitrogen. The third level of filtration contains a catalytic mixture of granulated manganese dioxide, copper oxide and a little silver oxide. The mixture is known widely as **hopcalite**. This catalyst converts carbon monoxide into carbon dioxide with close to 100% efficiency. It is, however, poisoned by water vapour. Indeed, the duration of the device is governed by the drying agent in the second level of filtration. The period of operation is, typically, about 1 h for saturated conditions and carbon monoxide concentrations of up to 1.5%.

Filter self-rescuers suffer from two disadvantages. First, they rely on there being at least 16% oxygen in the ambient atmosphere (section 11.2.2). Secondly, both the drying filter and the hopcalite involve exothermic reactions. Indeed, the temperature of the inhaled air is an indication of the level of carbon monoxide present. At high concentrations, the temperature of the filtered air may exceed 90 °C and cause great discomfort including blistering of the lips and mouth. However, removing the mouthpiece for temporary relief in such circumstances is likely to be fatal. The small heat exchanger close to the mouthpiece absorbs heat from the filtered air and rejects it to the exhaled air.

Self-contained self-rescuers (SCSRs)

As the name implies, these supply all the respiratory needs of the wearer and are independent of the gaseous constituents of the ambient atmosphere. A considerable variety of self-contained self-rescuers has been produced. The main disadvantage is their weight and bulk. Current generations of SCSRs vary from 2 to 5 kg and may not be sufficiently convenient for personal wear—hence caches of the units need to be available close to active working areas. Research continues in attempts to develop a self-contained self-rescuer that is sufficiently light and compact to be worn on the belt while giving the 60 or 90 min operating life that may be mandated by law. National and state legislation should be consulted for regulations that govern the requirements, types and operating durations of self-rescuers.

As with the FSRs, the self-contained versions are hermetically sealed in polypropylene or metal containers to ensure a long shelf life. However, if the seal is broken or damaged, the unit must be replaced. Periodic water bath tests should be carried out to ensure that the casings remain airtight. The units include noseclips, fitting straps and goggles. Two types of self-contained self-rescuers are in common use.

The compressed oxygen SCSR is a recirculating system. Expired air passes through a soda lime filter which removes carbon dioxide, then into a flexible 'breathing bag' where it is enriched with oxygen from a compact gas cylinder. The oxygen may be supplied at a base rate of some 1.2 to 2 l/min but will be increased automatically if deflation of the breathing bag indicates a rising demand for air. The air passes to and from the mouthpiece via a flexible breathing tube. The unit is worn on the chest and supported by neck and waist straps.

The chemical SCSRs are lighter devices and develop oxygen from the reaction that occurs when water vapour and carbon dioxide from exhaled breath pass through potassium superoxide (KO_2). The oxygen in this compound contains an additional electron compared with free gaseous oxygen. This gives a fairly weak bond and free oxygen is released readily according to the following reactions:

$$4KO_2 + CO_2 + H_2O \rightarrow K_2CO_3 + 2KOH + 3O_2 \tag{21.14}$$

$$KOH + CO_2 \rightarrow KHCO_3$$

Exhaled air from the mouthpiece passes down the flexible breathing tube and through a bed of granulated potassium superoxide where oxygen is added and carbon dioxide removed. It is then collected in a breathing bag for re-inhalation. A heat exchanger can be incorporated to maintain the temperature of the inhaled air at no more than 45 °C in most situations.

The chemical production of oxygen can be initiated simply by exhaling vigorously into the unit. However, there may be little or no time to accomplish this safely. Chlorate candles are incorporated into most SCSRs to overcome this difficulty. The chlorate candle is started by percussion or other device when a ripcord is pulled. This provides some 3 to 4 l of oxygen during the following few minutes and until the potassium superoxide reaction becomes effective.

21.9.4 Escapeways

In any subsurface facility, certain paths should be selected as preferred evacuation routes in the event of an emergency. These **escapeways** should be highlighted during training sessions and fire drills. There are three matters of importance:

1. airways selected as escapeways,
2. preparation and maintenance, and
3. protection and use of escapeways during an emergency.

The choice of escapeways commences by a pragmatic examination of the airflow routes, travel distances, geographical layout of airways, the physical state of those airways, directions of leakage through stoppings, and the locations of air crossings, doors and other ventilation controls that may be dislocated by a fire or explosion. The primary requirement is that escapeways should be maintained free from products of combustion for as long as possible following the outbreak of a fire.

At least two escapeways must be available from all areas of routine mineral extraction. This is usually mandated by law. In the case of single development headings or other special circumstances, legislation may limit the number of persons who can be allowed into those areas at any one time. For single-entry longwall systems, the section intake and return airways must serve as the escapeways. However, even in multi-entry systems it is prudent to maintain at least one return as a escapeway.

Incidents in the intakes or mineral extraction zone may prevent evacuation along intake escapeways. Depending on air velocities and the efficacy of any barricades that may be erected, it might be possible to outpace the airflow and stay ahead of products of combustion along a well-maintained return escapeway.

At least one (and preferably more) intake airway for each section of a mine should be maintained as an escapeway. Where two or more intakes are designated as escapeways then they should be truly independent ventilation routes as far back to the primary mine inlets as practicable. Two parallel airways, where one is ventilated by air that has passed through the other, are not independent ventilation routes and cannot be regarded as separate escapeways. Strata stresses should also be considered in selecting escapeways. Routes that are subject to crushing will become excessively expensive to maintain in a good travellable condition and should not be chosen as escapeways.

Network analysis programs with pollutant distribution modules (section 7.4.4) can be employed to check and improve the selection of escapeways for any given fire scenario. Computer packages that simulate fire situations are particularly valuable for this purpose (section 21.3.2). Even more specialized programs have been developed specifically for the identification of preferred escapeways (Barker-Read and Li, 1989). The reliability of escapeways can be subjected to objective tests through the techniques of fault-tree analysis through which the causes and consequences of events may be interlinked, tracked and analysed (Goodman and Kissell, 1989).

The preparation and maintenance of escapeways should be carried out with careful consideration given to the potential conditions in which those escapeways may be used, i.e. in zero visibility by persons wearing self-rescuers and who are in a state of anxiety. Obstructions, patches of poor roof, uneven or tracked floors and unfenced junctions may create little difficulty during routine travel but can make the difference between life and death when the airway is filled with smoke.

Escapeways must be maintained in a condition suitable for unimpeded foot travel at all times. Signs and coloured reflectors should be employed to indicate directions of escape. However, here again, these visual indicators may be rendered invisible by smoke. Reflectors should, preferably, be located on the sidewalls rather than at the roof of airways as smoke is liable to be thickest at roof level. Some form of lifeline should extend throughout the length of an escapeway. This may simply be a water or compressed air pipe, a cable, wire line or part of the structure of a conveyor, but placed at a convenient height to be followed by hand contact. Such lifelines should also have cones or other devices at intervals to give a tactile indication of direction. It is prudent to deposit extra caches of self-rescuers in well-identified locations within escapeways.

Persons who work in any section of a mine should be completely familiar with the relevant escapeways. This can be promoted by using alternative escapeways during fire drills and displaying maps which highlight escapeways within the section and at other locations, such as shaft stations, where miners may congregate.

During an actual emergency, every attempt should be made to maintain escapeways free from products of combustion for as long as possible. This might be accomplished by the placement of brattice cloths or other devices to control pressure differentials between airways (section 21.3.4). Again, training classes and fire drills give the opportunity of imparting familiarity with procedures involved in evacuation through escapeways. These include the advisability of travelling in groups for mutual assistance and the use of lifelines.

21.9.5 Refuge chambers

When personnel have become trapped within underground workings and all escape routes have become inaccessible then refuge chambers provide a last resort to preserve life. These are lengths of airway or prefabricated chambers within which miners may wait until they can be reached by rescue personnel. Blind headings or other single-ended zones are the preferred locations. However, totally enclosed prefabricated chambers may be sited in through-flow airways.

Refuge chambers should be sized according to the maximum number of persons who work in that area of the mine and for a period of at least two days. Assuming that each resting person requires 0.15 l of air per second (Table 11.2) gives $13\,m^3/day$. However, this should be multiplied by a safety factor of 3 to allow for increased concentrations of exhaled carbon dioxide. Hence, if 20 persons are to be accommodated, then the chamber should have a minimum volume of $13 \times 2 \times 3 \times 20 = 1560\,m^3$. An opening of cross-sectional area of $15\,m^2$ would, therefore, require a length of $1560/15 = 104\,m$.

Two types of refuge chamber have bought trapped miners vital waiting time. Pre-prepared and fully equipped refuge chambers should be considered for workings that are distant from surface connections, and in long development or exploration headings. They should be equipped with water, compressed air, telephones, extra self-rescuers, food in sealed containers, first aid supplies, oxygen cylinders and reading materials (Halasz, 1985). The entrances to pre-prepared refuge chambers should be well marked and all personnel should be aware of their existence and locations. Prefabricated chambers should be constructed from non-combustible and heat-resistant materials.

Trapped personnel may be able to construct improvised refuge chambers provided that brattice cloths, timber, hammers, saws and nails are available. Leakage points may be sealed by clothing, rags, boards or, indeed, anything else that is available. In such cases, an active compressed air line into the zone is invaluable in maintaining the pressure above that of the connecting and, perhaps, polluted airways. If such barricades have to be erected within a continuous airway then attempts should be made to minimize the pressure differential applied across the ends. This may entail holing stoppings or opening doors in the immediate vicinity.

21.10 EMERGENCY PROCEDURE AND DISASTER MANAGEMENT

The manner in which a major fire or an explosion in a mine is handled depends largely on the forethought and planning that has been expanded on such an eventuality.

The early stages of an emergency are often fraught with uncertainity, and chaos can easily occur. It is, therefore, vital that those is charge of subsurface operations should have established and documented a definitive procedure to be followed when a dangerous condition is discovered. All key personnel should be very familiar with that procedure. This final section discusses the three areas that should be addressed.

21.10.1 Immediate response

There are two sets of actions that should occur immediately and simultaneously when a major fire or an explosion takes place. First, all available warning systems for underground personnel should be activated; the numbers of persons underground and their work locations should be established, return airways should be evacuated, and responsible persons should be sent to inspect the affected area (as far as is possible). While this is in progress, telephone messages should be transmitted to key persons or agency representatives not already at the site. These include

1. management and senior supervisory staff,
2. heads of mine specialist departments (e.g. ventilation, safety, electrical, mechanical),
3. regional mine rescue centres and rescue departments at neighbouring mines,
4. government inspectorate or enforcement agencies,
5. union officials, and
6. local police, fire and medical facilities.

A list of these persons or agencies and their telephone numbers should be posted permanently on the wall of the surface operations centre of the mine and updated whenever necessary. At all times, when people are in the mine, there must be a designated person on surface who has the authority and responsibility to initiate these immediate actions.

21.10.2 Command centre

An emergency command centre should be established in a surface office as rapidly as possible. This should be manned by a few key officials, normally the senior management of the mine and others with detailed and expert knowledge of fire-fighting and rescue operations. Those persons should have authority to take part in decision making. However, they should be supported and advised by specialist engineers, government officials and union representatives. Office accommodations equipped with desks, telephones, mine maps and stationery should be made available for these supplementary groups, adjoining the command centre. Provision should also be made for rescue personnel and those involved in the analysis of gas samples.

It it important that the chain of command is firmly established. There should be one person in overall control. Decisions are usually arrived at after consultation and discussion. However, the final word lies with that one person. Instructions relating to firefighting, rescue and evacuation must be issued only from the command centre.

Direct communication should be maintained between the command centre and the mine telephone system as well as to surface fans, power centres and pumping stations.

21.10.3 Disaster management

In addition to the operations that are conducted underground during an emergency, there are numerous other facets that require detailed organization and management. It is to be expected that many other people will arrive at the mine site throughout the emergency, including the news media, family and friends of underground workers and volunteers varying from willing but unskilled laypersons to specialist consultants. Police authorities should be asked to assume control of traffic and all access roads to the mine property including surface buildings and mine entrances. Medical facilities may require to be set up on the mine surface and vehicles made readily available for transportation to hospitals. Medical personnel should be established on site and who must be able and willing to be escorted underground if necessary.

Accommodation may need to be found for immediate family members of trapped miners. This should, preferably, be close to but off the mine site. The presence of clergy and a specialist in trauma reactions can be of tremendous assistance in those facilities. Responsible persons should be put in charge of continuous catering facilities and janitorial services.

Every effort should be made to establish good liaison with media personnel and to accommodate them with arrangements for interviews and telephones. One spokesperson for the mining company should be responsible for all statements to the press. As part of disaster preparedness, that person should have been trained in communicating through the media; in particular, making accurate and relevant statements and responses to questions within a 30 or 60 second television interview. It assists in maintaining good relationships with media personnel to keep them informed of the time of forthcoming statements.

Both company personnel and reporters should be aware of their mutual responsibilities. The latter have a duty to inform the public while, at the same time, they should not engage in any activities that will interfere with rescue or firefighting operations. As televised interviews can be broadcast immediately or within seconds, it is of great importance that information relating to individuals be given to concerned family members before being made available to the media. While the great majority of news people act in a responsible manner, there is the occasional renegade whose enthusiasm exceeds his/her common sense. In one such incident, a young reporter managed to evade the check-in procedures at the top of a mine shaft and reached an underground fresh-air base during rescue operations. He was brought out on a stretcher, having fainted at the scene.

REFERENCES

Banerjee, S. C. (1985) *Spontaneous Combustion of Coal and Mine Fires, Central Mining Research Station, Dhanbad*, Balkema.

Barker-Read, G. R. and Li, H. (1989) Automatic selection of safe egress routes away from underground fires. *Min. Sci. Technol.* **9**, 289–308.

Both, W. (1981) Fighting mine fires with nitrogen in the German coal industry. *Min. Eng.*, (May).

Boulton, J. R. (1991) Wave trend crossing—a new tool for detecting fires in a mine employing diesel equipment. *Proc. 5th US Mine Ventilation Symp., West Virginia*, 9–17.

Browning, E. J. (1988) Frictional ignitions. *Proc. 4th Int. Ventilation Congr., Brisbane*, 319–26.

Brunner, D. J., Wallace, K. G. and Deen, J. B. (1991) The effects of natural ventilation pressure on the underground ventilation system at the Waste Isolation Pilot Plant. *Proc. 5th US Mine Ventilation Symp., West Virginia*, 593–604.

Chakravorty, R. N. and Woolf, R. L. (1980) Evaluation of systems for early detection of spontaneous combustion in coal mines. *Proc. 2nd Int. Mine Ventilation Congr., Reno, NV.*, 429–36.

Chakravorty, S. L. (1960) Auto-oxidation of Indian coals. *J. Min. Met. Fuels* (8, 9 and 11), 1, 10.

Cybulska, R. (1981) Examples of coal dust explosions. In *Ignitions, Explosions and Fires, Illiwarra Symp.* (ed. A. J. Hargraves), pp. 7.1–7.8 Australian Institute of Mining and Metallurgy.

Deliac, E. P., *et al.* (1985) Development of ventilation software on personal computers in France and the application to the simulation of mine fires. *Proc. 2nd US Mine Ventilation Symp., Reno, NV*, 19–27.

Dunn, M. F., *et al.* (1982) Main mine reverse performance characteristics. *Proc. 1st US Mine Ventilation Symp., Tuscaloosa, AL*, 23–8.

Dziurzynski, W., Tracz, J. and Trutwin, W. (1988) Simulation of mine fires. *Proc. 4th Int. Mine Ventilation Congr., Brisbane*, 357–63.

Eicker, H. and Kartenburg, H. J. (1984) Investigation of methods for early detection of fires due to spontaneous combustion. *Proc. 3rd Int. Ventilation Congr., Harrogate*, 411–16.

Eisner, H. S. and Smith, P. B. (1956) Firefighting in underground roadways: experiments with foam plugs. *SMRE Res. Rep. No. 130.*

Froger, C. E. (1985) Firefighting expertise in French underground mines. *Proc. 2nd US Mine Ventilating Symp., Reno., NV*, 3–10.

Goodman, G. V. R. and Kissell, F. N. (1989) Fault tree analysis of miner escape during mine fires. *Proc. 4th US Mine Ventilation Symp., Berkeley, CA*, 57–65.

Greuer, R. E. (1984) Transient-state simulation of ventilation systems in fire conditions. *Proc. 3rd Int. Mine Ventilation Congr., Harrogate*, 407–10.

Greuer, R. E. (1988) Computer models of underground mine ventilation and fires. *US Bureau of Mines, IC 9206*, 6–14.

Grieg, J. D., Lloyd, P. J. D. and Quail, R. W. (1975) Some aspects of the use of high-expansion foams in underground firefighting. *Proc. 1st Int. Mine Ventilation Congr., Johannesburg*, 239–45.

Halasz, L. (1985) *Establishment and Use of Refuge Bays at the Western Areas Gold Mining Company, Ltd.*, Vol. 4, No. 2, Chamber of Mines of S. Africa, Mine Safety Division.

Haldane, J. S. and Meachen, F. (1898) Oxidation and spontaneous combustion. *Trans. Inst. Min. Eng.* **16**, 457.

Hall, A. E., *et al.* (1989) Sulphide dust explosition studies at H-W mine of Westmin Resources, Ltd. *Proc. 4th US Mine Ventilation Symp., Berkeley, CA*, 532–9.

Herbert, M. J. (1988) The use of nitrogen inertization in the British Isles. *Proc. 4th Int. Mine Ventilation Congr., Brisbane*, 327–35.

Hjelmstad, K. E. and Pomroy, W. G. (1990) Novel fire warning system for underground mines. *Min. Eng. (USA)* (January), 107–12.

Hodges, D. J. and Hinsley, F. B. (1964) The influence of moisture on spontaneous heating of coal. *Trans. Inst. Min. Eng.* **123**(40), 211–24.
Holding, W. (1975) An approach to the potential problem of sulphide dust ignitions at Prieska Copper Mine. *Proc. 1st Int. Mine Ventilation Congr., Johannesburg*, 207–11.
Holding, W. (1982) Explosible dusts. *Environmental Engineering in South African Mines*, Chapter 28, Mine Ventilation Society of South Africa, 763–71.
Hwang, C. C., *et al.* (1991) Modeling the flow-assisted flame spread along conveyor belt surfaces. *Proc. 5th US Mine Ventilation Symp., West Virginia*, 39–45.
IME (1985) Sealing-off fires underground. *Institution of Mine Engineers Memo.*, 1–47, Doncaster.
Jensen, B., *et al.* (1989) An experimental approach to the determination of explosible lean limits for coal dust and methane mixtures. *Proc. 4th Mine Ventilation Symp., Berkeley, CA*, 549–57.
Kissell, F. N. and Timko, R. J. (1991) Pressurization of intake escapeways with parachute stoppings to reduce infiltration of smoke. *Proc. 5th US Mine Ventilation Symp., West Virginia*, 28–34.
Kolada, R. J. and Chakravorty, R. N. (1987) Controlling the hazard of methane explosions in coal storage facilities. *Proc. 3rd US Mine Ventilation Symp., Penn State*, 334–9.
Litton, C. D. (1988) Complexities of mine fire detection due to the use of diesel-powered equipment. *Proc. 4th US Mine Ventilation Symp., Berkeley, CA*, 162–6.
Litton, C. D., *et al.* (1987) Calculating fire-throttling of mine ventilation airflow. *US Bureau of Mines Rep., RI 9076.*
Mitchell, D. W. (1990) *Mine Fires.* Maclean Hunter, Chicago, IL, 167 pp.
Ouderkirk, S. J., *et al.* (1985) Mine stench fire warning computer model development and in-mine validation testing. *Proc. 2nd US Mine Ventilation Symp., Reno, NV*, 29–35.
PD–NCB Consultants (1978) A review of spontaneous combustion problems and controls with application to US coal mines. *Rep. ET*-77-C-01-8965 (US Dept. of Energy.)
Pomroy, W. H. (1988) Fire detection systems for non-coal underground mines. *US Bureau of Mines IC 9206*, 21–7.
Pomroy, W. H. and Laage, L. W. (1988) Real-time monitoring and simulation analysis of mine atmospheres to locate and characterize underground mine fires. *Proc. 4th Int. Mine Ventilation Congr., Brisbane*, 37–43.
Sheer, T. J. (1988) Safety precautions for the protection of miners against underground fires. *Proc. 4th Int. Mine Ventilation Congr., Brisbane*, 1–12.
Stefanov, T. P., *et al.* (1984) Unsteady-state processes during an open fire in a ventilation network. *Proc. 3rd Int. Mine Ventilation Congr., Harrogate*, 417–20.
Strang, J. and MacKenzie-Wood, P. (1985) *Mines Rescue, Safety and Gas Detection*, Weston, Kiama, 366 pp.
Sun, W. T. (1963) First industrial trial of fighting underground fires by combustion gases at Tchegan Colliery. *Int. Mining Congr., Salzburg.*
Timko, R. J., Derick, R. L. and Thimons, E. D. (1988) Analysis of a fire in a Colorado coal mine. *Proc. 4th Int. Mine Ventilation Congr., Brisbane*, 345–53.
Trutwin, W. (1975) Reversal of airflow by means of water sprays in upcast shafts. *Pol. Acad. Sci., Min. Arch. Q.* **20**(4), (in Polish).
Verakis, H. C. (1991) Reducing the fire hazard of mine conveyor belts. *Proc. 5th US Mine Ventilation Symp., West Virginia*, 69–73.
Verakis, H. C. and Dalzell, R. W. (1988) Impact of entry air velocity on the fire hazard of conveyor belts. *Proc. 4th Int. Mine Ventilation Congr., Brisbane*, 375–81.
Zabetakis, M. G., *et al.* (1959) Determining the explosibility of mine atmospheres. *US Bureau of Mines. IC 7901*, 11.

Zabrosky, C. E. and Klinefelter, G. (1988) Microcomputers assist in extinguishing a bituminous coal mine fire. *Proc. 4th Int. Mine Ventilation Congr., Brisbane*, 337–44.

FURTHER READING

Bhowmick, B. C., *et al.* (1991) Dynamic balancing of pressure—its relevance for control of fire in Indian coal mines. *Proc. Indo-Polish Workshop of Mining Research and Mining Methods, Central Mining Research Station, Danbad*, 1–15.

Graham, J. I. (1914) Absorption of oxygen by coal. *Trans. Inst. Min. Eng.* **48**, 521.

Willett, H. L. (1951) Gas analysis behind stoppings. *Trans. Inst. Min. Eng.* **111**, 629.

Index